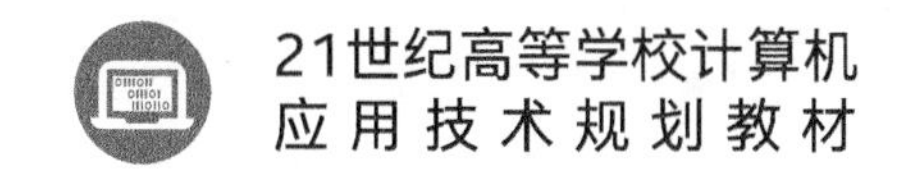

Android 应用开发从入门到精通

◎ 郑耿忠 庄桂东 编著

清華大学出版社
北京

内 容 简 介

本书采用模块化结构，以大量案例分析为主线，介绍了 Android 手机操作系统的开发与应用。全书分为 Android 操作系统与开发环境、UI 事件控制、Android 基本界面组件和 Android 高级界面组件、资源文件使用；Activity、Service 及 BroadcastReceiver 应用、数据存储以及文件读写、ContentProvider 应用、Android 网络编程、综合案例分析共 11 章。本书体系结构清晰，内容围绕 Android 手机操作系统开发与应用，对 Android 的功能按照特性进行分类，根据由浅入深的原则，以教学单元搭配步骤讲解，每个章节都包含精心设计和讲解的应用程序开发案例，使书的内容在广度和讲解的详细程度上达到最佳的平衡，另外，本书着重实际操作，辅以适当的理论讲解，让读者在理解 Android 手机技术的原理的同时掌握 Android 重要函数库的使用，然后再通过综合案例的方式将所学的开发技术融会贯通。

本书适合所有有志于从事 Android 手机操作系统开发并有一定 Java 程序设计基础的人员参考使用，也可以作为 Android 手机操作系统开发的培训教材。

图书在版编目(CIP)数据

Android 应用开发从入门到精通/郑耿忠，庄桂东编著. —北京：清华大学出版社，2018 (2020.1重印)
(21 世纪高等学校计算机应用技术规划教材)
ISBN 978-7-302-47928-4

Ⅰ. ①A…　Ⅱ. ①郑…　②庄…　Ⅲ. ①移动终端－应用程序－程序设计　Ⅳ. ①TN929.53

中国版本图书馆 CIP 数据核字(2017)第 193234 号

责任编辑：黄　芝　李　晔
封面设计：刘　键
责任校对：梁　毅
责任印制：李红英

出版发行：清华大学出版社
　　网　　址：http://www.tup.com.cn，http://www.wqbook.com
　　地　　址：北京清华大学学研大厦 A 座　　**邮　　编**：100084
　　社 总 机：010-62770175　　**邮　　购**：010-62786544
　　投稿与读者服务：010-62776969，c-service@tup.tsinghua.edu.cn
　　质量反馈：010-62772015，zhiliang@tup.tsinghua.edu.cn
　　课件下载：http://www.tup.com.cn，010-62795954
印 装 者：北京九州迅驰传媒文化有限公司
经　　销：全国新华书店
开　　本：185mm×260mm　　**印　　张**：27　　**字　　数**：656 千字
版　　次：2018 年 5 月第 1 版　　**印　　次**：2020 年 1 月第 3 次印刷
印　　数：1801～2100
定　　价：59.50 元

产品编号：072419-01

随着我国改革开放的进一步深化，高等教育也得到了快速发展，各地高校紧密结合地方经济建设发展需要，科学运用市场调节机制，加大了使用信息科学等现代科学技术提升、改造传统学科专业的投入力度，通过教育改革合理调整和配置了教育资源，优化了传统学科专业，积极为地方经济建设输送人才，为我国经济社会的快速、健康和可持续发展以及高等教育自身的改革发展做出了巨大贡献。但是，高等教育质量还需要进一步提高以适应经济社会发展的需要，不少高校的专业设置和结构不尽合理，教师队伍整体素质亟待提高，人才培养模式、教学内容和方法需要进一步转变，学生的实践能力和创新精神亟待加强。

教育部一直十分重视高等教育质量工作。2007 年 1 月，教育部下发了《关于实施高等学校本科教学质量与教学改革工程的意见》，计划实施“高等学校本科教学质量与教学改革工程(简称‘质量工程’)”，通过专业结构调整、课程教材建设、实践教学改革、教学团队建设等多项内容，进一步深化高等学校教学改革，提高人才培养的能力和水平，更好地满足经济社会发展对高素质人才的需要。在贯彻和落实教育部“质量工程”的过程中，各地高校发挥师资力量强、办学经验丰富、教学资源充裕等优势，对其特色专业及特色课程(群)加以规划、整理和总结，更新教学内容、改革课程体系，建设了一大批内容新、体系新、方法新、手段新的特色课程。在此基础上，经教育部相关教学指导委员会专家的指导和建议，清华大学出版社在多个领域精选各高校的特色课程，分别规划出版系列教材，以配合“质量工程”的实施，满足各高校教学质量和教学改革的需要。

本系列教材立足于计算机公共课程领域，以公共基础课为主、专业基础课为辅，横向满足高校多层次教学的需要。在规划过程中体现了如下一些基本原则和特点。

(1) 面向多层次、多学科专业，强调计算机在各专业中的应用。教材内容坚持基本理论适度，反映各层次对基本理论和原理的需求，同时加强实践和应用环节。

(2) 反映教学需要，促进教学发展。教材要适应多样化的教学需要，正确把握教学内容和课程体系的改革方向，在选择教材内容和编写体系时注意体现素质教育、创新能力与实践能力的培养，为学生的知识、能力、素质协调发展创造条件。

(3) 实施精品战略，突出重点，保证质量。规划教材把重点放在公共基础课和专业基础课的教材建设上；特别注意选择并安排一部分原来基础比较好的优秀教材或讲义修订再版，逐步形成精品教材；提倡并鼓励编写体现教学质量和教学改革成果的教材。

(4) 主张一纲多本，合理配套。基础课和专业基础课教材配套，同一门课程可以有针对不同层次、面向不同专业的多本具有各自内容特点的教材。处理好教材统一性与多样化，基本教材与辅助教材、教学参考书，文字教材与软件教材的关系，实现教材系列资源配套。

（5）依靠专家，择优选用。在制定教材规划时依靠各课程专家在调查研究本课程教材建设现状的基础上提出规划选题。在落实主编人选时，要引入竞争机制，通过申报、评审确定主题。书稿完成后要认真实行审稿程序，确保出书质量。

繁荣教材出版事业，提高教材质量的关键是教师。建立一支高水平教材编写梯队才能保证教材的编写质量和建设力度，希望有志于教材建设的教师能够加入到我们的编写队伍中来。

21世纪高等学校计算机应用技术规划教材
联系人：魏江江 weijj@tup.tsinghua.edu.cn

目前 Android 是一门新兴技术，无论是相关书籍还是教育体制都处于初级阶段，因此 Android 人才在短期之内将会呈现供不应求的状态。从长期来看，随着各种移动应用需求的增加，手机应用开发商对 Android 应用的开发力度也会不断加大，因此，随着安卓手机用户比例的增长，更加剧了市场对有关 Android 系统开发书籍的需求。

本书对 Android 的功能按照特性进行分类，根据由浅入深的原则，以教学单元搭配步骤讲解，每个章节都包含精心设计和讲解的应用程序开发案例，使书的内容在广度和深度上达到最佳的平衡。另外，本书着重实际操作，并辅以适当的理论讲解，让读者在理解 Android 手机技术的原理的同时掌握 Android 重要函数库的使用，然后再通过综合案例的方式将所学的开发技术融会贯通。

相对其他教材，本书具有如下特点：

（1）遵循一个基础知识点对应一个实例的原则：将实例置于知识点之前，然后剖析实例，阐述知识点。

（2）内容安排更加合理，用最基础的实例讲解知识点，让初学者更加容易接受，真正做到由浅入深。

（3）通过对基本案例和综合案例循序渐进的介绍分析，由浅入深地完成掌握基本操作、基本理论到完成综合案例的全部过程。

本书可作为本科或高职高专软件工程、计算机科学与技术等专业的教材，也可供其他专业学生和从事 Android 开发与应用的有关技术人员参考。课程标准学时为 72 学时或 54 学时，在教学过程中可根据具体情况选学本书内容。

本书由郑耿忠主编和统稿，其中第 1～6 章由郑耿忠编写，第 7～11 章由庄桂东编写，书中案例由庄桂东录制。

本书在编写和出版过程中，得到清华大学出版社编辑的指导和支持，在此对他们的辛勤劳动和无私奉献表示真挚的谢意。同时，对本书参考文献中的有关作者致以诚挚的感谢。

Android 开发内容丰富，应用广泛，技术处于不断发展进步中，限于编者自身的水平和学识，书中难免存在疏漏之处，诚望读者不吝赐教，以便修正。

作 者

2017 年 5 月

目 录

第1章 Android操作系统与开发环境

Android一词的本义指“机器人”，同时也是Google于2007年11月5日宣布的基于Linux平台的开源手机操作系统的名称，该平台由操作系统、中间件、用户界面和应用软件组成。Android系统已经成为全球应用并具有广泛影响力的手机操作系统，国内对Android开发人才的需求也在迅速增长，从趋势上来看，Android软件人才的需求会越来越大。

Android 2.2平台的模拟器更加稳定，故本书所带的Android案例也是在Android 2.2平台上调试运行的。本章将重点讲解如何搭建Android开发环境，创建和启动手机模拟器以及学习使用Android操作系统。

1.1 Android简介

智能手机软件平台有Symbian、Windows Mobile、RIM BlackBerry、Android、iPhone、Palm、Brew、Java/J2ME。2012年11月的数据显示，Android占据全球智能手机操作系统市场76%的份额，在中国市场的占有率为90%。接下来将重点介绍什么是Android平台。

1.1.1 什么是Android

Android是一种基于Linux的自由及开放源代码的操作系统，主要使用于移动设备，如智能手机和平板电脑，由Google公司和开放手机联盟领导及开发。它包括一个操作系统、中间件和一些重要的应用程序。它采用软件堆层(Software Stack，又名软件叠层)的架构，主要分为三部分。底层以Linux内核工作为基础，由C语言开发，只提供基本功能；中间层包括函数库Library和Dalvik虚拟机，用C++语言开发。最上层是各种应用软件，包括通话程序、短信程序等，我们要做的，就是以Java作为编程语言编写各种各样的Android应用软件。本书中，学习Android其实就是学习怎么开发适用于在Android操作系统上运行的软件。

在国内，Android的前景十分广阔，国内很多的厂商和运营商也纷纷加入了Android阵营，同时Android应用的范围不仅仅在手机，国内一些厂家也陆续推出了采用Android系统的MID产品，比较著名的包括由Rockchip和蓝魔推出的同时具备高清播放和智能系统的音悦汇W7，可以预见，Android也将会被广泛应用在国产智能上网设备上，并将进一步扩大Android系统的应用范围。

1.1.2 Android 平台的架构详解

Android 平台采用一种被称为软件叠层的方式进行构建，就像一个多层蛋糕，每一层都有自己的特性和用途。这种软件结构使得层与层之间相互分离，明确各层的分工。这种分工保证了层与层之间的低耦合，当下层的层内或层下发生改变时，上层应用程序无须任何改变，图 1.1 为 Android 系统架构。

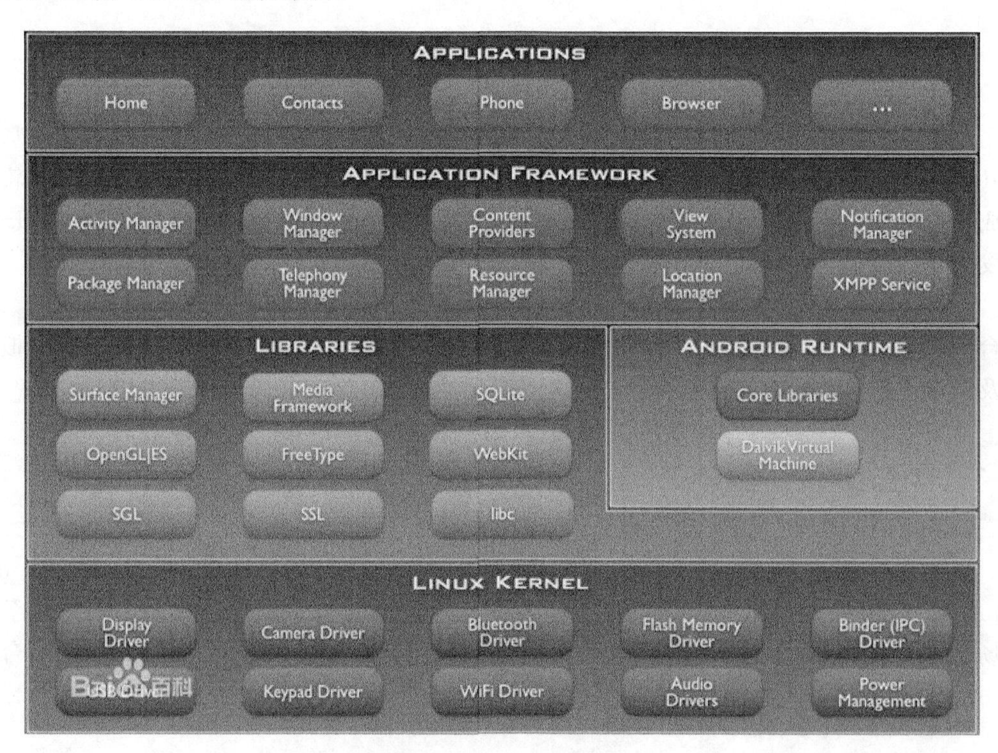

图 1.1 Android 系统架构

由图 1.1 可以很明显地看出，Android 系统架构由 5 部分组成，分别是 Applications（应用程序层）、Application Framework（应用程序框架）、Libraries（函数库）、Android Runtime（Android 运行时）、Linux Kernel（Linux 内核）。下面分别对这 5 部分进行简单介绍。

1. 应用程序层

Android 平台装配一个核心应用程序集合，这些程序包括电子邮件客户端、SMS 程序、日历、地图、浏览器、联系人和其他设置。所有应用程序都是用 Java 编程语言写的。更加丰富的应用程序有待我们去开发，本书介绍的内容则是如何编写 Android 系统上的应用程序。

2. 应用程序框架

通过提供开放的开发平台，Android 使开发者能够编写极其丰富和新颖的应用程序。开发者可以自由地利用设备硬件优势、访问位置信息、运行后台服务、设置闹钟、向状态栏添加通知等。

Android 应用程序框架提供了大量的 API 供开发者使用，关于这些 API 的具体功能和用法则是本书后面详细介绍的内容。

所有的应用程序其实是一组服务和系统，包括：

- 视图(View)——丰富的、可扩展的视图集合，可用于构建一个应用程序，包括列表、网格、文本框、按钮，甚至是内嵌的网页浏览器。
- 内容提供者(Content Providers)——使应用程序能访问其他应用程序(如通讯录)的数据，或共享自己的数据。
- 资源管理器(Resource Manager)——提供访问非代码资源，如本地化字符串、图形和布局文件。
- 通知管理器(Notification Manager)——使所有的应用程序能够在状态栏显示自定义警告。
- 活动管理器(Activity Manager)——管理应用程序生命周期，提供通用的导航回退功能。

3. 函数库

Android 包含一套 C/C++ 库的集合，供 Android 系统的各个组件使用。一般来说，Android 应用开发者不直接调用这套 C/C++ 库集，而是通过它上面的应用程序框架来调用这些库。下面为一些核心库：

- 系统 C 库——标准 C 系统库(libc)的 BSD 衍生，调整为基于嵌入式 Linux 设备。
- 媒体库——基于 PacketVideo 的 OpenCORE。这些库支持播放和录制许多流行的音频和视频格式，以及静态图像文件，包括 MPEG4、H. 264、MP3、AAC、AMR、JPG、PNG。
- 界面管理——管理访问显示子系统和无缝组合多个应用程序的二维和三维图形层。
- LibWebCore——新式的 Web 浏览器引擎，驱动 Android 浏览器和内嵌的 Web 视图。
- SGL——基本的 2D 图形引擎。
- 3D 库——基于 OpenGL ES 1.0 API 的实现。库使用硬件 3D 加速或包含高度优化的 3D 软件光栅。
- FreeType——位图和矢量字体渲染。
- SQLite——所有应用程序都可以使用的强大而轻量级的关系数据库引擎。

4. Android 运行时

Android 包含一个核心库的集合，提供大部分在 Java 编程语言核心类库中可用的功能。每一个 Android 应用程序是 Dalvik 虚拟机中的实例，运行在它们自己的进程中。Dalvik 虚拟机设计成在一个设备可以高效地运行多个虚拟机。Dalvik 虚拟机可执行的文件格式是. dex，. dex 格式是专为 Dalvik 设计的一种压缩格式，适合内存和处理器速度有限的系统。

大多数虚拟机包括 JVM 都是基于栈的，而 Dalvik 虚拟机则是基于寄存器的。两种架构各有优劣，一般而言，基于栈的机器需要更多指令，而基于寄存器的机器指令更大。dx 是一套工具,可以将 Java .class 转换成.dex 格式。一个.dex 文件通常会包括多个.class 文件。由于.dex 有时必须进行最佳化,会使文件大小增加 1～4 倍,并以.odex 结尾。

Dalvik 虚拟机依赖于 Linux 内核提供基本功能,如线程和底层内存管理。

5. Linux 内核

Android 系统建立在 Linux 2.6 之上,提供核心系统服务,例如,安全、内存管理、进程管理、网络堆栈、驱动模型。除此之外,Linux Kernel 也作为硬件和软件之间的抽象层,它隐藏具体硬件细节而为上层提供统一的服务。

如果只是做应用开发,就不需要深入了解 Linux Kernel 层。

1.2 搭建 Android 开发环境

在搭建 Android 开发环境之前,还需要 JDK(仅有 JRE 不够)、Eclipse IDE,而像 JDK 安装、环境变量设置之类的知识不在本书中进行讲解,若读者尚不明白这些操作,建议先掌握这些知识后再开始搭建 Android 开发环境。

1.2.1 如何下载和安装 ADT 插件

在企业开发中,很多程序员使用 Eclipse IDE 作为应用的开发环境,Android 推荐使用 Eclipse 来开发 Android 应用。为了使得 Android 应用的创建、运行和调试更加方便快捷,Android 的开发团队专门针对 Eclipse IDE 定制了一个插件:Android Development Tools (ADT)。

ADT 插件的安装有在线安装和离线安装两种方式。下面介绍离线安装 ADT 插件的步骤,图 1.2 为 ADT 插件下载链接。

① 登录 http://developer.android.com/sdk/installing/installing-adt.html#tmgr 站点,下载 ADT 插件的最新版本。

② 将页面往下拉,可以看到如图 1.2 所示的表格,单击 ADT-22.3.0.zip 链接直接下载 ADT 插件到本地。

Package	Size	MD5 Checksum
ADT-22.3.0.zip	14493723 bytes	0189080b23dfa0f866adafaaafcc34ab

图 1.2 ADT 插件下载链接

③ 启动 Eclipse,在 Eclipse 主菜单中选择 Help→Install New Software 命令,在出现的如图 1.3 所示的对话框中,单击 Add 按钮。

④ 在弹出的如图 1.4 所示的对话框的 Name 文本框中输入 ADT,然后单击 Archive 按钮,浏览和选择已经下载的 ADT 插件的压缩文件。

⑤ 单击 OK 按钮,返回如图 1.5 所示的可用软件的视图,选中 Developer Tools 复选框(即 ADT 插件),然后单击 Next 按钮,Eclipse 弹出一个对话框,该对话框会提示用户所有将要安装的插件详细清单,单击该对话框的 Next 按钮。

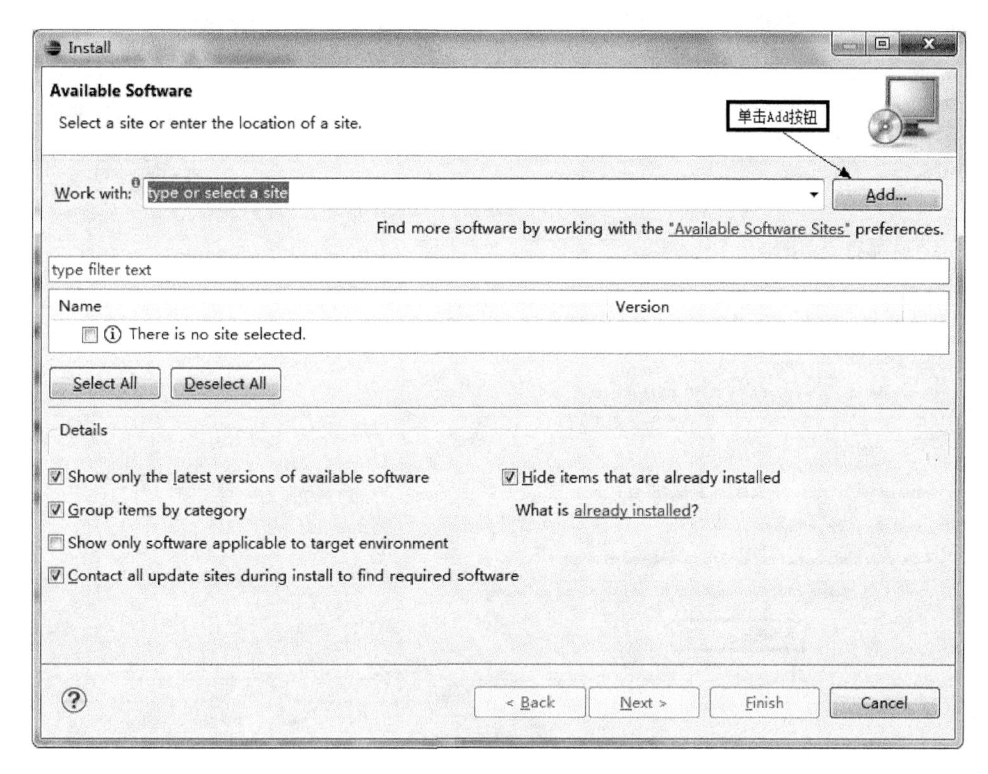

图 1.3 选择插件安装

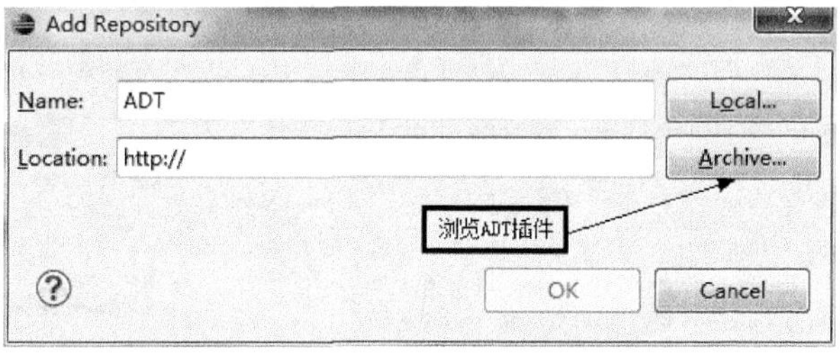

图 1.4 浏览 ADT 插件

⑥ 在 Eclipse 弹出的如图 1.6 所示的窗口中选择接受协议条款，单击 Finish 按钮，Eclipse 开始安装 ADT 插件。

ADT 插件的在线安装步骤跟离线安装的步骤基本一致，区别是在第 4 步时，在弹出的对话框中不再选择已经下载好的 ADT 插件，而是在 Location 文本框中直接输入"https://dl-ssl.google.com/android/eclipse/"，如图 1.7 所示，然后单击 OK 按钮，之后的步骤与离线安装完全一致。

1.2.2 如何下载和安装 Android SDK

Android SDK 包含了开发 Android 应用所依赖的 jar 文件、运行环境及相关工具，安装 Android SDK 请按下面步骤进行：

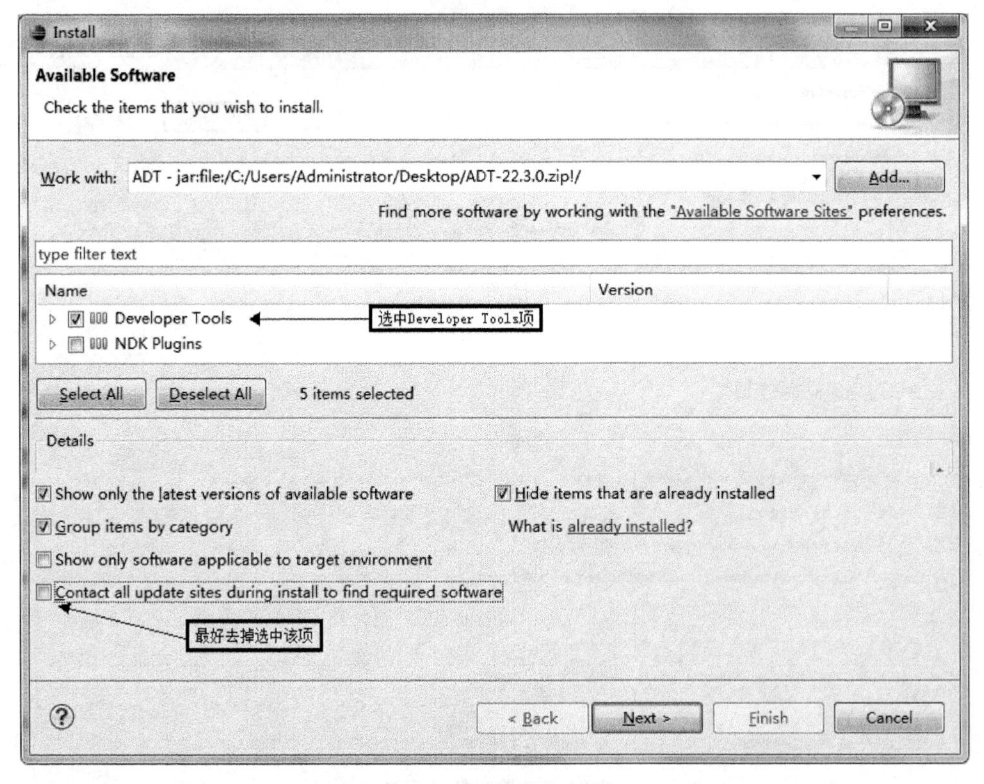

图 1.5　选中 ADT 插件

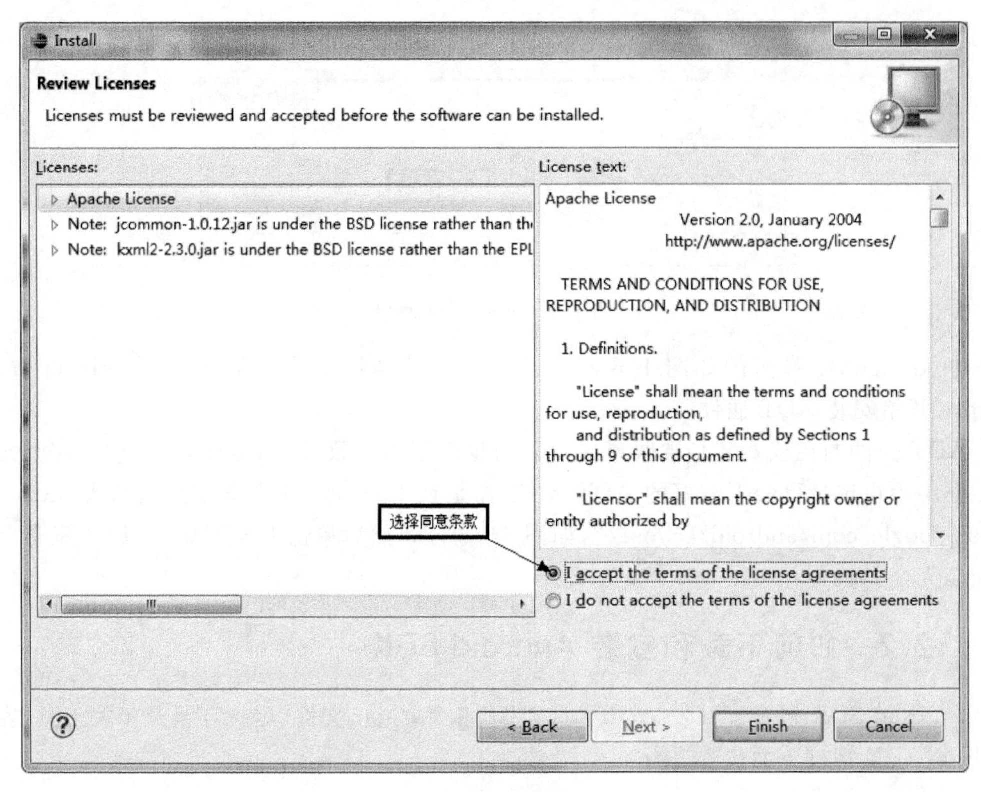

图 1.6　同意条款

图 1.7　离线安装 ADT

① Android SDK 的下载地址是 http://developer.android.com/sdk/index.html。输入该地址，将页面往下拉，单击 DOWNLOAD FOR OTHER PLATFORMS 链接，继续往下拖动页面，可看到 SDK 的下载链接，选择所需要的版本进行下载，一般下载解压版，如图 1.8 所示。

SDK Tools Only

单击所要下载的版本链接

Platform	Package	Size	MD5 Checksum
Windows 32 & 64-bit	android-sdk_r22.3-windows.zip	108847452 bytes	9f0fe8c8884d6aee2b298fee203c62dc
	installer_r22.3-windows.exe (Recommended)	88845794 bytes	ad50c4dd9e23cee65a1ed740ff3345fa
Mac OS X 32 & 64-bit	android-sdk_r22.3-macosx.zip	74893875 bytes	ecde88ca1f05955826697848fcb4a9e7
Linux 32 & 64-bit	android-sdk_r22.3-linux.tgz	100968558 bytes	6ae581a906d6420ad67176dff25a31cc

图 1.8　下载 SDK

② 将下载完成后得到的 android-sdk_r22.3-windows.zip 文件解压缩到任意路径下，例如“D:\”根目录。解压缩后得到一个名为 android-sdk-windows 的文件夹，该文件夹包含以下文件结构。

➢ add-ons：该目录下存放额外的附加软件，刚解压缩时该目录为空。
➢ platforms：该目录 F 存放不同版本的 Android 版本，刚解压缩时该目录为空。
➢ tools：该目录下存放了许多 Android 开发、调试的工具。
➢ AVD Manager.exe：AVD(Android 虚拟设备)管理器，通过该工具可以管理 AVD。
➢ SDK Manager.exe：Android SDK 管理器，通过该工具可以管理 Android SDK。

③ 单击 SDK Manager.exe，弹出如图 1.9 所示窗口，在窗口中选中需要安装的工具，其中 Android 平台工具是必选项，读者喜爱下载哪个版本的 SDK，则选中其版本的 SDK，可一次性选择所有版本，也可在以后需要的时候再对特定版本进行下载。选中后单击 Install 9 packages 按钮进行安装。

④ 在弹出的如图 1.10 所示的对话框中，列出了将要安装的 Android 工具包，选择接受所有许可内容，然后单击 Install 按钮，Android SDK 管理器就开始下载并安装读者所选的工具包了，如图 1.11 所示。等待一段时间即可完成，但该段时间的长短取决于读者的网络状态及所选中的工具包数量，有时候甚至会花费一两个小时。

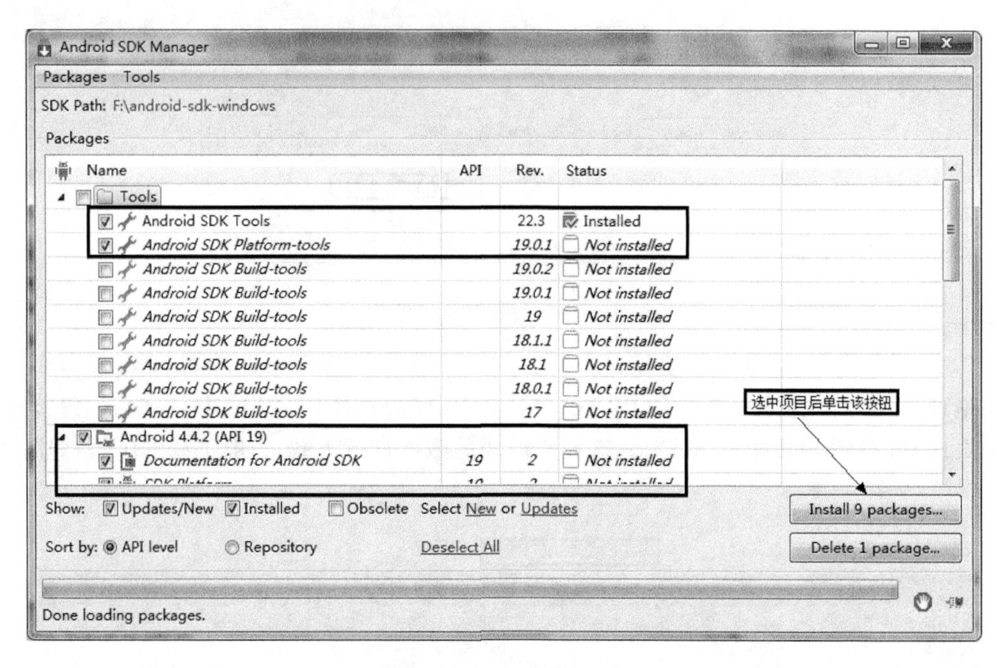

图 1.9 安装所需工具

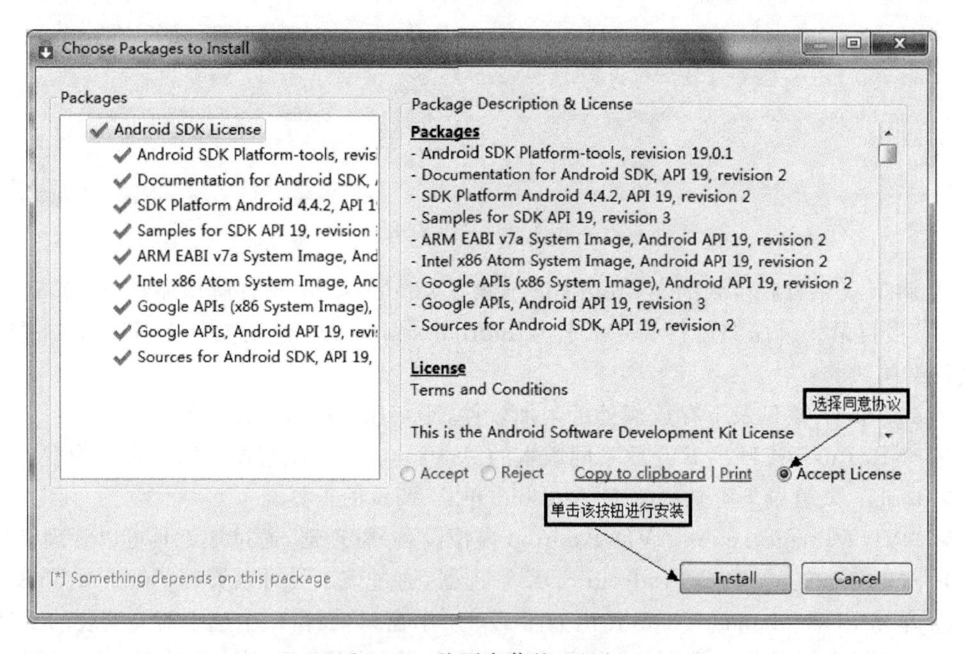

图 1.10 将要安装的工具包

⑤ 安装完成后，回到 Android SDK 文件夹界面，可以看到该目录下增加了如下几个文件夹。

➢ docs：该文件夹下存放了 Android SDK 开发文件和 API 文档等。

➢ platform-tools：该文件夹下存放了 Android 平台相关工具。

➢ samples：该文件夹下存放了不同 Android 平台的示例程序。

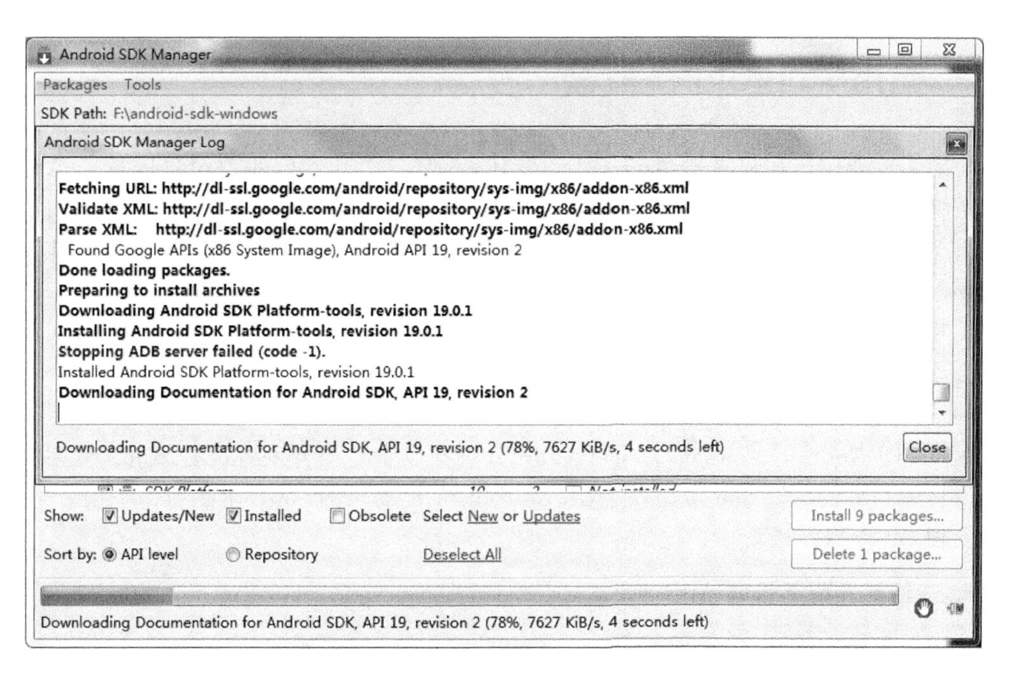

图 1.11 在线安装 Android 工具包

⑥ 启动 Eclipse，为 Eclipse 设置 Android SDK 的路径，选择 Eclipse 主菜单 Window→Preferences 菜单项，在打开的如图 1.12 所示的视图的左侧单击 Android 选项，在右侧的 SDK Location 文本框中输入 Android SDK 所在位置，单击 OK 按钮，完成 Android SDK 的路径设置。

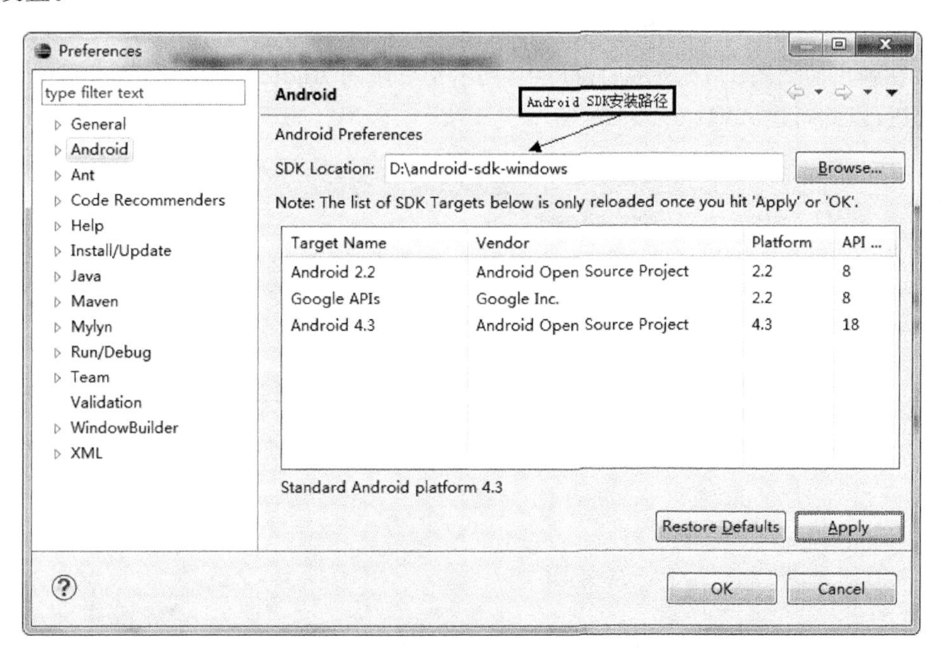

图 1.12 设置 SDK 路径

经过上面所介绍的过程，接下来就可以在 Eclipse 中开发 Android 应用了。

1.3 使用 Android 模拟器

前面主要介绍了如何搭建 Android 开发环境，但我们开发后的程序将运行于 Android 操作系统，不再像以前开发 Windows 软件一样运行于 Windows 平台。当然，我们不能要求每个开发者都去买一台搭建了 Android 平台的手机然后才开始学习，此时可以借助 Android 提供的"虚拟设备"工具来模拟 Android 手机。除此之外，Android SDK 还提供了大量工具来帮助我们进行开发、调试。

1.3.1 创建、删除和浏览 AVD

AVD，即 Android Virtual Device(Android 虚拟设备)，当开发者没有 Android 手机时，则可以将编写好的 Android 应用安装在 Android SDK 提供的 AVD 上运行。下面分别介绍两种管理 AVD 的方式。

1. 在图形界面下管理 AVD

在图形界面下管理 AVD 比较简单，可以借助 AVD 管理器来完成，完全在图形用户界面下操作，比较适合新上手的读者。

① 双击 Android SDK 安装目录下的 AVD Manager. exe 文件或者单击如图 1.13 所示的 Eclipse 工具栏上的 Android Virtual Device Manager 按钮，启动 AVD 管理器。

图 1.13　Eclipser 工具栏上的 AVD Manager 按钮

② 在弹出的如图 1.14 所示的窗口中，列出了可用的 Android 模拟器，单击窗口右边的 New 按钮，以此新建 AVD。

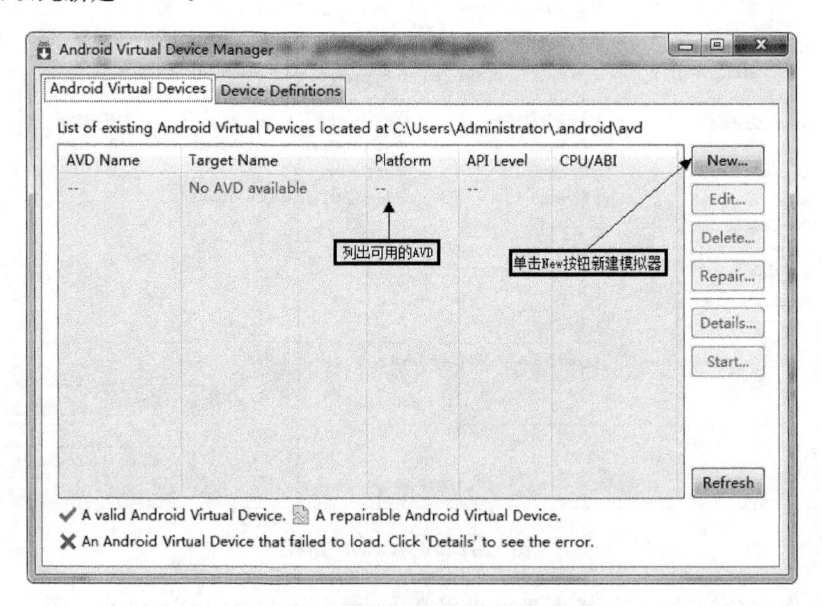

图 1.14　查看所有可用的 AVD 设备

③ 在弹出的如图 1.15 所示的对话框中，填写 AVD 设备的名称、选择 AVD 设备的分辨率以及运行的 Android 版本、填写虚拟 SD 卡的大小，然后单击该对话框下面的 OK 按钮，管理器则开始创建 AVD 设备，开发者只需稍作等待即可。

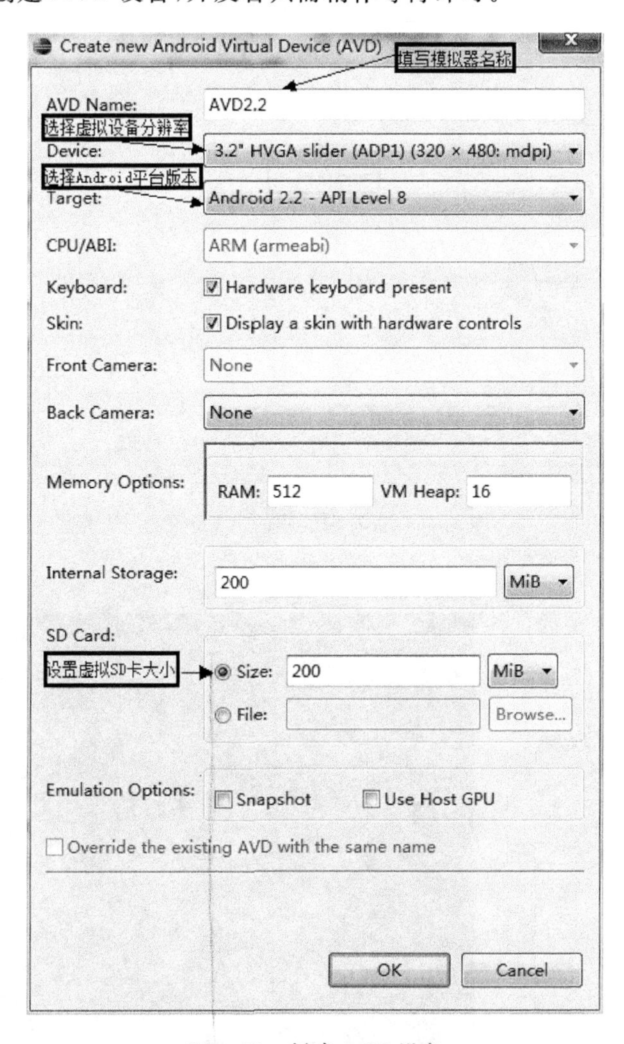

图 1.15　创建 AVD 设备

创建完成后，管理器返回如图 1.16 所示的窗口，该管理器将会列出当前所有可用的 AVD 设备，读者可以看到我们刚刚创建的 AVD 设备。如果开发者想删除某个 AVD 设备，只要在如图 1.16 所示的窗口中选择指定 AVD 设备，然后单击右边的 Delete 按钮即可。如果开发者想要浏览某个 AVD 设备，只要在如图 1.16 所示的窗口中选择指定 AVD 设备，然后单击右边的 Details 按钮，即会弹出一个 AVD 详情窗口，供开发者查看。

AVD 设备创建成功后，开发者即可运行该 AVD 了。借助如图 1.16 所示的 AVD 管理器来运行 AVD 设备非常简单：首先，在如图 1.16 所示的窗口左边选中所要启动的 AVD 设备；其次，在该窗口右边单击 Start 按钮，弹出如图 1.17 所示的窗口；最后，在图 1.17 所示窗口中，单击 Launch 按钮，模拟器即开始启动，启动过程如图 1.18 所示，启动完毕后的模拟器如图 1.19 所示。

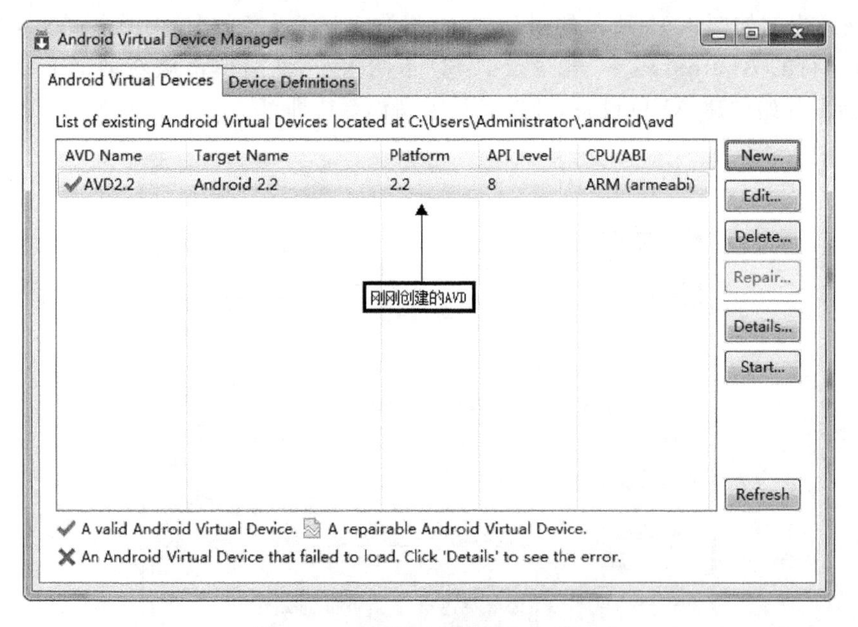

图 1.16　查看刚创建的 AVD 设备

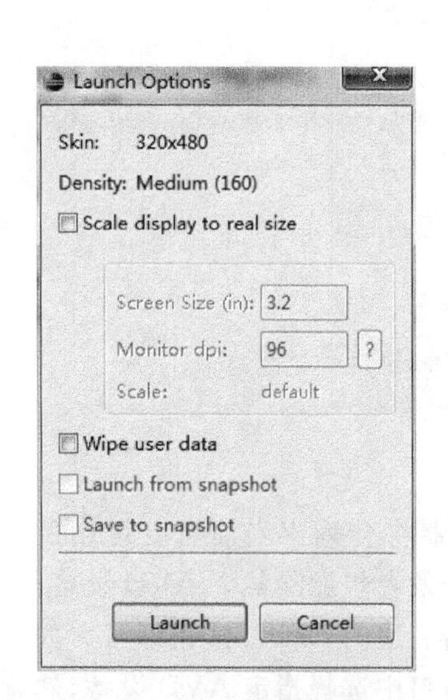

图 1.17　即将启动

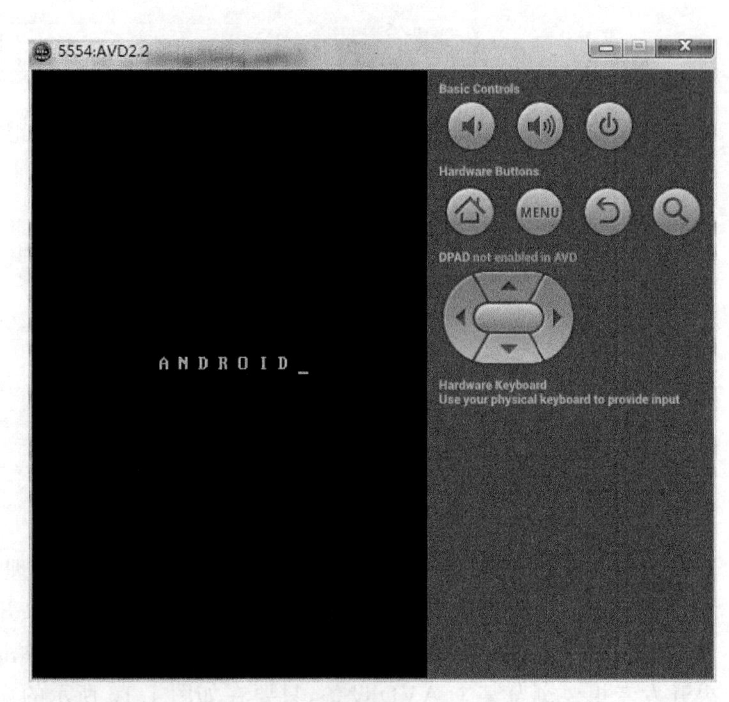

图 1.18　模拟器启动过程

2. 在命令行管理 AVD

在命令行中管理 AVD，首先要配置好 Android 环境变量，具体方法和配置 Java 环境变量一样。以 Windows XP 平台为例：右击“我的电脑”，在弹出的快捷菜单中选择“属性”命令，在“高级”选项卡中选择“环境变量”，然后新建一个 ANDROID_HOME，将本机中的

图 1.19　模拟器启动完毕

SDK 全路径粘贴进去，然后在 Path 下面加入“.；%ANDROID_HOME%/platform-tools；%ANDROID_HOME%/tools”即可。这里要注意标点符号要用英文，同时不要忘了最前面的“.”。在配置好环境变量后，就可以通过命令行管理 AVD 了。首先，如果直接在命令行执行 android 命令将会启动 Android SDK 管理器。其他命令如下：

- android list——列出计算机上所有已经安装的 Android 版本和 AVD 设备。
- android list avd——列出计算机上所有已经安装的 AVD 设备。
- android list target——列出计算机上所有已经安装的 Android 版本。
- android create avd——创建一个 AVD 设备。
- android move avd——移动或重命名一个 AVD 设备。
- android delete avd——删除一个 AVD 设备。
- android update avd——升级一个 AVD 设备使之符合新的 SDK 环境。
- android create project——创建一个新的 Android 项目。
- android update project——更新一个已有的 Android 项目。
- android create test-project——创建一个新的 Android 测试项目。
- android update test-project——更新一个已有的 Android 测试项目。

要创建一个 AVD，使用 android create avd 命令，给出几个参数：要创建的 AVD 的名称以及要创建的 AVD 搭载的 Android 版本。当然还可以指定其他参数，例如，AVD 设备的保存位置、虚拟 SD 卡的大小、模拟器的皮肤。例如，android create avd -n < avd 名称> -t < Android 版本> -p < AVD 设备保存位置> -s <选择 AVD 皮肤>。

在上面创建 AVD 命令中，只有 AVD 名称以及 Android 版本两个参数是必填的，而如果不设置 AVD 设备的保存位置，则默认保存在“%ANDROID_HOME/.avd%”路径下。

例如，创建一个名为 AVD2.2，搭载的安卓版本为 Android 2.2 的模拟器设备，由于 Android 2.2 的代号为"8"，所以输入如下命令即可：

```
android create avd - n AVD2.2  - t android - 8
```

执行上面的命令，系统会提醒用户是否需要定制 AVD 的硬件，开发者可以选择 yes 或 no。如果选择 no，即可直接开始创建 AVD 设备；如果选择 yes，即可开始定制 AVD 硬件的各种选项，定制完成后系统开始创建 AVD 设备。

1.3.2 使用 Android 模拟器

Android 模拟器就是一台运行在计算机上的"虚拟设备"，实际上前面我们已经使用过 Android 模拟器了，在 AVD 管理器中选中指定 AVD 设备，然后单击 Start 按钮就是启动模拟器来运行 Android 系统。

在 Android SDK 安装目录的 tools 子目录下有一个 emuLator. exe 文件，它就是 Android 模拟器。这个模拟器做得十分出色，几乎可以模拟真实手机的绝大部分功能，后面会陆续看到——当然它只是模拟，不要指望用模拟器与你现实中的朋友"煲电话粥"。

使用 emulator. exe 启动模拟器有两种用法：

- emulator -avd < AVD 名称>
- emulator -data 镜像文件名称

第一种用法是运行指定的 AVD 设备，例如如下命令：

```
emulator  - avd AVD2.2       //运行名为 AVD2.2 的 AVD 设备
```

第二种用法是直接使用指定镜像文件来运行 AVD，例如如下命令：

```
emulator  - data myfile      //以 myfile 作为镜像文件来运行 AVD 设备
```

1.4 开发第一个 Android 应用

前面已经介绍了 Android 系统的一些基本知识以及如何搭建开发 Android 应用的环境，包括 Android SDK 的安装和 ADT 插件安装，如何创建和使用模拟器。接下来，就可以开始进入第一个 Android 应用的开发了。其实，开发 Android 应用十分简单，Android 编程就是面向应用程序框架 API 编程，只要学习了 Android 的一般开发流程，就可以轻松开发具备丰富功能的应用了。

1.4.1 在 Eclipse 中开发第一个 Android 应用

Java 是 Android 官方推荐的用于开发 Android 应用的编程语言，而 Eclipse 是流行的编写 Java 代码的开发工具，也是最常用的软件开发工具，它可以很好地提高开发者的开发效率。使用 Eclipse 开发 Android 应用大致需要如下三步：

① 创建一个 Android 项目。

② 在 XML 布局文件中定义应用程序的用户界面。

③ 在 Java 代码中编写业务实现。

上面三个步骤是最基本的归纳，下面以一个 HelloWorld 级别的应用来介绍开发 Android 应用的一般流程。详细步骤如下：

① 在 Eclipse 主菜单单击 File→New→Other 菜单项，弹出如图 1.20 所示的窗口，在出现的列表中展开 Android 目录，选择 Android Application Project。

图 1.20　新建 Android 项目

② 单击 Next 按钮，出现如图 1.21 所示的对话框，在该对话框中填写应用名称、工程名称、默认包名，选择 Android 版本号，然后一直单击 Next 按钮，最后单击 Finish 按钮。

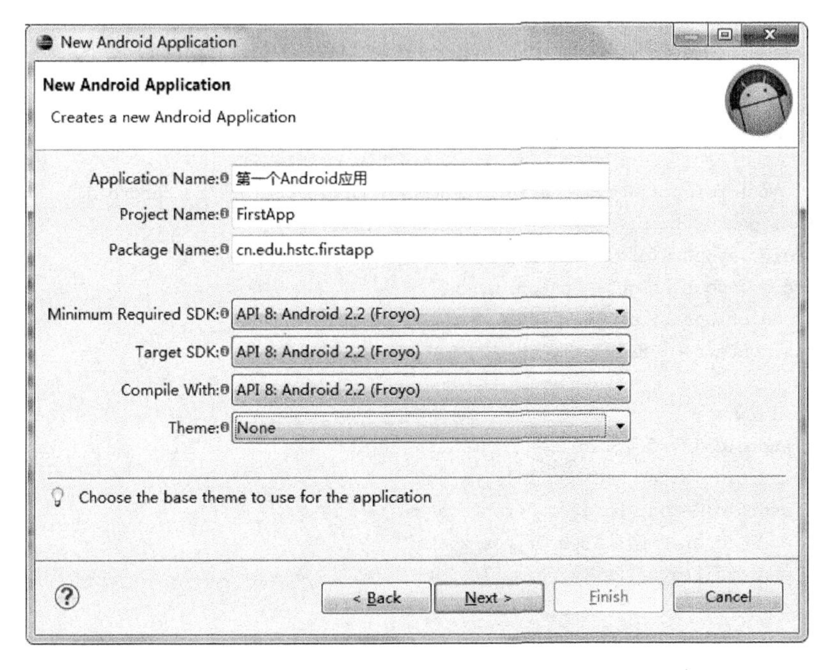

图 1.21　填写相关项

③ 单击 Finish 按钮，即成功创建了一个 Android 项目。创建完成后可以看到如图 1.22 所示的项目结构。

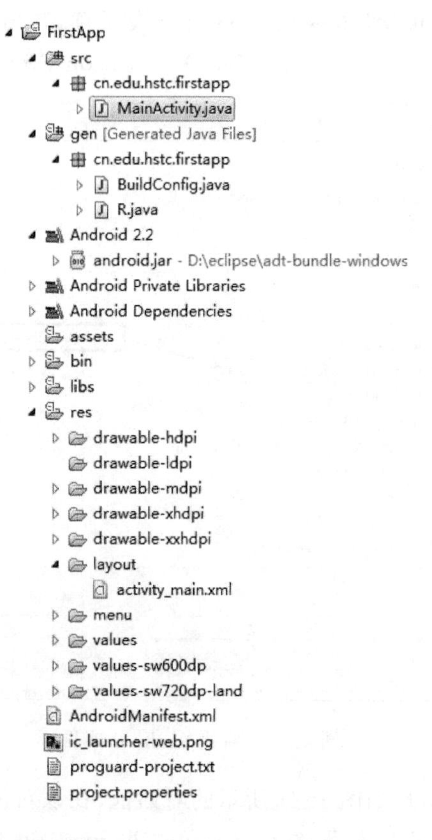

图 1.22　Android 项目结构

④ 可以看到，在如图 1.22 所示的 Android 项目结构中，layout 目录下有一个 activity_main.xml 文件，该文件用于定义 Android 应用的用户界面。双击该文件，将看到该文件中的代码，对代码稍作修改，修改后的文件内容如下所示。

```
<LinearLayout xmlns:android = "http://schemas.android.com/apk/res/android"
    xmlns:tools = "http://schemas.android.com/tools"
    android:layout_width = "match_parent"
    android:layout_height = "match_parent"
    android:orientation = "vertical"
    tools:context = ".MainActivity" >

    <TextView
        android:id = "@ + id/txt_title"
        android:layout_width = "fill_parent"
        android:layout_height = "wrap_content"
        android:gravity = "center_horizontal"
        android:text = "@string/hello_world" />

    <EditText
        android:id = "@ + id/edt_show"
```

```
        android:layout_width = "fill_parent"
        android:layout_height = "wrap_content" />

    < Button
        android:id = "@ + id/btn_confirm"
        android:layout_width = "fill_parent"
        android:layout_height = "wrap_content"
        android:text = "@string/confirm" />

</LinearLayout >
```

代码文件: codes\01\1.4\FirstApp\res\layout\activity_main.xml

这里对上面的 XML 文件作一个简单的介绍。

- LinearLayout：xml 文档的根元素，代表了一个线性布局，设置 orientation 属性为 vertical，将该界面布局里包含的 UI 控件按顺序从上至下排放。
- TextView：代表了一个文本框，用于显示文本，有点类似于 HTML 中的 label 标签。layout_width 属性与 layout_height 属性设置为 wrap_content，指定该控件的宽度与高度都是“包含内容”，取决于它所包裹的内容的大小，只要宽度与高度能包裹所含内容即可。text 属性则设置了该控件上所显示的文字。id 属性指定了该控件的唯一标识，在 Java 程序中可通过 findViewByld("id")来获取指定的 Android 界面组件。
- EditText：代表了一个文本编辑框，供用户输入文本，有点类似于 HTML 中的 text 标签。layout_width 属性设置为 fill_parent，指定该控件的宽度为占满了父容器所具有的宽度，这里为屏幕宽度。其他属性参考 TextView 控件。
- Button：代表了一个普通按钮，供用户单击操作，有点类似于 HTML 中的 button 标签。该控件所设置的 layout_width、layout_height、text、id 等通用属性所起作用参考上面三个组件，此处不再赘述。

在这里有一个小小的建议，就是当需要为一个控件指定 id 属性时，将该属性写在所有属性的后面，即放在该组件标签元素的最后一行，这样当你在 Java 代码中需要通过 findViewByld("id")方法来获取该控件，而又忘了该控件 id 时，可以方便快速地定位到该控件的 id 属性。

⑤ Android 项目结构中的 src 目录下存放着 Android 项目的源代码，该目录下的 cn\edu\hstc\firstapp 目录下有一个 MainActivity. java 文件，它就是 Android 项目的 Java 文件，用于控制 FirstApp 项目的业务实现。双击该文件，将其中的源代码如下所示进行修改。

```
package cn.edu.hstc.firstapp;

import android.app.Activity;
import android.os.Bundle;
import android.view.View;
import android.widget.Button;
import android.widget.EditText;
import android.widget.TextView;

public class MainActivity extends Activity {
```

```
    /**
     * 声明布局文件中的各个组件
     */
    private TextView title;
    private EditText show;
    private Button confirm;

    @Override
    protected void onCreate(Bundle savedInstanceState) {
        super.onCreate(savedInstanceState);
        setContentView(R.layout.activity_main);
        initView();
    }

    /**
     * 初始化界面控件
     */
    private void initView() {
    //通过 ID 获取 TextView 组件
    title = (TextView) this.findViewById(R.id.txt_title);
    //通过 ID 获取 EditText 组件
    show = (EditText) this.findViewById(R.id.edt_show);
    //通过 ID 获取 Button 组件
    confirm = (Button) this.findViewById(R.id.btn_confirm);
    //为 Button 控件绑定一个单击事件监听器
    confirm.setOnClickListener(new View.OnClickListener() {
            @Override
            public void onClick(View arg0) {
                //将 EditText 中所填入的内容显示在 TextView 中
                title.setText(show.getText().toString().trim());
            }
        });
    }
}
```

代码文件：codes\01\1.4\FirstApp\src\cn\edu\hstc\firstapp\MainActivity.java

至此，这个 HelloWorld 级别的 Android 应用已经开发完成。

在上面所介绍的步骤中，显然，Android 将用户界面交给 XML 文档来定义，而 Java 程序则专门负责业务实现，这样降低了程序的耦合性。这与读者熟悉的 MVC 设计模式也是有所相似的。其实我们可以将 XML 界面文件看成是一个 HTML 页面文件，只不过 Android 界面文件遵循 XML 文档格式，并使用 Android 标签，而 HTML 页面文件使用 HTML 的内置标签。

1.4.2 通过模拟器运行 Android 应用

想要将 Android 应用运行在模拟器上十分简单，只要按照如下步骤操作即可。

① 参照 1.3.2 节中的介绍，将指定 AVD 设备启动起来。

② 在 Eclipse 的包浏览器中选中需要运行的 Android 项目，然后右击，在弹出的菜单中

单击 Run As→Android Application 命令，如图 1.23 所示，等待片刻，即可看到该 Android 应用已经部署到模拟器上。

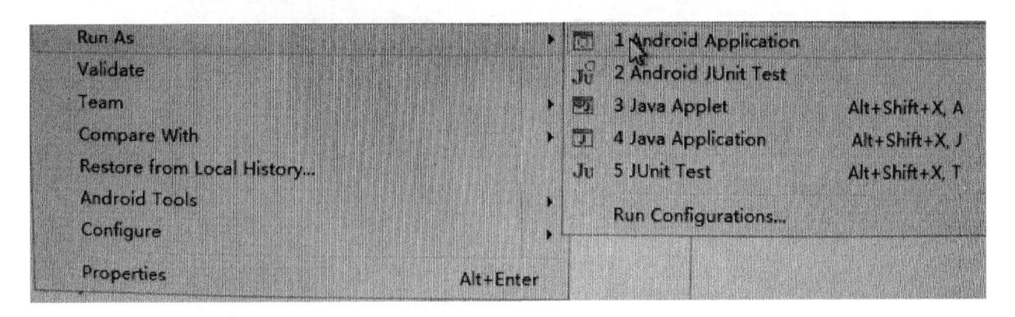

图 1.23　运行 Android 应用

完成第二步操作后，稍等片刻，就可以在原先启动的那台 AVD 设备中看到如图 1.24 所示的 Android 应用部署完成后的界面。

图 1.24　部署 Android 应用

在如图 1.24 所示界面的文本编辑框中输入“Hello Android”，然后单击“单击确定”按钮，即可看到图中的“我的第一个 Android 应用”变成了“Hello Android”，如图 1.25 所示。

如果想要在图 1.25 中的文本编辑框中输入中文，则可以进行以下步骤。

① 将鼠标指针放于图 1.25 中的文本编辑框上，长按鼠标左键，出现如图 1.26 所示的界面。

② 在如图 1.26 所示的界面中，单击 Input method 选项，出现如图 1.27 所示界面，选中“谷歌拼音输入法”选项，返回如图 1.25 所示的界面，此时就可以在文本编辑框中输入中文了。

图 1.25 运行 Android 应用

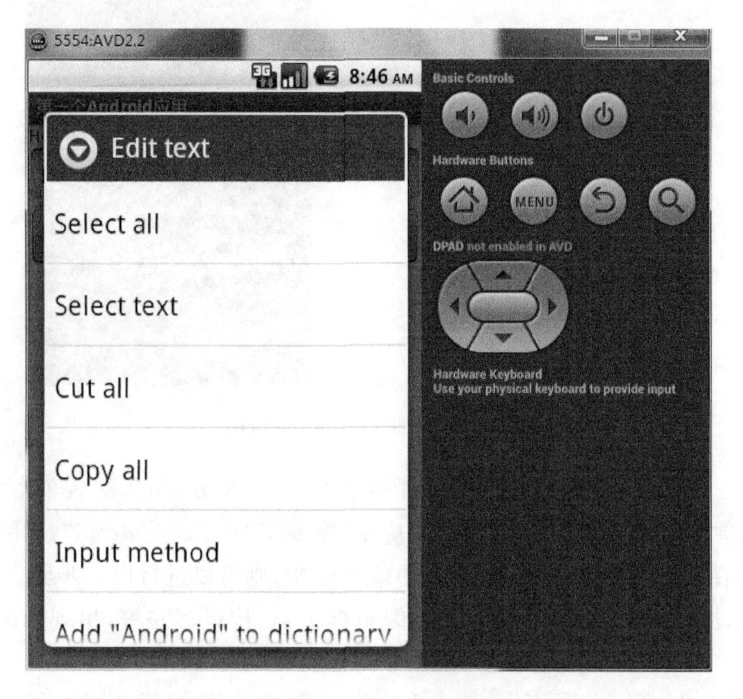

图 1.26 输入法设置

图 1.27　选择输入法

1.5 Android 应用程序架构

现在，读者应该已经学会如何新建一个 Android 项目以及开发 Android 项目的流程了，即在 XML 文件中定义用户界面，然后打开 Java 源代码编写业务实现。但是，读者也许现在对上面的 Java 源文件的代码还不是很理解，例如"findViewById(R. id. showTextView)；"代码中的 R. id. showTextView 参数是从何而来等等，对 Android 应用程序的项目结构也存在疑惑。

实际上，这些问题的答案并不太难，接下来对如图 1.22 所示的 Android 项目应用结构做一个简单介绍。

src 目录只是一个普通的、存放 Java 源文件的目录，这里不做详细介绍，下面介绍的是除 src 以外的其他目录结构。

1.5.1 自动生成的 gen 目录

gen 目录中存放所有由 Android 开发工具自动生成的文件。该目录中最重要的就是 R. java 文件。这个文件是由 Android 开发工具 AAPT 工具根据应用中的资源文件自动产生的。Android 开发工具会自动根据你放入 res 目录的 XML 界面文件、图标与常量，同步更新修改 R. java 文件。正因为 R. java 文件是由开发工具自动生成的，所以应避免手工修改 R. java。R. java 在应用中起到了资源字典的作用，它包含了界面、图标、常量等各种资源的 id，通过 R. java，应用可以很方便地找到对应资源。另外编译器也会检查 R. java 列表中的资源是否被使用到，没有被使用到的资源不会编译进软件中，这样可以减少应用在手机中占用的空间。打开 gen/cn/edu/hstc/firstapp 目录下的 R.java 文件，可以看到如下代码。

```
package cn.edu.hstc.firstapp;

public final class R {
    public static final class attr {
    }
    public static final class dimen {
        public static final int activity_horizontal_margin = 0x7f040000;
        public static final int activity_vertical_margin = 0x7f040001;
    }
    public static final class drawable {
        public static final int ic_launcher = 0x7f020000;
    }
    public static final class id {
        public static final int action_settings = 0x7f080003;
        public static final int btn_confirm = 0x7f080002;
        public static final int edt_show = 0x7f080001;
        public static final int txt_title = 0x7f080000;
    }
    public static final class layout {
        public static final int activity_main = 0x7f030000;
    }
    public static final class menu {
        public static final int main = 0x7f070000;
    }
    public static final class string {
        public static final int action_settings = 0x7f050001;
        public static final int app_name = 0x7f050000;
        public static final int confirm = 0x7f050003;
        public static final int hello_world = 0x7f050002;
    }
    public static final class style {
        public static final int AppBaseTheme = 0x7f060000;
        public static final int AppTheme = 0x7f060001;
    }
}
```

代码文件：codes\01\1.4\FirstApp\gen\cn\edu\hstc\firstapp\R.java

从上面的源代码文件可以看到：

- 在 R 类中，针对各种资源生成了对应的内部类。例如界面布局资源对应 layout 内部类；字符串资源对应 string 内部类；标识符资源对应 id 内部类。
- 每个内部类中以一个 public static final int 类型的 Field 来对应每个具体的资源项。例如前面在 activity_main.xml 文件中设置了 Button 控件的 id 属性为 confirmButton 以及设置了 EditText 控件的 id 属性为 showEditText，因此 R.id 类中就定义了这两个 Field；由于我们有一个名为 activity_main.xml 的布局文件，所以在 R.layout 类中就有了 activity_main 的 Field。

1.5.2 资源目录 res

Android 应用的 res 目录中存放了包括 XML 界面文件、图片资源、字符串资源、颜色资源、尺寸资源等在内的 Android 应用所用的全部资源。

通过展开 Android 项目的结构目录中的 res 目录，可以看到 layout、values、drawable 等资源文件夹，这些文件夹对不同的资源进行了分类，这样可以方便地让 AAPT 工具来扫描这些资源，并在 R.java 文件中为它们生成对应的内部类。下面对各个资源文件夹做一个简单的介绍。

- res/drawable：专门存放 png、jpg 等图片文件。在 Java 代码中使用 getResources().getDrawable(resourceId)获取该目录下的资源。
- res/layout：专门存放 XML 界面文件，XML 界面文件和 HTML 文件一样，主要用于显示用户操作界面。
- res/values：专门存放应用使用到的各种类型数据。不同类型的数据存放在不同的文件中，如下：

① strings.xml——定义字符串和数组，在 Activity 类中使用 getResources().getString(resourceId)或 getResources().getText(resourceId)来取得字符串资源。它的作用和 struts 中的国际化资源文件一样。一份 strings.xml 文件的内容如下：

```
<?xml version = "1.0" encoding = "utf - 8"?>
< resources >
    < string name = "app_name"> FirstApp </string >
    < string name = "action_settings"> Settings </string >
    < string name = "hello_world">我的第一个安卓应用</string >
    < string name = "confirm">确定</string >
</resources >
```

代码文件：codes\01\1.4\FirstApp\res\values\strings.xml

在 Java 代码中使用 getResources().getString(R.string.app_name)或 getResources().getText(R.string.app_name)来获取 key 为 app_name 的值。

在 XML 中使用@string/app_name 来获取 key 为 app_name 的值。

② arrays.xml：定义数组。一份 arrays.xml 文件的内容如下：

```
<?xml version = "1.0" encoding = "utf - 8"?>
< resources >
    < string - array name = "colors">
        < item > red </item >
        < item > yellow </item >
        < item > green </item >
        < item > blue </item >
    </string - array >
</resources >
```

代码文件：codes\01\1.4\FirstApp\res\values\arrays.xml

③ colors.xml：定义颜色和颜色字串数值，可以在 Activity 中使用 getResources().getDrawable(resourceId)以及 getResources().getColor(resourceId)取得这些资源。一份

colors.xml 文件的内容如下：

```
<?xml version="1.0" encoding="utf-8"?>
<resources>
    <color name="contents_text">#ff0000</color>
</resources>
```

代码文件：codes\01\1.4\FirstApp\res\values\colors.xml

④ dimens.xml：定义尺寸数据，在 Activity 中使用 getResources().getDimension(resourceId)取得这些资源。一份 dimens.xml 文件的内容如下：

```
<resources>
    <!-- Default screen margins, per the Android Design guidelines. -->
    <dimen name="activity_horizontal_margin">16dp</dimen>
    <dimen name="activity_vertical_margin">16dp</dimen>
</resources>
```

代码文件：codes\01\1.4\FirstApp\res\values\dimens.xml

⑤ styles.xml：定义样式。一份 styles.xml 文件的内容如下：

```
<resources>
    <style name="AppBaseTheme" parent="android:Theme.Light">
        <item name="android:textSize">18sp</item>
        <item name="android:textColor">#0066FF</item>
    </style>

    <style name="AppTheme" parent="AppBaseTheme">
    </style>
</resources>
```

代码文件：codes\01\1.4\FirstApp\res\values\styles.xml

- res/anim：存放定义动画的 XML 文件。
- res/xml：在 Activity 中使用 getResources().getXML()读取该目录下的 XML 资源文件。
- res/raw：该目录用于存放应用使用到的原始文件，如音效文件等。编译软件时，这些数据不会被编译，它们被直接加入到程序安装包里。为了在程序中使用这些资源，可以调用 getResources().openRawResource(resourcesId)来获取资源。

Android 除了提供 res 目录存放资源文件外，assets 目录也可以存放资源文件，而且 assets 目录下的资源文件不会在 R.java 中自动生成 ID，所以读取 assets 目录下的文件必须指定文件的路径，如：file:///android_asset/xxx.3gp。

1.5.3 项目清单文件：AndroidManifest.xml

AndroidManifest.xml 清单文件是每个 Android 项目都必需的，它是整个 Android 应用的全局描述文件。这个文件列出了应用程序所提供的功能，以后开发好的各种组件需要在该文件中进行配置，如果应用使用到了系统内置的应用(如电话服务、互联网服务、短信服务、GPS 服务等等)，还需在该文件中声明使用权限。

AndroidManifest.xml 清单文件说明了该应用的名称、所使用图标以及包含的组件等，一份 AndroidManifest. xml 文件的内容如下：

```
<?xml version = "1.0" encoding = "utf - 8"?>
<! -- 指定该 Android 应用的唯一包名,该包名可唯一表示该应用 -->
< manifest xmlns:android = "http://schemas.android.com/apk/res/android"
    package = "cn.edu.hstc.firstapp"
    android:versionCode = "1"
    android:versionName = "1.0" >

    < uses - sdk
        android:minSdkVersion = "18"
        android:targetSdkVersion = "18" />

    <! -- 指定 Android 应用标签、图标 -->
    < application
        android:allowBackup = "true"
        android:icon = "@drawable/ic_launcher"
        android:label = "@string/app_name"
        android:theme = "@style/AppTheme" >
        <! -- 为该应用定义一个 Activity 组件,并指定该 Activity 的标签 -->
        < activity
            android:name = "cn.edu.hstc.firstapp.MainActivity"
            android:label = "@string/app_name" >
            < intent - filter >
                <! -- 指定该 Activity 是程序的入口 -->
                < action android:name = "android.intent.action.MAIN" />
                <! -- 指定加载该应用时运行该 Activity -->
                < category android:name = "android.intent.category.LAUNCHER" />
            </intent - filter >
        </activity >
    </application >

</manifest >
```

代码文件：codes\01\1.4\FirstApp\AndroidManifest.xml

由以上文件内容可以看出，AndroidManifest. xml 清单文件通常包含如下信息：

➢ 应用程序的包名，该包名将会作为该应用的唯一标识。
➢ 应用程序所包含的组件，如 Activity、Service、BroadcastReceiver 和 ContentProvider 等。
➢ 应用程序兼容的最低版本。
➢ 应用程序使用系统所需的权限声明。
➢ 其他程序访问该程序所需的权限声明。

随着不断地进行开发，可能需要对 AndroidManifest. xml 清单文件进行适当的修改。

1.5.4 声明应用程序使用权限

上面提到，如果应用使用到了系统内置的应用(如电话服务、互联网服务、短信服务、GPS 服务等)，还需在该文件中声明使用权限；一个应用也可能被其他应用调用，因此也需要声明调用自身所需要的权限。

1. 声明该应用自身所拥有的权限

通过为< manifest.../>元素添加< uses-perrnission.../>子元素即可为自身声明权限。例如，在< manifest.../>元素中添加如下代码：

```
<! -- 声明该应用本身即有打电话的权限 -->
< uses - permission android:name = "android.permission.CALL_PHONe"/>
```

2. 声明调用该应用自身所需的权限

通过为应用的各组件元素，如< activity.../>元素添加< uses-permission.../>子元素即可声明调用该程序所需的权限。

例如，在< activity.../>元素中添加如下代码：

```
<! -- 声明该应用本身即有打电话的权限 -->
< uses - permission android:name = "android.permission.SEND_SMS"/>
```

下面给出一些 Android 系统常用权限，这些权限都可以通过 Android 官方文档查看到。

- 开机自动允许——android. permission. RECEIVE_BOOT_COMPLETED，允许程序开机自动运行。
- 电量统计——android. permission. BATTERY_STATS，获取电池电量统计信息。
- 使用蓝牙——android. permission. BLUETOOTH，允许程序连接配对过的蓝牙设备。
- 蓝牙管理——android. permission. BLUETOOTH_ADMIN，允许程序发现和配对新的蓝牙设备。
- 收到短信时广播——android. permission. BROADCAST_SMS，当收到短信时触发一个广播。
- 拨打电话——android. permission. CALL_PHONE，允许程序从非系统拨号器里输入电话号码。
- 拍照权限——android. permission. CAMERA，允许访问摄像头进行拍照。
- 安装应用程序——android. permission. INSTALL_PACKAGES，允许程序安装应用。
- 修改声音设置——android. permission. MODIFY_AUDIO_SETTINGS，修改声音设置信息。
- 修改电话状态——android. permission. MODIFY_PHONE_STATE，修改电话状态，如飞行模式，但不包含替换系统拨号器界面。
- 读取日程提醒——android. permission. READ_CALENDAR，允许程序读取用户的日程信息。
- 读取联系人——android. permission. READ_CONTACTS，允许应用访问联系人通讯录信息。
- 屏幕截图——android. permission. READ_FRAME_BUFFER，读取帧缓存用于屏幕截图。

- 读取短信内容——android. permission. READ_SMS，读取短信内容。
- 接收彩信——android. permission. RECEIVE_MMS，接收彩信。
- 接收短信——android. permission. RECEIVE_SMS，接收短信。
- 录音——android. permission. RECORD_AUDIO，录制声音通过手机或耳机的麦克。
- 发送短信——android. permission. SEND_SMS，发送短信。
- 设置闹铃提醒——com. android. alarm. permission. SET_ALARM，设置闹铃提醒。
- 设置系统时间——android. permission. SET_TIME，设置系统时间。
- 使用振动——android. permission. VIBRATE，允许振动。
- 写入联系人——android. permission. WRITE_CONTACTS，写入联系人，但不可读取。
- 写入外部存储——android. permission. WRITE_EXTERNAL_STORAGE，允许程序写入外部存储即SD卡上写文件。
- 编写短信——android. permission. WRITE_SMS，允许编写短信。
- 访问网络——android. permission. INTERNET，允许访问网络。
- 修改文件系统——android. permission. MOUNT_UNMOUNT_FILESYSTEMS，允许创建修改或删除文件。

1.6 Android应用的基本组件介绍

一个或多个基本组件组成了丰富多彩的Android应用。前面提到的Activity就是Android四大组件之一，其他三个为Service（服务）、ContentProvider（内容提供者）、BroadcastReceiver（广播接收器）。实际上，Android并不只有这四大组件，还有其他的一些组件，这些组件才构成了强大的功能丰富的Android应用。下面先对这些组件做一个简单的介绍，后面章节会进一步详细介绍每一个组件。

1.6.1 Activity

应用程序中，一个Activity通常就是一个单独的屏幕，作为与用户交互的组件，它上面可以显示一些控件，也可以监听和处理用户的事件并做出响应。Activity通过setContentView()显示指定的组件。该方法可以接收一个View对象作为参数，也可以接收一个布局资源id作为参数。但通常采用后者。图1.28分别采用了这两种方式设置了Activity中所显示的View。

```
//创建一个线性布局
LinearLayout layout = new LinearLayout(this);
//接收一个View对象，设置Activity显示该layout
Super.setContentView(layout);
//接收一个布局资源id，设置Activity显示activity_main.xml文件中定义的View
setContentView(R.layout.activity_main);
```

图1.28 显示Activity

Activity 组件继承 Activity 基类并且有自己的生命周期，这将在后面继续做深入的介绍。多个 Activity 构成了 Android 应用的 Activity 栈，当前活动的 Activity 位于栈顶。

1.6.2 Service

与 Activity 组件继承 Activity 基类相似，Service 组件需要继承 Service 基类。不同的是，Service 没有自己的用户界面，通常位于后台运行，用于开发监控类程序，为其他组件提供后台服务或监控其他组件的运行状态。

服务不能自己运行，需要通过 Contex. startService()或 Contex. bindService()启动服务，服务一旦启动，便有了自己独立的生命周期。

1.6.3 BroadcastReceiver

顾名思义，BroadcastReceiver 代表广播接收者。你的应用可以使用它对外部事件进行过滤——只对感兴趣的外部事件(如当电话呼入时，或者数据网络可用时)进行接收并做出响应。广播接收器没有用户界面，但它们可以启动一个 activity 或 serice 来响应它们收到的信息，或者用 NotificationManager 来通知用户。

开发自己的 BroadcastReceiver 的步骤如下：

① 写一个继承 BroadCastReceiver 的类，重写 onReceive(Context context, Intent intent)方法，广播接收器仅在它执行这个方法时处于活跃状态。当 onReceive()返回后，它即为失活状态。注意：为了保证用户交互过程的流畅，一些费时的操作要放到线程里，如 SMSBroadcastReceiver。

② 注册该广播接收者，注册有两种方法：程序动态注册和在 AndroidManifest 文件中进行静态注册(可理解为系统中注册)，如图 1.29 和图 1.30 所示。

```
<receiver android:name=".SMSBroadcastReceiver" >
    <intent-filter android:priority = "2147483647" >
        <action android:name="android.provider.Telephony.SMS_RECEIVED" />
    </intent-filter>
</receiver >
```

图 1.29　静态注册

```
IntentFilter intentFilter=new IntentFilter("android.provider.Telephony.SMS_RECEIVED");
registerReceiver(mBatteryInfoReceiver ,intentFilter);

//反注册
unregisterReceiver(receiver);
```

图 1.30　动态注册

- 静态注册，在 AndroidManifest. xml 文件中使用< receiver.../>元素完成注册。下面的 priority 属性表示接收广播的级别，"2147483647"为最高优先级。
- 动态注册，在 Java 代码中通过 Context. registReceiver()方法注册该 BroadcastReceiver。

广播有三种类型：普通广播，通过 Context. sendBroadcast()发送；有序广播，通过 Context. sendOrderedBroadcast()发送；异步广播，通过 Context. sendStickyBroadcast()发

送。不管是何种广播类型,如果 BroadcastReceiver 也对该消息"感兴趣"(通过 IntentFilter 配置),则 BroadcastReceiver 的 onReceive(Context context, Intent intent)方法将会被触发。

1.6.4　ContentProvider

Android 平台提供了 ContentProvider 使一个应用程序的指定数据集提供给其他应用程序。这些数据可以存储在文件系统、一个 SQLite 数据库中,或以任何其他合理的方式存储,其他应用可以通过 ContentResolver 类从该内容提供者中获取或存入数据。一个应用程序使用 ContentProvider 暴露自己的数据,而另一个应用程序则通过 ContentResolver 来访问数据。

只有需要在多个应用程序间共享数据时才需要内容提供者。例如,我们开发了一个发送短信的程序,当发送短信时需要从联系人管理应用中读取指定联系人的数据——这就需要多个应用程序之间进行实时的数据交换。

Android 系统为这种跨应用的数据交换提供了一个标准:ContentProvider。当用户实现自己的 ContentProvider 时,需要实现如下抽象方法。

➢ insert(Uri, ContentValues)——向 ContentProvider 插入数据。
➢ delete(Uri, ContentValues)——删除 ContentProvider 中指定数据。
➢ udpate(Uri, ContentValues, String, String[])——更新 ContentProvider 中指定数据。
➢ query(Uri, String[], String, String[], String)——从 ContentProvider 查询数据。

1.7　本章小结

本章简要介绍了 Android 应用开发的背景知识,包括什么是 Android、Android 的平台架构。读者阅读本章需要掌握的重点是如何搭建 Android 的开发环境,包括下载、安装、使用 ADT 工具;下载和安装 Android SDK、如何使用 Android 模拟器。这些内容是开发 Android 应用的基础。除此之外,本章还介绍了一个 Android 的 Hello World 级别的应用,通过该应用向读者介绍开发 Android 应用的一般流程并向读者分析了 Android 应用的目录结构。最后,本章简单介绍了 Android 的四大组件,让读者对这四大组件有了一个基本了解。通过阅读本章,读者应该已经具备了开发 Android 应用的一些基本知识了。

第2章 UI的各种事件控制

2.1 基于监听的事件响应

在认识各种 Android UI 控件之前，我们先来学习对程序界面上执行的各种操作作出响应，这些响应都是通过事件处理来完成的。实际上，在第 1 章介绍布局的时候，已经出现了部分 Android 的 UI 控件了。对于这些控件的详细介绍，将在第 3 章进行讲解。之所以将对 UI 的各种事件控制的介绍放在介绍 UI 控件之前，是因为在讲解各种控件的时候需要结合对具体控件的事件处理，如果不提前认识这些事件处理，就会加大学习的难度。学习事件处理时所出现的控件，大部分都是比较简单的控件，只起到显示界面的作用，难度并不大，读者在这一章只需知道有这个控件就可以了。

Android 提供了基于监听器的事件处理以及键盘事件、触摸屏事件等基于回调的事件处理模式。结合这两种事件处理机制，相信读者可以开发出更多丰富多彩的功能。下面首先介绍 Android 基于监听机制的事件处理模式。

2.1.1 第一种响应方法

这里将用一个实例来介绍第一种响应方法，先看以下程序界面代码。

```
<LinearLayout xmlns:android="http://schemas.android.com/apk/res/android"
    xmlns:tools="http://schemas.android.com/tools"
    android:layout_width="match_parent"
    android:layout_height="match_parent"
    android:background="@color/azure"
    android:gravity="center"
    android:orientation="vertical">

    <Button
        android:id="@+id/btn_add"
        android:layout_width="wrap_content"
        android:layout_height="wrap_content"
        android:text="+" />

    <TextView
```

```
        android:id = "@ + id/txt_show"
        android:layout_width = "wrap_content"
        android:layout_height = "wrap_content"
        android:text = "我会变大"
        android:textSize = "12sp"
        android:textStyle = "bold" />

    < Button
        android:id = "@ + id/btn_min"
        android:layout_width = "wrap_content"
        android:layout_height = "wrap_content"
        android:text = " - " />

</LinearLayout >
```

代码文件：codes\02\2.1\FirstMethod\res\layout\activity_main.xml

上面的界面文件非常简单，只是在一个线性布局中添加了一个 TextView 控件，用来显示文本，然后添加了两个 Button 按钮，用来与用户进行交互，三个控件自上而下排列。接下来的程序代码中，分别为两个 Button 按钮添加了单击事件处理。

```
package cn.edu.hstc.firstmethod;

import android.app.Activity;
import android.os.Bundle;
import android.view.View;
import android.widget.Button;
import android.widget.TextView;

public class MainActivity extends Activity {
    /**
     * 声明布局文件中的各个组件
     */
    private TextView show;
    private Button btnAdd, btnMin;

    @Override
    protected void onCreate(Bundle savedInstanceState) {
        super.onCreate(savedInstanceState);
        setContentView(R.layout.activity_main);
        initView();
    }

    private void initView() {
        /**
         * 加载布局文件中的各个组件
         */
        show = (TextView) this.findViewById(R.id.txt_show);
        btnAdd = (Button) this.findViewById(R.id.btn_add);
        btnMin = (Button) this.findViewById(R.id.btn_min);

        btnAdd.setOnClickListener(new View.OnClickListener() {
                @Override
                public void onClick(View arg0) {
```

```
                float fontSize = show.getTextSize();
                if (fontSize < 36) {
                    show.setText("我会变大或变小");
                    show.setTextSize(++fontSize);
                    btnMin.setEnabled(true);
                    if (fontSize == 36) {
                        show.setText("我不会再变大了");
                        btnAdd.setEnabled(false);
                    }
                }
            }
        });

        btnMin.setOnClickListener(new View.OnClickListener() {
            @Override
            public void onClick(View arg0) {
                float fontSize = show.getTextSize();
                if (fontSize > 12) {
                    show.setText("我会变大或变小");
                    show.setTextSize(--fontSize);
                    btnAdd.setEnabled(true);
                    if (fontSize == 12) {
                        show.setText("我不会再变小了");
                        btnMin.setEnabled(false);
                    }
                }
            }
        });
    }
}
```

代码文件: codes\02\2.1\FirstMethod\ src\cn\edu\hstc\firstmethod\activity\MainActivity.java

在上面的程序代码中,实现了单击"＋"按钮,界面中的文本字号变大;单击"-"按钮,界面中的文本字号变小。当字号大小大于12并小于36时,文本内容为"我会变大或变小",并且两个按钮皆可用;当字号等于12时,文本内容为"我不会再变小了",并且"-"按钮不可用;当字号大小为36时,文本内容为"我不会再变大了",并且"＋"按钮不可用。运行效果如图2.1所示。

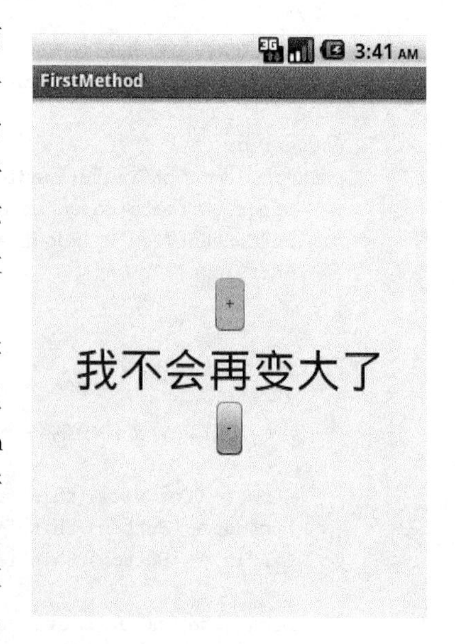

图2.1　控制单击屏幕字号大小

从代码文件MainActivity.java中可以看出,程序通过findViewById句柄获得界面中的各个控件,这里使用的是R.id.text、R.id.btnAdd以及R.id.btnMin作为该句柄的参数,即传入参数为界面中各个控件所对应的ID。

接着,程序通过setOnClickListener方法为界面文件中的两个Button按钮分别设置了单击事件监听器,这个方法的参数实际上是一个View.OnClickListener

类型的接口，这个接口需要 onClick 方法，在该方法中，则是实现事件处理的具体响应。为控件注册监听器其实是一种委托的机制。普通控件将整个事件处理委托给监听器，当指定事件发生时，就通知所委托的监听器去处理这个事件。

综上，为控件添加事件监听器，有以下两个编程要点：

➢ 使用 findViewById()获取 XML 布局文件中的控件。

➢ 使用 setOnXXXListener()为控件添加事件处理监听器。

在获取控件时需要强制转换成相应的控件类型，如上面 MainActivity. java 中的"btnAdd = (Button) this. findViewById(R. id. btnAdd);"将基础类型强制转换为 Button 类型。

实际上，在本节所介绍的这种为控件设置事件处理监听器的方式为采用匿名内部类作为事件监听器类的方式，即第一种响应方式。接下来的小节将为读者介绍第二种响应方式。

2.1.2 第二种响应方法

用户操作界面中的 UI 控件，除了上面所述的第一种响应方法之外，还有其他几种响应方法，本节介绍第二种响应方法，即 Activity 本身直接作为事件监听器。如下源代码中，在同样的布局文件和应用程序下实现同样的功能。

```
package cn.edu.hstc.secondmethod;

import android.app.Activity;
import android.os.Bundle;
import android.view.View;
import android.view.View.OnClickListener;
import android.widget.Button;
import android.widget.TextView;

public class MainActivity extends Activity implements OnClickListener {
    /**
     * 声明布局文件中的各个组件
     */
    private TextView show;
    private Button btnAdd, btnMin;

    @Override
    protected void onCreate(Bundle savedInstanceState) {
        super.onCreate(savedInstanceState);
        setContentView(R.layout.activity_main);
        //加载 TextView 组件
        show = (TextView) findViewById(R.id.txt_show);
        //加载增加按钮
        btnAdd = (Button) findViewById(R.id.btn_add);
        //加载减少按钮
        btnMin = (Button) findViewById(R.id.btn_min);
        //为增加按钮添加事件监听器
        btnAdd.setOnClickListener(this);
        //为减少按钮添加事件监听器
```

```
        btnMin.setOnClickListener(this);
    }

    @Override
    public void onClick(View v) {
        float fontSize = show.getTextSize();
        switch (v.getId()) {
        case R.id.btn_add:
            if (fontSize < 36) {
                show.setText("我会变大或变小");
                show.setTextSize(++fontSize);
                btnMin.setEnabled(true);
                if (fontSize == 36) {
                    show.setText("我不会再变大了");
                    btnAdd.setEnabled(false);
                }
            }
            break;
        case R.id.btn_min:
            if (fontSize > 12) {
                show.setText("我会变大或变小");
                show.setTextSize(--fontSize);
                btnAdd.setEnabled(true);
                if (fontSize == 12) {
                    show.setText("我不会再变小了");
                    btnMin.setEnabled(false);
                }
            }
            break;
        default:
            break;
        }
    }

}
```

代码文件：codes\02\2.1\SecondMethod\src\cn\edu\hstc\secondmethod\activity\MainActivity.java

上面的程序让Activity类实现了OnClickListener事件监听接口，并重写了事件处理器方法onClick(View v)，该方法内定义了具体的实现业务。当某个控件想要委托该实现了监听器接口的Activity作为事件处理器时，直接将this作为该控件注册监听器的方法的参数即可。如上面的程序代码“btnAdd.setOnClickListener(this);”，把当前Activity设置为“+”按钮的事件处理监听器。

为了保证不同的控件执行不同的操作，在实现onClick方法时，必须使用switch体，用各个控件的ID作为判断条件，实现不同的控件响应不同的业务。运行效果与图2.1一致，这里不再重复。

2.1.3 第三种响应方法

布局不变，本节采用第三种响应方法实现与前面实例同样的功能。实现代码如下：

```
package cn.edu.hstc.thirdmethod;

import android.app.Activity;
import android.os.Bundle;
import android.view.View;
import android.view.View.OnClickListener;
import android.widget.Button;
import android.widget.TextView;

public class MainActivity extends Activity {
    /**
     * 声明布局文件中的各个组件
     */
    private TextView show;
    private Button btnAdd, btnMin;

    @Override
    protected void onCreate(Bundle savedInstanceState) {
        super.onCreate(savedInstanceState);
        setContentView(R.layout.activity_main);
        //加载 TextView 组件
        show = (TextView) findViewById(R.id.txt_show);
        //加载增加按钮
        btnAdd = (Button) findViewById(R.id.btn_add);
        //加载减少按钮
        btnMin = (Button) findViewById(R.id.btn_min);
        //为增加按钮设置事件监听器
        btnAdd.setOnClickListener(new BtnAddOnClickLinstener());
        //为减少按钮设置事件监听器
        btnMin.setOnClickListener(new BtnMinOnClickLinstener());
    }

    private final class BtnAddOnClickLinstener implements OnClickListener {
        @Override
        public void onClick(View v) {
            float fontSize = show.getTextSize();
            if (fontSize < 36) {
                show.setText("我会变大或变小");
                show.setTextSize(++fontSize);
                btnMin.setEnabled(true);
                if (fontSize == 36) {
                    show.setText("我不会再变大了");
                    btnAdd.setEnabled(false);
                }
            }
        }
    }

    private final class BtnMinOnClickLinstener implements OnClickListener {
        @Override
        public void onClick(View v) {
```

```
                float fontSize = show.getTextSize();
                if (fontSize > 12) {
                    show.setText("我会变大或变小");
                    show.setTextSize( -- fontSize);
                    btnAdd.setEnabled(true);
                    if (fontSize == 12) {
                        show.setText("我不会再变小了");
                        btnMin.setEnabled(false);
                    }
                }
            }
        }

}
```

代码文件：codes\02\2.1\ThirdMethod\ src\cn\edu\hstc\thirdmethod\activity\MainActivity.java

在上述的代码中，定义了两个内部类，这两个内部类都实现了 View. OnClickListener 接口，并在 onClick 方法中定义了具体的业务。接着，为界面中的两个 Button 控件分别注册了事件处理监听器。如程序代码"btnAdd. setOnClickListener(new btnAddOnClickListener());"，将 btnAddOnClickListener 类的对象设置为 btnAdd 按钮的事件处理监听器。运行效果与第一种响应方法中的例子一致，此处不再赘述。

这种将内部类作为事件处理监听器类的方法，有两个好处：可以在当前类中复用该监听器类；该监听器类可以自由访问外部类中的所有界面组件。这两个优势其实也同样是上面的第一种响应方法（使用匿名内部类作为事件监听器类）以及第二种响应方法（使用当前 Activity 作为事件监听器类）所具备的。只是接下来第四种响应方法——使用外部类作为事件监听器类就没有同时具备这两个优势了。

2.1.4 第四种响应方法

在同样布局以及同一应用程序下，实现同样功能，如下代码中使用的是第四种事件响应处理方法。由于采用外部类作为事件监听器类，所以必须有一个实现了事件监听器接口 View. OnClickListener 的类。

```
package cn.edu.hstc.fourthmethod;

import android.view.View;
import android.view.View.OnClickListener;
import android.widget.Button;
import android.widget.TextView;

public class BtnOnClickLinstener implements OnClickListener {
    /**
     * 声明布局文件中的各个控件
     */
    private TextView show;
    private Button btnAdd, btnMin;
```

```
    BtnOnClickLinstener(TextView show, Button btnAdd, Button btnMin) {
        this.show = show;
        this.btnAdd = btnAdd;
        this.btnMin = btnMin;
    }

    @Override
    public void onClick(View v) {
        float fontSize = show.getTextSize();
        switch (v.getId()) {
        case R.id.btn_add:
            if (fontSize < 36) {
                show.setText("我会变大或变小");
                show.setTextSize(++fontSize);
                btnMin.setEnabled(true);
                if (fontSize == 36) {
                    show.setText("我不会再变大了");
                    btnAdd.setEnabled(false);
                }
            }
            break;
        case R.id.btn_min:
            if (fontSize > 12) {
                show.setText("我会变大或变小");
                show.setTextSize(--fontSize);
                btnAdd.setEnabled(true);
                if (fontSize == 12) {
                    show.setText("我不会再变小了");
                    btnMin.setEnabled(false);
                }
            }
            break;
        default:
            break;
        }
    }

}
```

代码文件：codes\02\2.1\FourthMethod\src\cn\edu\hstc\fourthmethod\listener\BtnOnClickListener.java

```
package cn.edu.hstc.fourthmethod;

import android.app.Activity;
import android.os.Bundle;
import android.view.Menu;
import android.widget.Button;
import android.widget.TextView;

public class MainActivity extends Activity {
    /**
```

```
     * 声明布局文件中的各个控件
     */
    private TextView show;
    private Button btnAdd, btnMin;

    @Override
    protected void onCreate(Bundle savedInstanceState) {
        super.onCreate(savedInstanceState);
        setContentView(R.layout.activity_main);
        //加载 TextView 组件
        show = (TextView) findViewById(R.id.txt_show);
        //加载增加按钮
        btnAdd = (Button) findViewById(R.id.btn_add);
        //加载减少按钮
        btnMin = (Button) findViewById(R.id.btn_min);
        //为增加按钮添加事件监听器
        btnAdd.setOnClickListener(new BtnOnClickLinstener(show, btnAdd, btnMin));
        //为减少按钮添加事件监听器
        btnMin.setOnClickListener(new BtnOnClickLinstener(show, btnAdd, btnMin));
    }

    @Override
    public boolean onCreateOptionsMenu(Menu menu) {
        //Inflate the menu; this adds items to the action bar if it is present.
        getMenuInflater().inflate(R.menu.main, menu);
        return true;
    }

}
```

代码文件：codes\02\2.1\FourthMethod\ src\cn\edu\hstc\fourthmethod\activity\MainActivity.java

上面的程序中，首先定义了一个实现了 View.OnClickListener 接口的外部类 BtnOnClickListener，在该外部类中重写了 onClick 方法来处理具体业务。但由于该外部类不能直接访问 Activity 中的控件，所以需要通过传参方式来间接访问。接着在 Activity 类中为两个 Button 控件注册了事件处理监听器，如代码行"btnAdd.setOnClickListener(new BtnOnClickListener(text, btnAdd, btnMin));"，为"+"按钮注册了监听器，委托该监听器响应具体操作，而该监听器则为程序定义的外部类 BtnOnClickListener。

运行效果与第一种响应方法中的例子一致，此处不再赘述。

实际上，这种使用顶级类来定义事件监听器类的方法比较少用，因为这样不能在事件监听器类中自由访问 Activity 中的控件，除非确实有多个界面共享同一个事件监听器的情况。

2.1.5 在 XML 界面文件中指定事件处理方法

上面讲到的四种响应方法都是在程序中为控件注册事件处理监听器，接下来要介绍的这种响应方法是在 XML 界面文件中为组件绑定事件处理方法。如下 XML 代码文件以及 .java 源文件，在同样的布局以及相同的应用程序下，实现同样的功能。

```
<LinearLayout xmlns:android = "http://schemas.android.com/apk/res/android"
    xmlns:tools = "http://schemas.android.com/tools"
    android:layout_width = "match_parent"
    android:layout_height = "match_parent"
    android:background = "@color/azure"
    android:gravity = "center"
    android:orientation = "vertical" >

    <! -- 为控件指定 onClick 属性,绑定事件处理方法 btnAddClick -->
    <Button
        android:id = "@ + id/btn_add"
        android:layout_width = "wrap_content"
        android:layout_height = "wrap_content"
        android:text = " + "
        android:onClick = "btnAddClick" />

    <TextView
        android:id = "@ + id/txt_show"
        android:layout_width = "wrap_content"
        android:layout_height = "wrap_content"
        android:text = "我会变大"
        android:textSize = "12sp"
        android:textStyle = "bold" />

    <! -- 为控件指定 onClick 属性,绑定事件处理方法 btnMinClick -->
    <Button
        android:id = "@ + id/btn_min"
        android:layout_width = "wrap_content"
        android:layout_height = "wrap_content"
        android:text = " - "
        android:onClick = "btnMinClick" />

</LinearLayout>
```

代码文件: codes\02\2.1\XMLMethod\res\layout\activity_main.xml

```
package cn.edu.hstc.xmlmethod;

import android.app.Activity;
import android.os.Bundle;
import android.view.View;
import android.widget.Button;
import android.widget.TextView;

public class MainActivity extends Activity {
    private TextView show;
    private Button btnAdd, btnMin;

    @Override
    protected void onCreate(Bundle savedInstanceState) {
```

```
        super.onCreate(savedInstanceState);
        setContentView(R.layout.activity_main);
        //加载 TextView 组件
        show = (TextView) findViewById(R.id.txt_show);
        //加载增加按钮
        btnAdd = (Button) findViewById(R.id.btn_add);
        //加载减少按钮
        btnMin = (Button) findViewById(R.id.btn_min);
    }

    public void btnAddClick(View v) {
        float fontSize = show.getTextSize();
        if (fontSize < 36) {
            show.setText("我会变大或变小");
            show.setTextSize(++fontSize);
            btnMin.setEnabled(true);
            if (fontSize == 36) {
                show.setText("我不会再变大了");
                btnAdd.setEnabled(false);
            }
        }
    }

    public void btnMinClick(View v) {
        float fontSize = show.getTextSize();
        if (fontSize > 12) {
            show.setText("我会变大或变小");
            show.setTextSize(--fontSize);
            btnAdd.setEnabled(true);
            if (fontSize == 12) {
                show.setText("我不会再变小了");
                btnMin.setEnabled(false);
            }
        }
    }
}
```

代码文件：codes\02\2.1\XMLMethod\ src\cn\edu\hstc\xmlmethod\activity\MainActivity.java

上面的 XML 代码与介绍四种响应方法时所举例子的 XML 界面文件的代码基本一致，唯一的区别在于本例的 XML 界面文件中的两个 Button 控件都指定了 android:onClick 属性，并给出属性值，而属性值则是在 MainActivity 类中分别定义的事件处理方法的方法名。如代码行 android:onClick="btnAddClick"为"+"按钮绑定了事件处理方法 btnAddClick。而在 MainActivity 类中定义了 btnAddClick 方法，该方法内则是具体的响应内容。当界面中的"+"按钮被单击时，该方法将会被激发并处理"+"按钮上的单击事件。运行效果与前面介绍的四种响应方法时所举例子的运行效果一致，此处不再赘述。

其实，大多数 Android 标签都支持如 onClick、onLongClick 等属性，这种属性的属性值则是 Activity 中所对应定义的事件处理方法的方法名。

2.2　键盘事件

2.1节的内容主要为读者介绍了UI控件基于监听机制的事件响应方法，这些响应方法也是Android UI最常见的响应方法。读者掌握了这些响应方法就已经基本掌握了如何对操作Android UI执行响应了。但是，有时候用户也需要直接操作Android手机上的键盘，这个时候就需要直接对键盘事件进行响应了。本节将介绍基于回调机制的键盘事件的响应方法。

```
<LinearLayout xmlns:android="http://schemas.android.com/apk/res/android"
    android:layout_width="match_parent"
    android:layout_height="match_parent"
    android:background="@color/azure"
    android:gravity="center" >

    <TextView
        android:id="@+id/show"
        android:layout_width="wrap_content"
        android:layout_height="wrap_content"
        android:textSize="24sp" />

</LinearLayout>
```

代码文件：codes\02\2.2\KeyboardEvent\res\layout\activity_main.xml

上面的XML界面文件非常简单，在一个线性布局里添加了一个TextView控件，用来在后面显示文本。接下来的程序代码控制用户按下键盘上的返回键，在界面输出文字。

```
package cn.edu.hstc.keyboardevent.acitivity;

import android.app.Activity;
import android.os.Bundle;
import android.view.KeyEvent;
import android.widget.TextView;

public class MainActivity extends Activity {
    private TextView show;

    protected void onCreate(Bundle savedInstanceState) {
        super.onCreate(savedInstanceState);
        setContentView(R.layout.activity_main);
        show = (TextView) findViewById(R.id.show);
    }

    @Overrid
    public boolean onKeyDown(int keyCode, KeyEvent event) {
        //按下返回键,同时没有重复按,event.getRepeatCount() == 0 表示只按了一次
        if (keyCode == KeyEvent.KEYCODE_BACK && event.getRepeatCount() == 0) {
            show.setText("您刚刚按下了返回键");
```

```
        }
        return false;
    }
}
```

代码文件：codes\02\2.2\KeyboardEvent\src\cn\edu\hstc\keyboardevent\activity\MainActivity.java

上面的程序代码中，在 Activity 中重写了 onKeyDown(int keyCode, KeyEvent event)方法，在该方法中，实现了键盘事件。参数 keyCode 为按键码，event 表示按键事件，其中包含更详细的内容。在本例中，按键码 keyCode== KeyEvent.KEYCODE_BACK 表示用户按下的是返回键。该方法会在用户按下返回键时被回调，运行效果如图 2.2 所示。

图 2.2 响应键盘事件

实际上，为了实现回调机制的事件处理，Android 为所有 GUI 组件都提供了一些可重写的事件处理方法，以 View 为例，该类包含如下方法。

- boolean onKeyDown(int keyCode, KeyEvent event)：当用户在该组件上按下某个按键时触发该方法。
- boolean onKeyLongPress(int keyCode, KeyEvent event)：当用户在该组件上长按某个按键时触发该方法。
- boolean onKeyShortcut(int keyCode, KeyEvent event)：当一个键盘快捷键事件发生时触发该方法。
- boolean onKeyUp(int keyCode, KeyEvent event)：当用户在该组件上松开某个按键时触发该方法。
- boolean onTouchEvent(int keyCode, KeyEvent event)：当用户在该组件上触发触摸屏事件时触发该方法。
- boolean onTrackballEvent(int keyCode, KeyEvent event)：当用户在该组件上触发轨迹球事件时触发该方法。

上面的函数 boolean onTouchEvent(int keyCode, KeyEvent event)为 Android 添加触摸屏事件时需要重写的方法。接下来介绍 Android 的触摸屏事件处理。

2.3 触摸屏事件

上一节讲的是 Android 键盘事件，本节将介绍同样基于回调机制的 Android 触摸屏事件。本节的例子实现在 Android 设备屏幕上描绘一个红色的小圆球，这个小圆球随着手指在屏幕上的移动而移动。

```
package cn.edu.hstc.touchevent.activity;
```

```
import android.content.Context;
import android.graphics.Canvas;
import android.graphics.Color;
import android.graphics.Paint;
import android.util.AttributeSet;
import android.view.MotionEvent;
import android.view.View;

public class DrawView extends View {
    public float currentX = 40;
    public float currentY = 50;

    public DrawView(Context context, AttributeSet atts) {
        super(context, atts);
    }

    @Override
    protected void onDraw(Canvas canvas) {
        super.onDraw(canvas);
        Paint p = new Paint();
        p.setColor(Color.RED);
        canvas.drawCircle(currentX, currentY, 15, p);
    }

    @Override
    public boolean onTouchEvent(MotionEvent event) {
        this.currentX = event.getX();
        this.currentY = event.getY();
        //通知组件重绘
        this.invalidate();
        //返回 true 表明处理方法已经处理该事件
        return true;
    }
}
```

代码文件：codes\02\2.3\TouchEvent\src\cn\edu\hstc\touchevent\activity\DrawView.java

上面的程序自定义了一个 View 类，重写了 View 组件的 onTouchEvent(MotionEvent event)方法，使该组件能够处理触摸屏事件。本例的 onTouchEvent 方法中获取了当前手指的 x、y 坐标，然后在当前位置上用该 View 重绘小球。

MotionEvent 是用于处理运动事件的类，这个类可以获取动作的类型、坐标。在 Android 2.0 版本之后，MotionEvent 中还包含了多点触摸的信息，当有多个触点同时起作用的时候，可以获得触点的数目和每一个触点的坐标。

接下来就是直接在界面中使用该 View 组件了。

```
<LinearLayout xmlns:android = "http://schemas.android.com/apk/res/android"
    android:layout_width = "match_parent"
    android:layout_height = "match_parent"
```

```
    android:background = "@color/azure"
    android:orientation = "vertical" >

    < cn.edu.hstc.touchevent.activity.DrawView
        android:layout_width = "fill_parent"
        android:layout_height = "fill_parent"
        android:orientation = "vertical" />

</LinearLayout >
```

代码文件：codes\02\2.3\TouchEvent\res\layout\activity_main.xml

图 2.3　跟随手指的小球

接下来在 Activity 类中无须为这个 View 绑定任何事件监听器，因为这个 View 自己就可以处理它的触摸屏事件了。运行效果如图 2.3 所示。

通过为 View 提供事件处理的回调方法，可以很好地将时间处理方法封装在 View 内部，从而提高程序的内聚性。

基于回调的事件处理更适合那种事件处理逻辑比较固定的 View。

2.4　Handler 消息传递机制

假如你的应用需要联网读取数据，这是个比较耗时的操作，你不能把这些耗时的操作放在主线程里，因为这样会出现界面假死现象，所以必须把这些耗时的操作放在子线程中。但由于子线程又涉及 UI 的更新操作，而 UI 必须由主线程访问。这时，接收子线程发送过来的数据，并配合主线程更新 UI 的 Handler 便出现了。本节将介绍 Handler 的消息传递机制。

2.4.1　认识 Handler

对于 Handler，它的主要任务就是接收从子线程传递过来的数据、消息，并在主线程中处理获取到的消息。

Handler 运行在主线程中，它与子线程通过 Message 对象来传递数据。分两步：子线程通过 sendMessage()方法传递 Message 对象，里面包含数据；主线程中的 Handler 接收数据，并配合主线程更新 UI。

Handler 是通过回调的方式来处理子线程传递过来的数据的。开发者只要重写 Handler 类中的处理消息的方法，当子线程中的消息被发送时，Handler 类中处理消息的方法将被自动调用。

用于发送、处理消息的方法如下：

- void handleMessage(Message msg)——处理消息的方法。该方法通常用于被重写。
- final boolean hasMessages(int what)——检查消息队列中是否包含 what 属性为指

定值的消息。

- final boolean hasMessages(int what, Object object)——检查消息队列中是否包含 what 属性为指定值且 object 属性为指定对象的消息。
- 多个重载的 Message obtainMessage()——获取消息。
- sendEmptyMessage(int what)——发送空消息。
- final boolean sendEmptyMessageDelayed(int what, long delayMillis)——指定多少毫秒之后发送空消息。
- final boolean sendMessage(Message msg)——立即发送消息。
- final boolean sendMessageDelayed(Message msg, long delayMillis)——指定多少毫秒之后发送消息。

2.4.2　使用 Handler

上节已经简单介绍了 Android Handler 的作用以及 Handler 类中常用的回调方法。本节将通过一个简单的例子来进一步学习 Android Handler 的使用。

```
<LinearLayout xmlns:android = "http://schemas.android.com/apk/res/android"
    android:layout_width = "match_parent"
    android:layout_height = "match_parent"
    android:background = "@color/azure"
    android:orientation = "vertical" >

    <EditText
        android:id = "@ + id/show"
        android:layout_width = "fill_parent"
        android:layout_height = "wrap_content"
        android:hint = "触碰读取 SD 卡中的文件内容"
        android:lines = "10" />

</LinearLayout>
```

代码文件：codes\02\2.4\HandlerDemo\res\layout\activity_main.xml

上面的 XML 界面文件很简单，只是在一个空白的界面布局中添加了一个文本框，用于显示从 SD 卡里读取到的内容。

```
package cn.edu.hstc.handlerdemo.activity;

import cn.edu.hstc.handlerdemo.util.FileHelper;
import android.app.Activity;
import android.os.Bundle;
import android.os.Handler;
import android.os.Message;
import android.view.MotionEvent;
import android.view.View;
import android.view.View.OnTouchListener;
import android.widget.EditText;

public class MainActivity extends Activity {
```

```
    private EditText show;
    private FileHelper fileHelper = new FileHelper();
    private String result;

    @Override
    protected void onCreate(Bundle savedInstanceState) {
        super.onCreate(savedInstanceState);
        setContentView(R.layout.activity_main);
        show = (EditText) findViewById(R.id.show);
        //自定义 Handler
        final Handler handler = new Handler() {
         @Override
         public void handleMessage(Message msg) {
             if (msg.what == 0x123) {
                 //当消息表示为 0x123 时,将消息中的数据填写到 EditText 中
                 show.setText(msg.obj.toString());
             }
         }
        };

        show.setOnTouchListener(new OnTouchListener() {
            @Override
            public boolean onTouch(View arg0, MotionEvent arg1) {
                new Thread(new Runnable() {
                    @Override
                    public void run() {
                        result = fileHelper.readSDFile("test.txt");
                        Message msg = new Message();
                        msg.what = 0x123;
                        msg.obj = result;
                        handler.sendMessage(msg);
                    }
                }).start();
                return false;
            }
        });
    }
}
```

代码文件：codes\02\2.3\HandlerDemo\src\cn\edu\hstc\handlerdemo\activity\MainActivity.java

```
package cn.edu.hstc.handlerdemo.util;

import java.io.File;
import java.io.FileInputStream;

import org.apache.http.util.EncodingUtils;

import android.os.Environment;
```

```
public class FileHelper {
    //SD 卡的路径
    private String SDPATH;

    public FileHelper() {
        SDPATH = Environment.getExternalStorageDirectory().getPath();
    }

    public String readSDFile(String fileName) {
        File file = new File(SDPATH + "//" + fileName);
        String res = "";

        try {
            FileInputStream fis = new FileInputStream(file);
            int length = fis.available();
            byte[] buffer = new byte[length];

            fis.read(buffer);
            res = EncodingUtils.getString(buffer, "UTF-8");
            fis.close();
        } catch (Exception e) {
            e.printStackTrace();
        }
        return res;
    }
}
```

代码文件：codes\02\2.3\HandlerDemo\src\cn\edu\hstc\handlerdemo\activity\FileHelper.java

上面的代码中，定义了一个 Handler 类，重写了 handleMessage 回调方法。该方法实现当接收到的消息标识为 0x123 时，将接收到的消息作为界面中的 EditText 的显示内容。这里使用的是回调机制，当子线程发送消息时，handleMessage 方法将会被自动回调。之所以在 handleMessage 方法中能够访问 Activity 中的组件，是因为该方法是在主线程中被调用的。

程序中为 EditText 注册了触摸屏监听器，该监听器实现启动一条子线程，用于读取 SDCard 中的 aa.txt 文本并将读取到的内容封装在消息对象中，如代码行 msg.obj = res。然后通过 handler.sendMessage(msg)发送消息，此处使用的 handler 即为之前所定义的 Handler 类的对象。这样，在该线程中所发送的消息将会被这个 handler 所接收。

运行效果如图 2.4 所示。

图 2.4 借助 Handler 更新 UI

2.5 本章小结

本章的学习重点是掌握 Android 的两种事件处理机制：基于回调的事件处理和基于监听的事件处理。对于基于监听的事件处理模式来说，需要读者了解的有五种不同的响应方法：使用匿名内部类作为事件监听器类；使用内部类作为事件监听器类；使用外部类作为事件监听器类；使用 Activity 类本身作为事件监听器类；将事件处理方法绑定到界面文件中。对于基于回调的事件处理模式来说，开发者需要掌握不同事件对应的回调方法。本章介绍了键盘事件处理、触摸屏事件处理以及使用 Handler 消息传递机制处理耗时事件。这三种事件处理都是基于 Android 回调机制。学习本章后，相信读者已经了解了各种常见的 Android 事件处理方法。

第3章 Android基本界面组件

我们判定一个应用的好与坏，用户体验起着至关重要的作用，而一个具有良好的用户体验的应用也必将是一个具有友好的图形用户界面的应用，否则将很难吸引最终用户。实际上，很多优秀的、市面上流行的 Android 软件，首先都会给用户提供友好的图形用户界面，这样的程序才会被接受而流行起来。

Android 提供了大量功能丰富的 UI 组件，开发者只要按一定规律把这些 UI 组件搭建在一起就可以开发出优秀的图形用户界面。Android 界面开发是 Android 应用开发的基础，也是 Android 开发非常重要的组成部分。

3.1 Android 五大布局管理器

Android 的界面是由布局和组件协同完成的，布局好比是建筑里的框架，而组件则相当于建筑里的砖瓦。组件按照布局的要求依次排列，就组成了用户所看见的界面。通过使用布局管理器，Android 应用的界面组件可在不同分辨率的手机上得到良好的控制。因此通常推荐使用布局管理器来管理组件的分布、大小，而不是设置组件的位置和大小。如果在程序中设置了组件的大小和位置，那么这个应用一般只能运行于特定的分辨率的手机，不能做到手机自适应。而 Android 布局管理器可以根据运行的手机来调整组件的大小与位置，我们只需为容器选择合适的布局管理器。

Android 的五大布局分别是 LinearLayout(线性布局)、FrameLayout(单帧布局)、RelativeLayout(相对布局)、AbsoluteLayout(绝对布局)和 TableLayout(表格布局)。

3.1.1 线性布局

线性布局管理器在 XML 文件中为一个 LinearLayout 标签，在该标签下的子标签所代表的组件将会按照从左到右或从上到下的顺序依次排列起来。LinearLayout 通过设置 android:orientation 属性来控制容器里各组件的排列方向，可横向排列，亦可纵向排列。

提示：当组件太多时，排列到手机屏幕的尽头时，Android 线性布局不会自动将余下的组件换到下一行显示，剩下的组件将看不到。

LinearLayout 中的子元素有一个重要的属性是 android:layout_weight，它用于描述该子元素在该 LinearLayout 中的权重。如一行只有一个文本框，那么该属性的默认值就为 0。如果一行中有两个等长的文本框，那么它们该属性的值可以同为 1 。如果想让同一行的两个文本框其中的一个长占三分之二，另一个长占三分之一，那么它们该属性的值就为 1 和 2。总之记住，android:layout_weight 的数值越小，权重越高。

如下 XML 代码文件定义了一个线性布局。

```
<LinearLayout xmlns:android = "http://schemas.android.com/apk/res/android"
    android:layout_width = "match_parent"
    android:layout_height = "match_parent"
    android:gravity = "bottom|center_horizontal"
    android:orientation = "vertical" >

    <Button
        android:layout_width = "wrap_content"
        android:layout_height = "wrap_content"
        android:text = "@string/first" >

    <Button
        android:layout_width = "wrap_content"
        android:layout_height = "wrap_content"
        android:text = "@string/second" >

    <Button
        android:layout_width = "wrap_content"
        android:layout_height = "wrap_content"
        android:text = "@string/third" >

</LinearLayout>
```

代码文件：codes\03\3.1\LineayLayoutDemo\res\layout\activity_main.xml

上面的布局文件定义了一个简单的线性布局，布局中的三个按钮从上到下垂直排列，并且整体底部居中。运行上面的程序，可以看到如图 3.1 所示的界面。

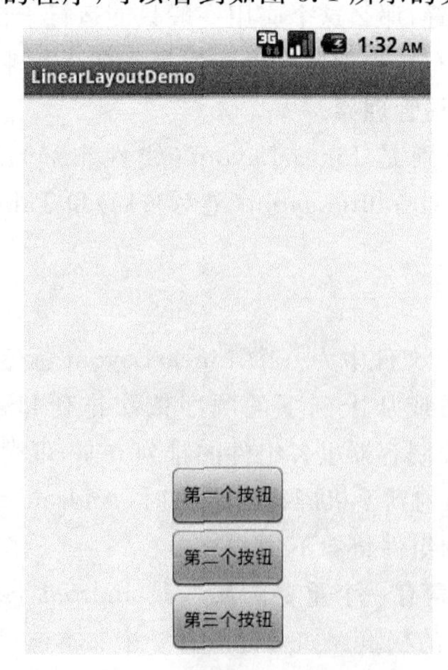

图 3.1　垂直排列，底部居中

如果把上面代码文件中的 android:gravity 的属性值改为 center,则运行结果变为如图 3.2 所示。

如果要将这三个按钮设置成从左到右顺序排列,那么只要将 android:orientation 属性值改为 horizontal 即可。运行结果如图 3.3 所示。

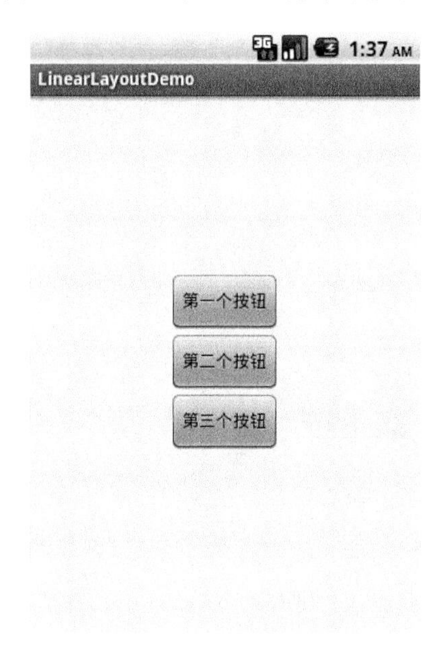

图 3.2 垂直排列,整体居中

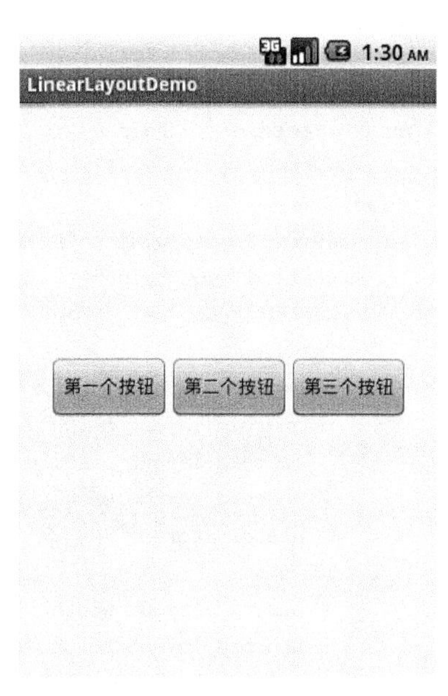

图 3.3 横向排列,整体居中

3.1.2 表格布局

表格布局在 XML 代码文件中用一个 TableLayout 表示,可以在一个 Activity 中添加多行,每一行又可以设置多个列,横竖交叉,形成表格,所以叫表格布局。

表格布局用添加一个 TableRow 标签来表示新增加一行,然后在这个 TableRow 中添加一个组件,表示该 TableRow 新增了一列。列的宽度由该列中最宽的那个单元格决定,整个 TableLayout 的宽度则取决于父容器的宽度(如手机屏幕宽度)。

如果直接向 TableLayout 添加组件,那么这个组件就直接占用一行。

在表格布局中,可以为单元格设置如下属性。

android:collapseColumns="1",隐藏该 TableLayout 里的 TableRow 的列 1,即第 2 列(从 0 开始算起),若有多列要隐藏,则列数之间用“,”隔开。

android:shrinkColumns="1",将 TableLayout 里的 TableRow 的列 1 设置收缩,即第 2 列(从 0 开始算起),若有多列要收缩,则列数之间用“,”隔开。

android:stretchColumns="1",将 TableLayout 里的 TableRow 的列 1 设置伸张,即第 2 列(从 0 开始算起),若有多列要伸张,则列数之间用“,”隔开。

注意: TableLayout 的行数由开发人员制定,即有多少个 TableRow 对象(或 view 控件),就有多少行。列的宽度: 由该列中最宽的单元格决定,整个表格布局的宽度取决于父容器的宽度(默认是占满父容器本身)。如第一个 TableRow 含 2 个控件,第二个 TableRow

含3个控件，那么该TableLayout的列数为3。

如下XML代码文件定义了一个TableLayout。

```
<TableLayout xmlns:android = "http://schemas.android.com/apk/res/android"
    android:layout_width = "match_parent"
    android:layout_height = "match_parent"
    android:background = "@color/azure"
    android:stretchColumns = "1"
    android:padding = "20dip">

    <TableRow
        android:layout_width = "fill_parent"
        android:layout_height = "wrap_content"
        android:layout_marginTop = "10dip">
        <TextView
            android:layout_width = "wrap_content"
            android:layout_height = "wrap_content"
            android:layout_gravity = "right|center_vertical"
            android:textColor = "@color/black"
            android:text = "用户名: "/>
        <EditText
            android:layout_width = "wrap_content"
            android:layout_height = "wrap_content"
            android:background = "@drawable/shurukuang"
            android:hint = "请输入用户名"/>
    </TableRow>

    <TableRow
        android:layout_width = "fill_parent"
        android:layout_height = "wrap_content"
        android:layout_marginTop = "5dip">
        <TextView
            android:layout_width = "wrap_content"
            android:layout_height = "wrap_content"
            android:layout_gravity = "right|center_vertical"
            android:textColor = "@color/black"
            android:text = "密码: "/>
        <EditText
            android:layout_width = "wrap_content"
            android:layout_height = "wrap_content"
            android:background = "@drawable/shurukuang"
            android:inputType = "textPassword"
            android:hint = "请设定密码"/>
    </TableRow>

    <TableRow
        android:layout_width = "fill_parent"
        android:layout_height = "wrap_content"
        android:layout_marginTop = "5dip">
        <LinearLayout
```

```
            android:layout_width="wrap_content"
            android:layout_height="wrap_content"/>
        <LinearLayout
            android:layout_width="wrap_content"
            android:layout_height="wrap_content"
            android:layout_gravity="right">
            <Button
                android:layout_width="60dip"
                android:layout_height="40dip"
                android:background="@drawable/bt2"
                android:text="登录"/>
            <Button
                android:layout_width="60dip"
                android:layout_height="40dip"
                android:background="@drawable/bt2"
                android:layout_marginLeft="5dip"
                android:text="注册"/>
        </LinearLayout>
    </TableRow>

</TableLayout>
```

代码文件：codes\03\3.1\TableLayoutDemo\res\layout\activity_main.xml

运行结果如图3.4所示。

图3.4 定义TableLayout

3.1.3 相对布局

Android相对布局在XML文件的元素为RealativeLayout，在相对布局中，一个控件的位置决定于它和其他控件的关系。这样做的好处是比较灵活，缺点是比较复杂。接下来将相对布局常用的属性分成四组进行讲解。

表3.1给出了四个属性设置控件之间的关系和位置。

表3.1 位置属性

属性名称	描述	备注
android:layout_above	将该控件置于给定ID控件之上	该属性值为某个控件的ID 如：android:layout_above="@id/entry"
android:layout_below	将该控件置于给定ID控件之下	
android:layout_toLeftOf	将该控件置于给定ID控件的左边	
android:layout_toRightOf	将该控件置于给定ID控件的右边	

上面四个属性并没有设置各个控件之间是否对齐。例如使用android:layout_below属性将B控件置于A控件之下，那么可能出现的情况是B控件确实在A控件之下，但B控件跟A控件并没有对齐，另一种情况是B控件在A控件之下的同时两控件对齐。

如下定义的XML界面文件，将B控件置于A控件之下，但两者并没有对齐。

```
<RelativeLayout xmlns:android="http://schemas.android.com/apk/res/android"
    android:layout_width="match_parent"
    android:layout_height="match_parent"
```

```
    android:padding = "10dip"
    android:background = "@color/azure">

    < Button
        android:layout_width = "wrap_content"
        android:layout_height = "50dip"
        android:text = "我是 A 控件"
        android:layout_alignParentRight = "true"
        android:background = "@drawable/bt2"
        android:id = "@ + id/btn_A"/>

    < Button
        android:layout_width = "wrap_content"
        android:layout_height = "50dip"
        android:text = "我是 B 控件"
        android:background = "@drawable/bt2"
        android:layout_below = "@id/btn_A"/>

</RelativeLayout >
```

代码文件: codes\03\3.1\RelativeLayoutDemo\res\layout\activity_main.xml

运行效果如图 3.5 所示。

将上面代码文件中的 android:layout_alignParentRight="true"去掉,B 控件与 A 控件将对齐,运行效果如图 3.6 所示。

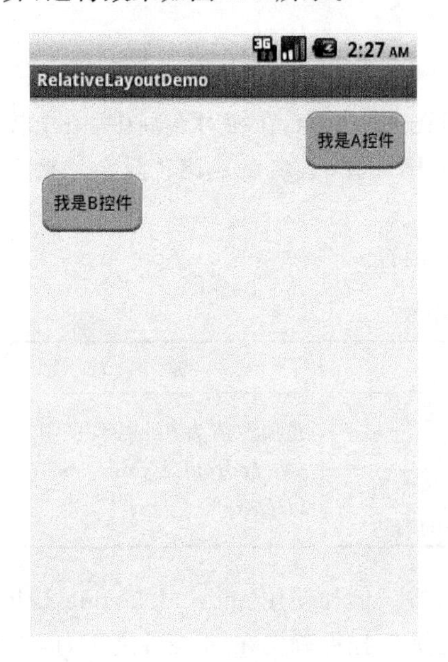

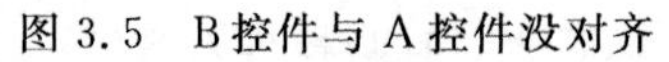
图 3.5　B 控件与 A 控件没对齐

图 3.6　B 控件与 A 控件对齐

如表 3.2 所示的五个属性,设置的是控件与控件之间的对齐方式。

表 3.2　控件之间对齐方式属性

属性名称	描　　述	备　　注
android:layout_alignBaseline	将该控件的基线(baseline)与给定 ID 控件的基线对齐	该属性值为某个控件的 ID 如: android:layout_alignTop="@id/entry"
android:layout_alignTop	将该控件的顶部与给定 ID 控件的顶部对齐	
android:layout_alignBottom	将该控件的底部与给定 ID 控件的底部对齐	
android:layout_alignLeft	将该控件的左边边缘与给定 ID 控件的左边边缘对齐	
android:layout_alignRight	将该控件的右边边缘与给定 ID 控件的右边边缘对齐	

给代码文件：codes\O3\3.1\RelativeLayoutDemo\res\layout\activity_main.xml 中的 B 控件加上 android:layout_alignRight="@id/btn_A"这样一句代码，即可控制 A 控件跟 B 控件的右边缘对齐。运行效果如图 3.7 所示。

图 3.7　B 控件跟 A 控件的右边边缘对齐

表 3.3 给出了四个属性设置控件与父控件之间对齐的方式。

表 3.3　控件与父控件之间对齐的方式属性

属性名称	描　　述	备　　注
android:layout_alignParentTop	将该控件的顶部与父控件的顶部对齐	
android:layout_alignParentBottom	将该控件的顶部与父控件的底部对齐	可选值为 true 和 false
android:layout_alignParentLeft	将该控件的顶部与父控件的左边边缘对齐	
android:layout_alignParentRight	将该控件的顶部与父控件的右边边缘对齐	

如下面的 XML 代码文件，将 A 控件与 B 控件上下置放，并且 B 控件右边边缘与父控件对齐。

```
<RelativeLayout xmlns:android = "http://schemas.android.com/apk/res/android"
    android:layout_width = "match_parent"
    android:layout_height = "match_parent"
    android:padding = "10dip"
    android:background = "@color/azure">

    <Button
        android:layout_width = "wrap_content"
        android:layout_height = "50dip"
        android:text = "我是 A 控件"
        android:background = "@drawable/bt2"
        android:id = "@ + id/btn_A"/>

    <Button
        android:layout_width = "wrap_content"
        android:layout_height = "50dip"
        android:text = "我是 B 控件"
        android:background = "@drawable/bt2"
        android:layout_below = "@id/btn_A"
         android:layout_alignParentRight = "true" />

</RelativeLayout>
```

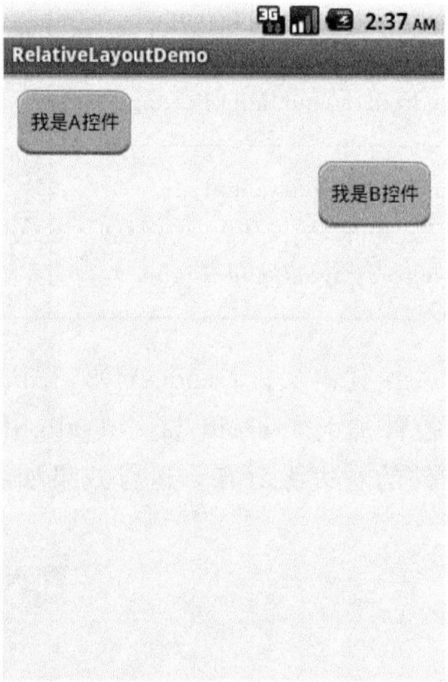

图 3.8　B 控件右边边缘跟父控件对齐

运行效果如图 3.8 所示。

表 3.4 给出了三个属性控制控件的方向。

表 3.4　控件方向属性

属性名称	描　述	备　注
android:layout_centerHorizontal	将该控件左右居中	可选值为 true 和 false
android:layout_centerVertical	将该控件上下居中	
android:layout_centerInParent	将该控件上下左右居中	

下面的 XML 代码文件中，B 控件在父控件上下左右方向居中。

```
<RelativeLayout xmlns:android = "http://schemas.android.com/apk/res/android"
    android:layout_width = "match_parent"
    android:layout_height = "match_parent"
    android:padding = "10dip"
    android:background = "@color/azure">

    <Button
        android:layout_width = "wrap_content"
        android:layout_height = "50dip"
        android:text = "我是 A 控件"
        android:background = "@drawable/bt2"
        android:id = "@ + id/btn_A"/>
```

```
    <Button
        android:layout_width = "wrap_content"
        android:layout_height = "50dip"
        android:text = "我是 B 控件"
        android:background = "@drawable/bt2"
        android:layout_centerInParent = "true"/>

</RelativeLayout>
```

运行效果如图 3.9 所示。

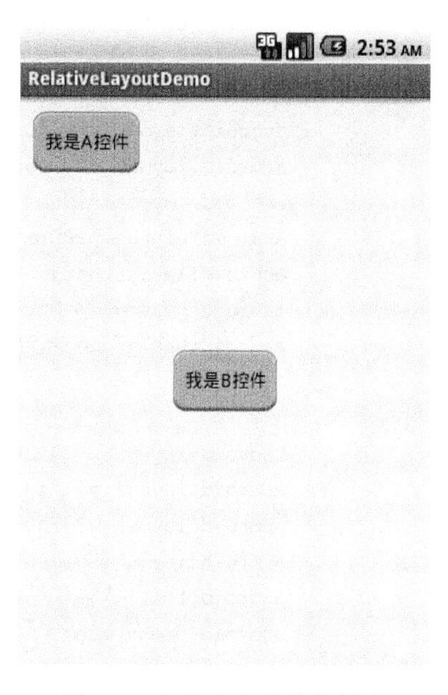

图 3.9 B 控件在父控件居中

3.1.4 绝对布局

绝对布局在 Android 中用 AbsoluteLayout 表示。绝对布局中的组件大小需要开发人员自己来控制，甚至组件的位置都需要开发人员通过坐标来控制。

在相对布局中，通过设置每个控件的 layout_x 以及 layout_y 属性来设置控件的 x 坐标和 y 坐标，以此来控制控件的位置。

注意：使用绝对布局的方法在直接拖控件的时候显得比较方便，但是不利于程序的后期调整，而且也不能做到界面的自适应性，很难做到兼顾不同手机的屏幕大小以及分辨率。

如下 XML 代码文件中，用绝对布局定义了一个登录界面。

```
<AbsoluteLayout xmlns:android = "http://schemas.android.com/apk/res/android"
    android:layout_width = "match_parent"
    android:layout_height = "match_parent"
    android:background = "@color/azure">

    <TextView
        android:layout_width = "wrap_content"
        android:layout_height = "wrap_content"
        android:text = "用户名: "
        android:layout_x = "20dip"
        android:layout_y = "20dip"/>

    <EditText
        android:layout_x = "80dip"
        android:layout_y = "15dip"
        android:layout_width = "wrap_content"
        android:width = "200px"
        android:layout_height = "wrap_content"
        android:background = "@drawable/shurukuang"/>

    <TextView
        android:layout_x = "20dip"
        android:layout_y = "80dip"
        android:layout_width = "wrap_content"
        android:layout_height = "wrap_content"
```

```
        android:text = "密　码: "/>

    <EditText
        android:layout_x = "80dip"
        android:layout_y = "75dip"
        android:layout_width = "wrap_content"
        android:width = "200px"
        android:layout_height = "wrap_content"
        android:password = "true"
        android:background = "@drawable/shurukuang"/>

    <Button
        android:layout_x = "130dip"
        android:layout_y = "135dip"
        android:layout_width = "wrap_content"
        android:layout_height = "40dip"
        android:text = "登　录"
        android:background = "@drawable/bt2"/>

</AbsoluteLayout>
```

代码文件：codes\03\3.1\AbsoluteLayoutDemo\res\layout\activity_main.xml

运行效果如图 3.10 所示。

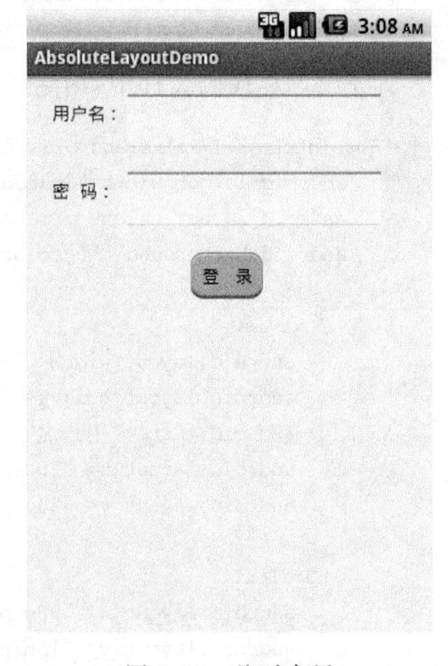

图 3.10　绝对布局

上面在设置组件的 x 坐标、y 坐标中用了 android:layout_x="20dip"这样的语句，dip 是一个距离单位。Android 支持如下常用的距离单位。

- px：像素，每个像素对应屏幕中的一个点。
- dip 或 dp：设备独立像素，一种基于屏幕密度的抽象单位。1dip＝1px，但随着屏幕密度的改变，dip 与 px 的换算会发生改变。
- sp：比例像素，主要处理字体的大小。

3.1.5　帧布局

FrameLayout 帧布局在屏幕上开辟出一块区域，在这块区域中可以添加多个子控件，但是所有的子控件都被对齐到屏幕的左上角。帧布局的大小由子控件中尺寸最大的那个子控件来决定。如果子控件一样大，那么同一时刻只能看到最上面的子控件。

在帧布局中，子控件是通过栈来绘制的，所以后添加的子控件会被控制在上层。如下 XML 代码文件中，在一个帧布局中定义了三个线性布局，层层相叠。

```
<FrameLayout xmlns:android = "http://schemas.android.com/apk/res/android"
    android:layout_width = "match_parent"
    android:layout_height = "match_parent"
    android:background = "@color/azure">
```

```
    <LinearLayout
        android:layout_width = "100dip"
        android:layout_height = "100dip"
        android:background = "@color/red"/>

    <LinearLayout
        android:layout_width = "80dip"
        android:layout_height = "80dip"
        android:background = "@color/seagreen"/>

    <LinearLayout
        android:layout_width = "60dip"
        android:layout_height = "60dip"
        android:background = "@color/yellow"/>

</FrameLayout>
```

代码文件：codes\03\3.1\FrameLayoutDemo\res\layout\activity_main.xml

运行效果如图 3.11 所示。

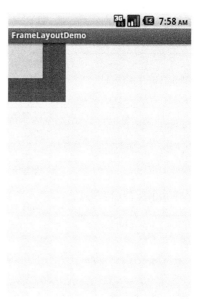

图 3.11　帧布局

3.2　Android 基本界面组件

实际上，无论是看起来多么美观的操作界面，都是由一个又一个的界面组件堆砌而成的，这些组件就放在一个容器(ViewGroup)里面。而单独掌握这些基本界面组件也是学习 Android 编程的一个必不可少的环节。接下来将重点介绍 Android 的基本界面组件。

3.2.1　文本框和编辑框

文本框(android.widget.TextView)是 android.view.View 的直接子类，同时也是 Button、CheckedTextView、Chronometer、DigitalClock、EditText 的直接父类，它的间接子类是 AutoCompleteTextView、CheckBox、CompoundButton、ExtractEditText、MultiAutoCompleteTextView、RadioButton、ToggleButton。

TextView 的作用就是在界面上显示文本，但是它没有文本编辑功能，如果开发者需要一个可以编辑内容的文本框，那么可以使用 TextView 的子类 EditText(编辑框)，EditText 允许用户在文本框中编辑内容。

TextView 提供大量的 XML 属性，这些属性大部分不仅适用于 TextView，还适用于 EditText。表 3.5 显示了 TextView 支持的属性及其描述。

表 3.5　TextView 支持的属性及其描述

属性名称	描　述
android:autoLink	设置是否当文本为 URL 链接/email/电话号码/map 时，文本显示为可单击的链接。可选值为 none/web/email/phone/map/all
android:autoText	如果设置，将自动执行输入值的拼写纠正。此处无效果，在显示输入法并输入的时候起作用

续表

属性名称	描　述
android:bufferType	指定 getText()方式取得的文本类别。选项 editable 类似于 StringBuilder 可追加字符,也就是说,在 getText 后可调用 append 方法设置文本内容。spannable 则可在给定的字符区域使用样式
android:capitalize	设置英文字母大写类型。此处无效果,需要弹出输入法才能看得到,参见 EditText 此属性说明
android:cursorVisible	设定光标为显示/隐藏,默认显示
android:digits	设置允许输入哪些字符。如"1234567890.+-*/%\n()"
android:drawableBottom	在 text 的下方输出一个 drawable,如图片。如果指定一个颜色,会把 text 的背景设为该颜色,并且同时和 background 使用时覆盖后者
android:drawableLeft	在 text 的左边输出一个 drawable
android:drawablePadding	设置 text 与 drawable(图片)的间隔,与 drawableLeft、drawableRight、drawableTop、drawableBottom 一起使用,可设置为负数,单独使用没有效果
android:drawableRight	在 text 的右边输出一个 drawable
android:drawableTop	在 text 的正上方输出一个 drawable
android:editable	设置是否可编辑
android:editorExtras	设置文本的额外的输入数据,在 EditView 中再讨论
android:ellipsize	设置当文字过长时,该控件该如何显示。有如下值设置:start——省略号显示在开头;end——省略号显示在结尾;middle——省略号显示在中间;marquee——以跑马灯的方式显示(动画横向移动)
android:freezesText	设置保存文本的内容以及光标的位置
android:gravity	设置文本位置,如设置成 center,则文本将居中显示
android:hint	Text 为空时显示的文字提示信息,可通过 textColorHint 设置提示信息的颜色。此属性在 EditView 中使用,但是这里也可以用
android:imeOptions	附加功能,设置右下角 IME 动作与编辑框相关的动作,如 actionDone 右下角将显示一个"完成",而不设置默认是一个回车符号。这个在 EditText 中再详细说明,此处无用
android:imeActionId	设置 IME 动作 ID,在 EditText 中再做说明
android:imeActionLabel	设置 IME 动作标签,在 EditText 中再做说明
android:includeFontPadding	设置文本是否包含顶部和底部额外空白,默认为 true
android:inputMethod	为文本指定输入法,需要完全限定名(完整的包名)。例如,com.google.android.inputmethod.pinyin,但是这里报错找不到
android:inputType	设置文本的类型,用于帮助输入法显示合适的键盘类型。在 EditText 中再详细说明,这里无效果
android:linksClickable	设置链接是否单击连接,即使设置了 autoLink
android:marqueeRepeatLimit	在 ellipsize 指定 marquee 的情况下,设置重复滚动的次数,当设置为 marquee_forever 时表示无限次
android:ems	设置 TextView 的宽度为 N 个字符的宽度。这里测试为一个汉字字符宽度,效果如右:
android:maxEms	设置 TextView 的宽度为最长为 N 个字符的宽度。与 ems 同时使用时覆盖 ems 选项

续表

属性名称	描述
android:minEms	设置 TextView 的宽度为最短为 N 个字符的宽度。与 ems 同时使用时覆盖 ems 选项
android:maxLength	限制显示的文本长度,超出部分不显示
android:lines	设置文本的行数,设置两行就显示两行,即使第二行没有数据
android:maxLines	设置文本的最大显示行数,与 width 或者 layout_width 结合使用,超出部分自动换行,超出行数将不显示
android:minLines	设置文本的最小行数,与 lines 类似
android:lineSpacingExtra	设置行间距
android:lineSpacingMultiplier	设置行间距的倍数。如 1.2
android:numeric	如果被设置,则该 TextView 有一个数字输入法。此处无用,设置后唯一效果是 TextView 有单击效果,此属性在 EditText 中将详细说明
android:password	以小点"."显示文本
android:phoneNumber	设置为电话号码的输入方式
android:privateImeOptions	设置输入法选项,此处无用,在 EditText 中将进一步讨论
android:scrollHorizontally	设置文本超出 TextView 的宽度的情况下,是否出现横拉条
android:selectAllOnFocus	如果文本是可选择的,则让它获取焦点而不是将光标移动到文本的开始位置或者末尾位置。EditText 中设置后无效果
android:shadowColor	指定文本阴影的颜色,需要与 shadowRadius 一起使用。效果:Hello World, TestActivity!
android:shadowDx	设置阴影横向坐标开始位置
android:shadowDy	设置阴影纵向坐标开始位置
android:shadowRadius	设置阴影的半径。设置为 0.1 就变成字体的颜色了,一般设置为 3.0 的效果比较好
android:singleLine	设置单行显示。如果和 layout_width 一起使用,当文本不能全部显示时,后面用"…"来表示。如 android:text="test_ singleLine " android:singleLine="true" android:layout_width="20dp"将只显示"t…"。如果不设置 singleLine 或者设置为 false,文本将自动换行
android:text	设置显示文本
android:textAppearance	设置文字外观。如"? android:attr/textAppearanceLargeInverse"这里引用的是系统自带的一个外观,? 表示系统是否有这种外观,否则使用默认的外观。可设置的值如下:textAppearanceButton/textAppearanceInverse/textAppearanceLarge/textAppearanceLargeInverse/textAppearanceMedium/textAppearanceMediumInverse/textAppearanceSmall/textAppearanceSmallInverse
android:textColor	设置文本颜色
android:textColorHighlight	被选中文字的底色,默认为蓝色
android:textColorHint	设置提示信息文字的颜色,默认为灰色。与 hint 一起使用
android:textColorLink	文字链接的颜色
android:textScaleX	设置文字缩放,默认为 1.0f。分别设置 0.5f/1.0f/1.5f/2.0f,效果如下: abcdef 0.5f abcdef 1.0f abcdef 1.5f abcdef 2.0f

续表

属 性 名 称	描　　述
android:textSize	设置文字大小,推荐度量单位 sp,如 15sp
android:textStyle	设置字形[bold(粗体) 0, italic(斜体) 1, bolditalic(又粗又斜) 2]可以设置一个或多个,用"|"隔开
android:typeface	设置文本字体,必须是以下常量值之一: normal 0, sans 1, serif 2, monospace(等宽字体) 3
android:height	设置文本区域的高度,支持度量单位: px/dp/sp/in/mm
android:maxHeight	设置文本区域的最大高度
android:minHeight	设置文本区域的最小高度
android:width	设置文本区域的宽度,支持度量单位: px/dp/sp/in/mm
android:maxWidth	设置文本区域的最大宽度
android:minWidth	设置文本区域的最小宽度

实例:使用 TextView 设置不同大小、颜色、作用的文本。

```
<LinearLayout xmlns:android = "http://schemas.android.com/apk/res/android"
    android:layout_width = "match_parent"
    android:layout_height = "match_parent"
    android:orientation = "vertical">

    <TextView
        android:layout_width = "wrap_content"
        android:layout_height = "wrap_content"
        android:text = "我的字体大小为 20pt"
        android:textSize = "20pt"/>

    <TextView
        android:layout_width = "wrap_content"
        android:layout_height = "wrap_content"
        android:text = "我的字体颜色是红色"
        android:textColor = "@color/red"/>

    <TextView
        android:layout_width = "wrap_content"
        android:layout_height = "wrap_content"
        android:singleLine = "true"
        android:text = "百度一下: http://www.baidu.com"
        android:autoLink = "web"/>

    <TextView
        android:layout_width = "wrap_content"
        android:layout_height = "wrap_content"
        android:text = "我是一行密码"
        android:password = "true"/>

</LinearLayout>
```

代码文件: codes\03\3.2\TextViewDemo\res\layout\activity_main.xml

将上面这个 XML 文件作为 Activity 的界面文件，并将应用部署在 Android 模拟器上，运行效果如图 3.12 所示。

图 3.12　TextView 实例

实例：使用 EditText 定义一个友好的输入界面。

```
<TableLayout xmlns:android="http://schemas.android.com/apk/res/android"
    android:layout_width="match_parent"
    android:layout_height="match_parent"
    android:background="@color/azure"
    android:padding="10dip"
    android:stretchColumns="1">


    <TableRow>
        <TextView
            android:layout_width="wrap_content"
            android:layout_height="wrap_content"
            android:gravity="right"
            android:text="账号："/>
        <EditText
            android:layout_width="wrap_content"
            android:layout_height="wrap_content"
            android:hint="请输入账号..."
            android:background="@drawable/shurukuang"/>
    </TableRow>

    <TableRow
        android:layout_marginTop="10dip">
        <TextView
            android:layout_width="wrap_content"
            android:layout_height="wrap_content"
            android:gravity="right"
            android:text="密码："/>
        <EditText
            android:layout_width="wrap_content"
            android:layout_height="wrap_content"
            android:hint="请输入密码..."
            android:password="true"
            android:background="@drawable/shurukuang"/>
    </TableRow>

    <TableRow
        android:layout_marginTop="10dip">
        <TextView
            android:layout_width="wrap_content"
            android:layout_height="wrap_content"
            android:text="手机号码："/>
        <EditText
            android:layout_width="wrap_content"
```

```
            android:layout_height = "wrap_content"
            android:hint = "请输入您的手机号码..."
            android:phoneNumber = "true"
            android:background = "@drawable/shurukuang"/>
    </TableRow>

</TableLayout>
```

代码文件：codes\03\3.2\EditTextDemo\res\layout\activity_main.xml

运行效果如图 3.13 所示。

图 3.13　EditText 实例

3.2.2　按钮与图片按钮

上面提到过，Button（按钮）为 TextView 的直接子类，而 ImageButton 继承了 Button。两者都是为了在 UI 界面上生成一个按钮，供用户单击或双击。当用户单击或双击按钮时，按钮就会触发一个单击或双击事件，完成某些动作。

Button 按钮与 ImageButton 按钮之间的区别在于：Button 按钮上显示文字，而 ImageButton 上显示图片。

注意：为 ImageButton 按钮指定 android:text 属性是没有用的，即使指定了该属性，图片按钮上也不会显示任何文字。

实例：按钮与图片按钮的一般用法。

```
<LinearLayout xmlns:android = "http://schemas.android.com/apk/res/android"
    android:layout_width = "match_parent"
    android:layout_height = "match_parent"
    android:orientation = "vertical"
    android:padding = "10dip">

    <Button
        android:layout_width = "wrap_content"
        android:layout_height = "wrap_content"
        android:text = "我是一个普通按钮"
        android:textSize = "10pt"
        android:background = "@color/red"/>

    <ImageButton
        android:layout_width = "wrap_content"
        android:layout_height = "wrap_content"
        android:layout_marginTop = "10dip"
        android:src = "@drawable/bt2"
        android:background = "@color/yellow"/>

    <Button
```

```
        android:layout_width = "wrap_content"
        android:layout_height = "wrap_content"
        android:layout_marginTop = "10dip"
        android:text = "带文字的图片按钮"
        android:textSize = "10pt"
        android:background = "@drawable/bt2"/>

</LinearLayout>
```

代码文件：codes\03\3.2\ButtonDemo\res\layout\activity_main.xml

运行效果如图 3.14 所示。

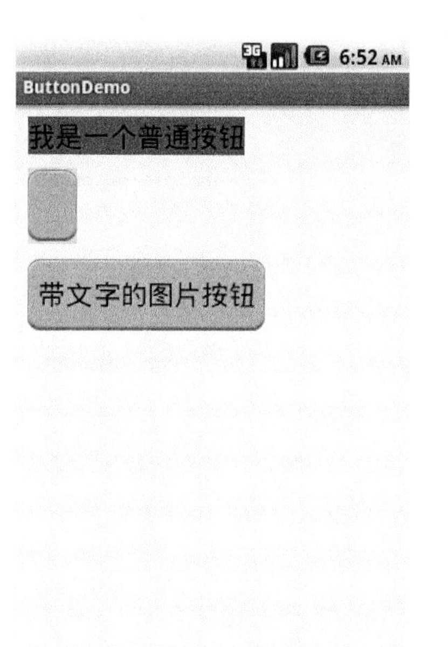

图 3.14　Button 与 ImageButton 的一般用法

3.2.3　单选按钮与复选框

单选按钮(RadioButton)和复选框(CheckBox)都继承自 Button,因此它们可以直接使用 Button 支持的各种属性。

RadioButton 与 CheckBox 都可以指定 android:checked 属性,该属性用来设置 RadioButton 与 CheckBox 的选中状态。

RadioButton 和 CheckBox 的区别如下：

- 单个 RadioButton 在选中后,通过单击无法变为未选中；单个 CheckBox 在选中后,通过单击可以变为未选中。
- 一组 RadioButton,只能同时选中一个；一组 CheckBox,能同时选中多个。
- RadioButton 在大部分 UI 框架中默认都以圆形表示；CheckBox 在大部分 UI 框架中默认都以矩形表示。

实例：使用 RadioButton 和 CheckBox 显示个人信息。

```
<LinearLayout xmlns:android = "http://schemas.android.com/apk/res/android"
    android:layout_width = "match_parent"
    android:layout_height = "match_parent"
    android:padding = "10dip"
    android:background = "@color/azure"
    android:orientation = "vertical">

    <LinearLayout
        android:layout_width = "fill_parent"
        android:layout_height = "wrap_content"
        android:orientation = "horizontal">

        <TextView
            android:layout_width = "wrap_content"
            android:layout_height = "wrap_content"
            android:text = "请选择您的性别: "
            android:textSize = "9pt"
            android:layout_gravity = "center_vertical"/>
```

```
        <RadioGroup
            android:layout_width="wrap_content"
            android:layout_height="wrap_content"
            android:orientation="horizontal">

            <RadioButton
                android:layout_width="wrap_content"
                android:layout_height="wrap_content"
                android:text="男"/>

            <RadioButton
                android:layout_width="wrap_content"
                android:layout_height="wrap_content"
                android:layout_marginLeft="10dip"
                android:text="女"/>

        </RadioGroup>

    </LinearLayout>

    <LinearLayout
        android:layout_width="fill_parent"
        android:layout_height="wrap_content"
        android:orientation="horizontal">

        <TextView
            android:layout_width="wrap_content"
            android:layout_height="wrap_content"
            android:text="请选择您的爱好："
            android:textSize="9pt"/>

        <LinearLayout
            android:layout_width="fill_parent"
            android:layout_height="wrap_content"
            android:orientation="vertical">

            <CheckBox
                android:layout_width="wrap_content"
                android:layout_height="wrap_content"
                android:text="看书"/>

            <CheckBox
                android:layout_width="wrap_content"
                android:layout_height="wrap_content"
                android:text="听歌"/>

            <CheckBox
```

```
                android:layout_width = "wrap_content"
                android:layout_height = "wrap_content"
                android:text = "看电影"/>

        </LinearLayout >

    </LinearLayout >

</LinearLayout >
```

代码文件：codes\03\3.2\RadioBttonCheckBoxDemo\res\layout\activity_main.xml

运行效果如图 3.15 所示。

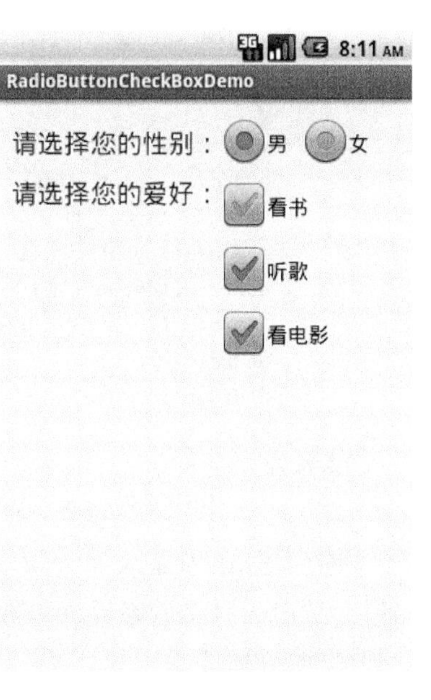

图 3.15　使用 RadioButton 和 CheckBox 显示个人信息

3.2.4　开关按钮

android.widget.ToggleButton 译为开关按钮，ToggleButton 通过一个带有亮度指示同时默认文本为 ON 或 OFF 的按钮显示选中或未选中状态，ToggleButton 也只有这两种状态，在这两种状态间切换的同时可以修改开关按钮上的默认文本。

ToggleButton 支持如表 3.6 所示的 XML 属性。

表 3.6　ToggleButton 支持的 XML 属性

属性名称	描述
android:disabledAlpha	设置按钮在禁用时透明度 disabledAlpha:未指定 OFF disabledAlpha:0 OFF disabledAlpha:0.1 OFF disabledAlpha:0.5 OFF disabledAlpha:1.0 OFF disabledAlpha:1.1 OFF disabledAlpha:2.0 OFF
android:checked	设置该按钮是否被选中
android:textOff	未选中时按钮的文本
android:textOn	选中时按钮的文本

实例：用开关按钮来控制白天与黑夜之间的切换。

```
<LinearLayout xmlns:android = "http://schemas.android.com/apk/res/android"
    android:layout_width = "match_parent"
    android:layout_height = "match_parent"
```

```
    android:background = "@drawable/day"
    android:id = "@ + id/layout">

    < ToggleButton
        android:layout_width = "wrap_content"
        android:layout_height = "wrap_content"
        android:layout_gravity = "center_vertical"
        android:textOn = "白天"
        android:textOff = "黑夜"
        android:checked = "true"
        android:id = "@ + id/toggle"/>

</LinearLayout >
```

代码文件：codes\03\3.2\ToggleButtonDemo\res\layout\activity_main.xml

上面的 XML 代码文件定义了一个界面，界面的背景为一张资源 id 为 day 的图片，该图片景色为“白天”，接着在界面上放置了一个状态开关按钮，下面的程序代码用来切换开关状态同时更改界面背景图片。

```
package cn.edu.hstc.togglebutton.activity;

import android.app.Activity;
import android.os.Bundle;
import android.widget.CompoundButton;
import android.widget.CompoundButton.OnCheckedChangeListener;
import android.widget.LinearLayout;
import android.widget.ToggleButton;

public class MainActivity extends Activity {
    //声明界面布局
    private LinearLayout layout;
    //声明开关按钮 ToggleButton
    private ToggleButton toggle;
    @Override
    protected void onCreate(Bundle savedInstanceState) {
        super.onCreate(savedInstanceState);
        setContentView(R.layout.activity_main);
        initWidget();
    }

    private void initWidget() {
        //通过资源 id 获取界面布局以及界面文件中的 ToggleButton
        layout = (LinearLayout) this.findViewById(R.id.layout);
        toggle = (ToggleButton) this.findViewById(R.id.toggle);

        toggle.setOnCheckedChangeListener(new OnCheckedChangeListener() {
            @Override
            public void onCheckedChanged(CompoundButton buttonView, boolean isChecked) {
                if (isChecked) {      //当开关打开时，界面背景图片为白天
                    layout.setBackgroundResource(R.drawable.day);
                } else {              //当开关关闭时，界面背景图片为黑夜
```

```
                layout.setBackgroundResource(R.drawable.night);
            }
        }
    });
  }
}
```

codes\03\3.2\ToggleButtonDemo\src\cn\edu\hstc\togglebutton\activity\MainActivity.java

将程序部署在 Android 模拟器上，运行效果如图 3.16 所示。

当用户单击开关按钮，运行效果如图 3.17 所示。

图 3.16 开关打开，白天背景

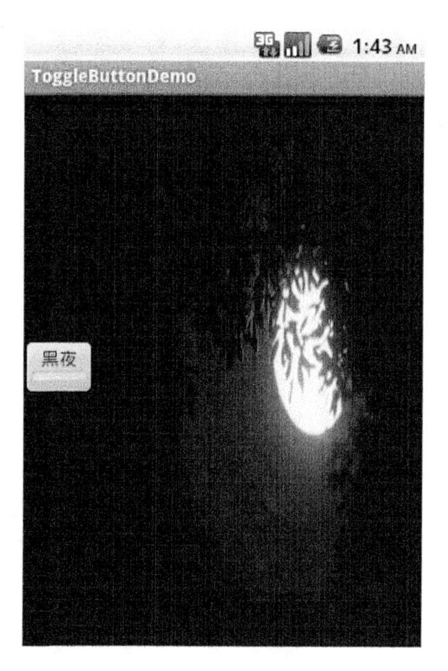

图 3.17 开关关闭，黑夜背景

3.2.5 时钟

在 Android 中有两个组件可以用来显示时间：一个是 AnalogClock，该组件模拟了现实中的时钟界面；另一个是 DigitalClock，该组件只是简单地显示当前时间，本身可看作是一个现实内容为当前时间的 TextView。这两个组件的使用都比较简单，只需要在布局文件中写入这两个组件并设置其位置即可。

实例：在界面中使用 AnalogClock 和 DigitalClock 来显示当前时间。

```
<LinearLayout xmlns:android="http://schemas.android.com/apk/res/android"
    android:layout_width="match_parent"
    android:layout_height="match_parent"
    android:background="@color/azure"
    android:gravity="center"
    android:orientation="vertical">

    <AnalogClock
```

```
            android:layout_width="wrap_content"
            android:layout_height="wrap_content"/>

    <DigitalClock
        android:layout_width="wrap_content"
        android:layout_height="wrap_content"
        android:textSize="14pt"/>

</LinearLayout>
```

代码文件：codes\03\3.2\ AnalogClockDigitalClockDemo \res\layout\activity_main.xml

运行效果如图 3.18 所示。

图 3.18　显示当前时间

3.2.6　图像视图

android.widget.ImageView 译为图像视图，直接继承自 android.view.View，显示任意图像，例如图标。ImageView 类可以加载各种来源的图片(如资源或图片库)，需要计算图像的尺寸，比便它可以在其他布局中使用，并提供例如缩放和着色(渲染)各种显示选项。

ImageView 支持表 3.7 所示的 XML 属性。

表 3.7　ImageView 支持的 XML 属性

属性名称	描　述
android:adjustViewBounds	设置该属性为真，可以在 ImageView 调整边界时保持图片的纵横比例
android:baseline	视图内基线的偏移量
android:baselineAlignBottom	如果为 true，图像视图将基线与父控件底部边缘对齐
android:cropToPadding	如果为 true，则剪切图片以适应内边距的大小

续表

<table>
<tr><th>属性名称</th><th>描述</th></tr>
<tr><td>android:maxHeight</td><td>为视图提供最大高度的可选参数。(单独使用无效，需要与setAdjustViewBounds一起使用。如果想设置图片固定大小，又想保持图片宽高比，需要如下设置：
(1) 设置setAdjustViewBounds为true;
(2) 设置maxWidth、MaxHeight;
(3) 设置layout_width和layout_height为wrap_content)</td></tr>
<tr><td>android:maxWidth</td><td>为视图提供最大宽度的可选参数</td></tr>
<tr><td>android:scaleType</td><td>控制为了使图片适合ImageView的大小，应该如何变更图片大小或移动图片。一定是下列常量之一：
<table>
<tr><th>常量</th><th>值</th><th>描述</th></tr>
<tr><td>matrix</td><td>0</td><td>用矩阵来绘图</td></tr>
<tr><td>fitXY</td><td>1</td><td>拉伸图片(不按比例)以填充View的宽高</td></tr>
<tr><td>fitStart</td><td>2</td><td>按比例拉伸图片，拉伸后图片的高度为View的高度，且显示在View的左边</td></tr>
<tr><td>fitCenter</td><td>3</td><td>按比例拉伸图片，拉伸后图片的高度为View的高度，且显示在View的中间</td></tr>
<tr><td>fitEnd</td><td>4</td><td>按比例拉伸图片，拉伸后图片的高度为View的高度，且显示在View的右边</td></tr>
<tr><td>center</td><td>5</td><td>按原图大小显示图片，但图片宽高大于View的宽高时，截图图片中间部分显示</td></tr>
<tr><td>centerCrop</td><td>6</td><td>按比例放大原图直至等于某边View的宽高显示</td></tr>
<tr><td>centerInside</td><td>7</td><td>当原图宽高或等于View的宽高时，按原图大小居中显示；反之将原图缩放至View的宽高居中显示</td></tr>
</table></td></tr>
<tr><td>android:src</td><td>设置可绘制对象作为ImageView显示的内容</td></tr>
<tr><td>android:tint</td><td>为图片设置着色颜色</td></tr>
</table>

实例：使用ImageView制作一个图片查看器。

```
<RelativeLayout xmlns:android = "http://schemas.android.com/apk/res/android"
    android:layout_width = "match_parent"
    android:layout_height = "match_parent"
    android:background = "@color/azure">

    <TableLayout
        android:layout_width = "fill_parent"
        android:layout_height = "wrap_content"
        android:stretchColumns = "0,1"
        android:layout_alignParentBottom = "true"
        android:id = "@ + id/layout">
        <TableRow>
            <Button
                android:layout_width = "wrap_content"
```

```
                android:layout_height = "wrap_content"
                android:text = "上一张"
                android:enabled = "false"
                android:id = "@ + id/btnPrev"/>
            < Button
                android:layout_width = "wrap_content"
                android:layout_height = "wrap_content"
                android:text = "下一张"
                android:id = "@ + id/btnNext"/>
        </TableRow >
    </TableLayout >

    < ImageView
        android:layout_width = "fill_parent"
        android:layout_height = "fill_parent"
        android:src = "@drawable/a"
        android:scaleType = "fitXY"
        android:layout_above = "@id/layout"
        android:id = "@ + id/imageView"/>
</RelativeLayout >
```

代码文件: codes\03\3.2\ ImageViewDemo\res\layout\activity_main.xml

上面的 XML 界面文件在界面底部左右排放了两个按钮：一个用于查看上一张图片，一个用于查看下一张图片，接着在这两个按钮的上方定义了一个 ImageView 用于盛放图片，ImageView 的宽高填充了除了两个按钮所占的其他界面的宽高。下面的程序代码实现单击界面中的两个按钮以便按顺序查看上一张或者下一张图片的功能。

```
package cn.edu.hstc.imageviewdemo.activity;

import android.app.Activity;
import android.graphics.BitmapFactory;
import android.graphics.drawable.BitmapDrawable;
import android.os.Bundle;
import android.view.View;
import android.widget.Button;
import android.widget.ImageView;

public class MainActivity extends Activity {
    private Button btnPrev, btnNex;
    private ImageView imageView;
    int[] images = new int[] {R.drawable.a, R.drawable.b, R.drawable.c};
    int currentImage = 0;
    @Override
    protected void onCreate(Bundle savedInstanceState) {
        super.onCreate(savedInstanceState);
        setContentView(R.layout.activity_main);
        initWidget();
    }
```

```
    private void initWidget() {
        btnPrev = (Button) this.findViewById(R.id.btnPrev);
        btnNex = (Button) this.findViewById(R.id.btnNext);
        imageView = (ImageView) this.findViewById(R.id.imageView);
        btnPrev.setOnClickListener(new TheOnClickListener());
        btnNex.setOnClickListener(new TheOnClickListener());
    }

    private class TheOnClickListener implements View.OnClickListener {
        @Override
        public void onClick(View v) {
            BitmapDrawable bitmapDrawable = (BitmapDrawable) imageView.getDrawable();
            if (!bitmapDrawable.getBitmap().isRecycled()) { //如果未回收图片,强制回收
                bitmapDrawable.getBitmap().recycle();
            }
            switch (v.getId()) {
            case R.id.btnPrev:
                    imageView.setImageBitmap(BitmapFactory.decodeResource(getResources(),
images[--currentImage]));
                if (currentImage <= 0) {
                    currentImage = 0;
                    btnPrev.setEnabled(false);
                } else if (currentImage < 2) {
                    btnNex.setEnabled(true);
                }
                break;
            case R.id.btnNext:
                    imageView.setImageBitmap(BitmapFactory.decodeResource(getResources(),
images[++currentImage]));
                if (currentImage >= 2) {
                    currentImage = 2;
                    btnNex.setEnabled(false);
                } else if (currentImage > 0) {
                    btnPrev.setEnabled(true);
                }
                break;
            default:
                break;
            }
        }
    }
}
```

codes\03\3.2\ImageViewDemo\src\cn\edu\hstc\imageviewdemo\activity\MainActivity.java

将应用部署到模拟器上，运行程序，若当前图片为第一张图片时，“上一张”按钮不可用，若当前图片为最后一张图片时，“下一张”按钮不可用。运行效果如图3.19所示。

图 3.19 ImageView 照片查看器

3.3 本章小结

本章重点介绍了 Android 五大布局，其中读者必须重点关注的是线性布局、表格布局以及相对布局，因为这三个布局在具体的 Android 程序开发中是较常用的，而这三个布局之中又属线性布局尤为常用。接着，介绍了 Android 基本界面组件，因为应用的最终用户是一些不太懂软件的人，那么与用户交互的软件界面是否友好就在某个程度决定了用户对该应用的喜爱和依赖程度，掌握好界面编程的重要性可见一斑。在接下来的第 4 章中，我们将对 Android 的高级组件编程进行系统的学习。

第4章 Android高级界面组件

第 3 章介绍了 Android 的基本界面组件，在实际项目开发中，我们也会使用到 Android 一些比较高级的控件，这些控件的使用使我们开发出来的程序更显“高大上”，更加智能化，也正因为这些高级控件的使用，才丰富了我们的应用。

4.1 Android 高级界面组件的组成

Android 的高级界面组件包括自动完成文本框、下拉列表框 spinner、日期时间选择器、进度条、拖动条、星级评分条、选项卡、滚动视图、列表视图等，本节将重点介绍这几种组件。

4.1.1 自动完成文本框

假如现在正在编制一个地图的应用，我们在界面中放置了一个文本框，用于输入地址，然后希望在输入“韩”这个字的时候就会以下拉框的形式联想显示类似于“韩文公祠”“韩山师范学院”这样的一些地址供我们直接选择，完成地址文本框的输入。这个时候会用到的一个 Android 控件便是自动完成文本框 AutoCompleteTextView。表 4.1 介绍了 AutoCompleteTextView 的一些常用属性及其对应功能。

表 4.1 AutoCompleteTextView 常用属性

属　性	说　明
android:completionHint	设置出现在下拉菜单中的提示标题
android:completionThreshold	设置用户至少输入多少个字符才会显示提示
android:dropDownHorizontalOffset	下拉菜单于文本框之间水平偏移。默认与文本框左对齐
android:dropDownHeight	下拉框的高度
android:dropDownWidth	下拉框的宽度
android:singleLine	单行显示
android:dropDownVerticalOffset	垂直偏移量
android:popupBackground	设置下拉框的背景

使用 AutoCompleteTextView 需要为它设置一个 Adapter 适配器，该 Adapter 封装了 AutoCompleteTextView 预设的提示文本。

实例：自动完成地址输入框信息输入。

```
<LinearLayout xmlns:android = "http://schemas.android.com/apk/res/android"
```

```
    android:layout_width = "match_parent"
    android:layout_height = "match_parent"
    android:orientation = "horizontal" >

    < AutoCompleteTextView
        android:id = "@ + id/txt_auto_address"
        android:layout_width = "fill_parent"
        android:layout_height = "wrap_content"
        android:completionThreshold = "1"
        android:hint = "请输入地址..."
        android:popupBackground = " # 444444"
        android:textColor = "@android:color/secondary_text_dark" />

</LinearLayout >
```

代码文件: codes\04\4.1\AutoCompleteTextViewDemo\res\layout\activity_main.xml

上面界面文件实现将一个自动完成文本框放置于界面中,并设置了自动完成文本框的下拉框的背景颜色为黑色,文本框的字体颜色为灰色,并且设置当用户输入一个字符的时候就会自动提示文本。

```
package cn.edu.hstc.autocompletetextviewdemo.activity;

import android.app.Activity;
import android.os.Bundle;
import android.widget.ArrayAdapter;
import android.widget.AutoCompleteTextView;

public class MainActivity extends Activity {
    String[] address = new String[] { "广东潮州", "广东汕头", "广东揭阳", "广东深圳", "广
东广州", "广东珠海" };

    @Override
    protected void onCreate(Bundle savedInstanceState) {
        super.onCreate(savedInstanceState);
        setContentView(R.layout.activity_main);
        //创建适配器
        ArrayAdapter < String > add = new ArrayAdapter < String >(MainActivity.this, android.
R.layout.simple_dropdown_item_1line, address);
        AutoCompleteTextView act = (AutoCompleteTextView) this .findViewById(R.id.txt_auto
_address);
        //为自动完成输入框设置设配器
        act.setAdapter(add);
    }

}
```

代码文件: codes\04\4.1\AutoCompleteTextViewDemo\src\cn\edu\hstc\autocompletetextview\activity\MainActivity.java

将程序部署在 Android 模拟器上,运行效果如图 4.1 所示。

如图 4.1 所示,用户只需选中任何一个选项,便可自动完成地址文本框的信息输入。

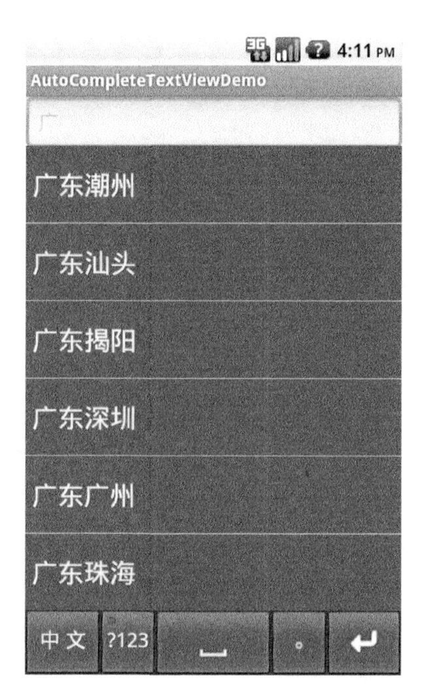

图 4.1 自动完成地址输入

4.1.2 下拉列表框的功能和用法

下拉列表框 Spinner 提供一种下拉列表框选择的输入方式，在手机应用界面上非常常见，使用 Spinner 可以节省有限的屏幕空间占用。

有两种方法可以配置下拉列表控件的下拉内容项：一种是在 XML 文件中预先定义数据，另一种是在编程的时候载入数据。如果下拉列表框中的列表项是确定的，那么建议采用 XML 预设方式。接下来首先介绍这种配置方式。

实例：下拉列表框 Spinner 的使用。

```
<LinearLayout xmlns:android = "http://schemas.android.com/apk/res/android"
    android:layout_width = "match_parent"
    android:layout_height = "match_parent"
    android:orientation = "vertical" >

    <TextView
        android:layout_width = "wrap_content"
        android:layout_height = "wrap_content"
        android:text = "Spinner 数据源来自 XML 文件" />

    <Spinner
        android:id = "@ + id/spinner_xml"
        android:layout_width = "fill_parent"
        android:layout_height = "wrap_content"
        android:entries = "@array/education"
        android:prompt = "@string/hint" />
```

```
    <TextView
        android:layout_width = "wrap_content"
        android:layout_height = "wrap_content"
        android:text = "在 java 文件中加载 Spinner 数据源" />

    <Spinner
        android:id = "@ + id/spinner_java"
        android:layout_width = "fill_parent"
        android:layout_height = "wrap_content"
        android:prompt = "@string/hint1" />

</LinearLayout>
```

代码文件：codes\04\4.1\SpinnerDemo\res\layout\activity_main.xml

由于上面的界面代码中为 Spinner 控件指定了一个 XML 文件中的数组作为其数据源，所以这里需要新建一个文件，代码如下：

```
<?xml version = "1.0" encoding = "UTF - 8"?>
<resources>
    <string - array name = "education">
        <item>博士研究生</item>
        <item>硕士研究生</item>
        <item>本科</item>
        <item>专科</item>
    </string - array>
</resources>
```

代码文件：codes\04\4.1\SpinnerDemo\res\value\education.xml

使用 XML 作为 Spinner 的数据源这种方式的特点是代码简单，将程序部署在模拟器上，运行效果如图 4.2 所示。

图 4.2　Spinner 的用法(一)

接下来继续介绍在 Java 程序中为 Spinner 控件加载数据源的方式，这是一种可以配置 Spinner 下拉列表项内容的方式。

这种方式与将 XML 文件中的数组配置为 Spinner 的数据源的方式的主要区别在于 MainActivity.java，该类文件的代码如下：

```
package cn.edu.hstc.spinnerdemo.activity;

import android.app.Activity;
import android.os.Bundle;
import android.widget.ArrayAdapter;
import android.widget.Spinner;

public class MainActivity extends Activity {
    //定义需要载入的数据
    private static final String[ ] EXPRESS_STRINGS  =
```

```
{"顺丰速递", "申通快递", "圆通快递", "中通快递"};
    //声明 Spinner 对象
    private Spinner expressSpinner;
    //声明适配器对象
    private ArrayAdapter<String> adater;

    @Override
    protected void onCreate(Bundle savedInstanceState) {
        super.onCreate(savedInstanceState);
        setContentView(R.layout.activity_main);
        //从布局文件中加载 Spinner 对象
        expressSpinner = (Spinner) MainActivity.this.findViewById(R.id.spinner_java);
        //创建适配器
        adater = new ArrayAdapter<String>(MainActivity.this, android.R.layout.simple_
spinner_item, EXPRESS_STRINGS);
        //将适配器设置到 Spinner 中
        expressSpinner.setAdapter(adater);
    }

}
```

代码文件：codes\04\4.1\SpinnerDemo\cn\edu\hstc\activity\MainActivity.java

将应用再次部署到模拟器上，运行效果如图 4.3 所示。

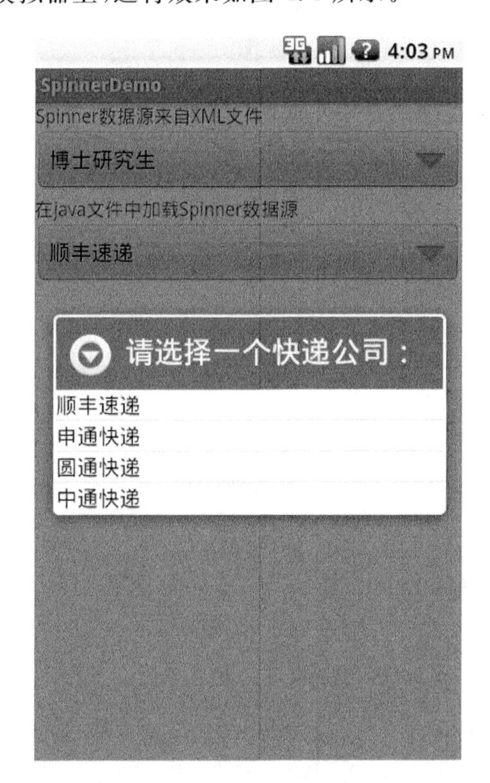

图 4.3　Spinner 的用法(二)

4.1.3 日期、时间选择器

日期选择控件 DatePicker 继承自 FrameLayout 类，其主要功能是向用户提供包含年、月、日的日期数据并允许用户对其进行修改。如果要捕获用户修改日期选择控件中的数据事件，需要为 DatePicker 添加 OnDateChangedListener 监听器。

时间选择器 TimePicker 也继承自 FrameLayout 类，其主要功能是向用户显示一天中的时间，可以为 24 小时的格式，也可以为 AM/PM 制，并允许用户进行选择。如果要捕获用户修改时间数据事件，需要为 TimePicker 添加 OnTimeChangedListener 监听器。

以下通过一个让用户选择日期、时间的实例来示范 DatePicker 和 TimePicker 的功能和用户。界面代码如下：

```
<LinearLayout xmlns:android = "http://schemas.android.com/apk/res/android"
    android:layout_width = "match_parent"
    android:layout_height = "match_parent"
    android:background = "@android:color/white"
    android:orientation = "vertical" >

    <TextView
        android:id = "@ + id/txt_show"
        android:layout_width = "fill_parent"
        android:layout_height = "wrap_content"
        android:paddingLeft = "5dp"
        android:paddingTop = "10dp"
        android:textColor = "@android:color/black" />

    <DatePicker
        android:id = "@ + id/datePicker"
        android:layout_width = "wrap_content"
        android:layout_height = "wrap_content"
        android:layout_gravity = "center_horizontal"
        android:paddingTop = "5dp" />

    <TimePicker
        android:id = "@ + id/timePicker"
        android:layout_width = "wrap_content"
        android:layout_height = "wrap_content"
        android:layout_gravity = "center_horizontal"
        android:paddingTop = "5dp" />

</LinearLayout>
```

代码文件：codes\04\4.1\DateTimePickerDemo\res\layout\activity_main.xml

该代码实现在界面中从上至下放置一个 TextView，用于显示日期与时间；一个 DatePicker 日期控件；一个 TimePicker 时间控件，用于供用户设置日期和时间。

```
package cn.edu.hstc.datetimepickerdemo.activity;

import java.util.Calendar;
```

```
import android.app.Activity;
import android.os.Bundle;
import android.widget.DatePicker;
import android.widget.DatePicker.OnDateChangedListener;
import android.widget.TextView;
import android.widget.TimePicker;
import android.widget.TimePicker.OnTimeChangedListener;

public class MainActivity extends Activity {
    //声明五个记录当前时间的变量
    private int year, month, day, hour, minute;
    private DatePicker datePicker;
    private TimePicker timePicker;

    @Override
    protected void onCreate(Bundle savedInstanceState) {
        super.onCreate(savedInstanceState);
        setContentView(R.layout.activity_main);

        datePicker = (DatePicker) findViewById(R.id.datePicker);
        timePicker = (TimePicker) findViewById(R.id.timePicker);

        //获取当前年、月、日、小时、分钟
        Calendar c  = Calendar.getInstance();
        year = c.get(Calendar.YEAR);
        month = c.get(Calendar.MONTH);
        day = c.get(Calendar.DAY_OF_MONTH);
        hour = c.get(Calendar.HOUR);
        minute = c.get(Calendar.MINUTE);

        datePicker.init(year, month, day, new OnDateChangedListener() {
            @Override
            public void onDateChanged(DatePicker view, int year, int month, int day) {
                MainActivity.this.year = year;
                MainActivity.this.month = month;
                MainActivity.this.day = day;
                show(year, month, day, hour, minute);
            }
        });

        timePicker.setIs24HourView(true);
        timePicker.setCurrentHour(hour);
        timePicker.setCurrentMinute(minute);
        timePicker.setOnTimeChangedListener(new OnTimeChangedListener() {
            @Override
            public void onTimeChanged(TimePicker view, int hour, int minute) {
                MainActivity.this.hour = hour;
                MainActivity.this.minute = minute;
                show(year, month, day, hour, minute);
            }
```

```
        });
    }

    private void show(int year, int month, int day, int hour, int minute) {
        TextView show = (TextView) findViewById(R.id.txt_show);
        show.setText("选择的日期和时间为: " + year + "年" + (month + 1) + "月" + day +
"日" + hour + "时" + minute + "分");
    }
}
```

代码文件: codes\04\4.1\DateTimePickerDemo\cn\edu\hstc\activity\MainActivity.java

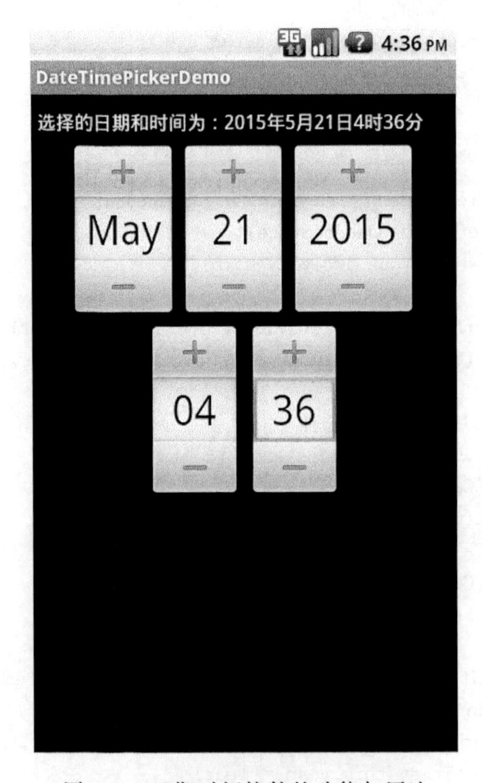

图 4.4　日期时间控件的功能与用法

4.1.4　进度条的介绍与应用

进度条 ProgressBar 是个非常实用的组件，最直观的作用就是提供进度的显示，通常向用户展示某个耗时操作完成的百分比。例如，流媒体播放的缓冲区进度、加载网络内容的进度，进度条的显示避免让用户感觉应用程序失去了响应，改善了用户体验。

Android 对进度条提供了几种不同风格的样式，通过 style 属性可以设置 ProgressBar 的风格，该属性支持如下属性值：

- @android:style/Widget.ProgressBar.Horizontal——水平进度条。
- @android:style/Widget.ProgressBar.Inverse——不断跳跃、旋转画面的进度条。
- @android:style/Widget.ProgressBar.Large——大进度条。
- @android:style/Widget.ProgressBar.Large.Inverse——不断跳跃、旋转画面的大

进度条。

- @android:style/Widget.ProgressBar.Small——小进度条。
- @android:style/Widget.ProgressBar.Small.Inverse——不断跳跃、旋转画面的小进度条。

ProgressBar 也支持如表 4.2 所示的常用 XML 属性。

表 4.2 ProgressBar 常用 XML 属性

XML 属性	说 明
android:max	设置进度条的最大值
android:progress	设置进度条的已完成进度值
android:progressDrawable	设置进度条的轨道绘制形式
android:progressBarStyle	默认进度条样式
android:progressBarStyleHorizontal	水平进度条样式
android:progressBarStyleLarge	大进度条样式
android:progressBarStyleSmall	小进度条样式

最常见的 ProgressBar 是水平进度条与圆形进度条，下面通过一个实例来介绍这两种进度条。

实例：进度条的功能和用法。界面代码如下：

```
<LinearLayout xmlns:android="http://schemas.android.com/apk/res/android"
    xmlns:tools="http://schemas.android.com/tools"
    android:layout_width="match_parent"
    android:layout_height="match_parent"
    android:paddingLeft="10dp"
    android:orientation="vertical">

    <LinearLayout
        android:layout_width="fill_parent"
        android:layout_height="wrap_content"
        android:paddingTop="10dp"
        android:orientation="horizontal">

        <TextView
            android:layout_width="wrap_content"
            android:layout_height="wrap_content"
            android:text="水平进度条当前进度值: " />

        <TextView
            android:id="@+id/txt_horizontal"
            android:layout_width="wrap_content"
            android:layout_height="wrap_content" />

    </LinearLayout>

    <ProgressBar
        android:id="@+id/bar_horizontal"
        android:layout_width="170dp"
```

```
        android:layout_height = "wrap_content"
        android:paddingTop = "5dp"
        android:max = "100"
        style = "@android:style/Widget.ProgressBar.Horizontal" />

    < LinearLayout
        android:layout_width = "fill_parent"
        android:layout_height = "wrap_content"
        android:paddingTop = "10dp"
        android:orientation = "horizontal">

        < TextView
            android:layout_width = "wrap_content"
            android:layout_height = "wrap_content"
            android:text = "圆形进度条当前进度值: " />

        < TextView
            android:id = "@ + id/txt_default"
            android:layout_width = "wrap_content"
            android:layout_height = "wrap_content" />

    </LinearLayout >

    < ProgressBar
        android:id = "@ + id/bar_default"
        android:layout_width = "wrap_content"
        android:layout_height = "wrap_content"
        android:paddingTop = "5dp" />

</LinearLayout >
```

代码文件: codes\04\4.1\ProgressBarDemo\res\layout\activity_main.xml

在上述界面代码中,从上而下放置了两个进度条:一个为水平进度条,一个为默认的圆形进度条。同时放置了两个 TextView 用于显示进度条当前进度值,这两个进度值是随着进度条的进度百分比不断变化的,实现实时更新进度值。核心的业务代码如下:

```
package cn.edu.hstc.progressbardemo.activity;

import android.app.Activity;
import android.os.Bundle;
import android.os.Handler;
import android.os.Message;
import android.view.View;
import android.widget.ProgressBar;
import android.widget.TextView;

public class MainActivity extends Activity {
    //声明界面中的进度条
    private ProgressBar horizontalBar, defaultBar;
    private TextView horizontalTxt, defaultTxt;
    //声明一个 Handler
```

```
private Handler handler;
//声明一个变量用于存放进度条的进度值
private int progress;

@Override
protected void onCreate(Bundle savedInstanceState) {
    super.onCreate(savedInstanceState);
    setContentView(R.layout.activity_main);
    //通过 Id 获取界面中的进度条
    horizontalBar = (ProgressBar) MainActivity.this.findViewById(R.id.bar_horizontal);
    defaultBar = (ProgressBar) MainActivity.this.findViewById(R.id.bar_default);
    horizontalTxt = (TextView) MainActivity.this.findViewById(R.id.txt_horizontal);
    defaultTxt = (TextView) MainActivity.this.findViewById(R.id.txt_default);

    //创建一个 Handler
    handler = new Handler() {
        @Override
        public void handleMessage(Message msg) {
            super.handleMessage(msg);
            switch (msg.what) {
            case 0X11:
                //设置进度条的当前进度以及 EditText 中的值
                horizontalBar.setProgress(progress);
                horizontalTxt.setText(String.valueOf(progress));
                defaultBar.setProgress(progress);
                defaultTxt.setText(String.valueOf(progress));
                if (progress == 100) {
                    defaultBar.setVisibility(View.INVISIBLE);
                }
                break;
            default:
                break;
            }
        }
    };

    //创建一个线程用于改变进度值
    new Thread(new Runnable() {
        @Override
        public void run() {
            int max = horizontalBar.getMax();
            try {
                while (progress < max) {
                    progress += 10;
                    Message msg = new Message();
                    msg.what = 0X11;
                    handler.sendMessage(msg);
                    Thread.sleep(1000);
                }
            } catch (Exception e) {
                e.printStackTrace();
```

```
                    }
                }
            }).start();
        }
    }
```

代码文件：codes\04\4.1\ProgressBarDemo\cn\edu\hstc\activity\MainActivity.java

在MainActivity类中，开启一条子线程不断将进度条的进度值加10，直到该值等于100，并在子线程中发送消息给一个handler，通知其更新界面中的进度条当前进度值，实现实时更新进度条。将程序部署在模拟器上，运行效果如图4.5所示。

水平进度条当前进度值：45

圆形进度条当前进度值：45

图4.5　ProgressBar的功能和用法

4.1.5　拖动条的介绍与应用

拖动条SeekBar，顾名思义，是一种允许用户手动拖动来改变值的组件，类似水平的ProgressBar，只是ProgressBar用颜色的填充来表示进度条的完成百分比，而SeekBar通过滑块的位置来标识数值。由于SeekBar允许用户手动拖动，因此常被用于调节音量、亮度、对比度、饱和度等，旨在对系统某个数值进行调节。

SeekBar可以被用户控制，因此需要对其进行事件监听，这就需要实现SeekBar.OnSeekBarChangeListener接口。在SeekBar中需要监听三个事件，分别是数值的改变(onProgressChanged)、开始拖动(onStartTrackingTouch)、停止拖动(onStopTrackingTouch)。在onProgressChanged中可以得到当前数值的大小。

SeekBar还允许用户自定义拖动条的滑块外观，用户可以通过android:thumb属性来设置，该属性指定一个Drawable对象，将该对象作为自定义滑块。下面通过一个实例来演示拖动条的开发。

实例：使用拖动条SeekBar控制图片大小。

```
<LinearLayout xmlns:android="http://schemas.android.com/apk/res/android"
    xmlns:tools="http://schemas.android.com/tools"
    android:layout_width="match_parent"
    android:layout_height="match_parent"
    android:background="@android:color/black"
    android:orientation="vertical" >

    <SeekBar
        android:id="@+id/seekBar"
        android:layout_width="fill_parent"
        android:layout_height="wrap_content"
        android:max="200"
        android:progress="200"
```

```
        android:thumb = "@drawable/ic_launcher" />

    < ImageView
        android:id = "@ + id/imageView"
        android:layout_width = "200dp"
        android:layout_height = "200dp"
        android:src = "@drawable/ic_launcher"
        android:layout_gravity = "center"
        android:scaleType = "fitXY" />

</LinearLayout >
```

代码文件：codes\04\4.1\SeekBarDemo\res\layout\activity_main.xml

以上界面代码非常简单，只是在界面中从上而下放置了一个拖动条以及一个用于显示图片的 ImageView，主程序的代码如下：

```
package cn.edu.hstc.seekbardemo.activity;

import android.app.Activity;
import android.content.Context;
import android.os.Bundle;
import android.view.WindowManager;
import android.widget.ImageView;
import android.widget.LinearLayout;
import android.widget.SeekBar;
import android.widget.SeekBar.OnSeekBarChangeListener;

public class MainActivity extends Activity {
    private SeekBar seekBar;              //声明一个 SeekBar 对象
    private ImageView imageView;          //声明一个 ImageView 对象
    private int width;                    //声明一个 int 类型的变量,用于存放当前手机屏幕宽度

    @Override
    protected void onCreate(Bundle savedInstanceState) {
        super.onCreate(savedInstanceState);
        setContentView(R.layout.activity_main);
        //接下来两行代码获取了手机屏幕的宽度
         WindowManager wm = (WindowManager) MainActivity.this.getSystemService(Context.
WINDOW_SERVICE);
        width = wm.getDefaultDisplay().getWidth();
        //加载界面代码中的 SeekBar 和 ImageView 对象
        seekBar = (SeekBar) MainActivity.this.findViewById(R.id.seekBar);
        imageView = (ImageView) MainActivity.this.findViewById(R.id.imageView);
        //为 SeekBar 对象设置 OnSeekBarChangeListener 监听器并重写 onProgressChanged 方法
        seekBar.setOnSeekBarChangeListener(new OnSeekBarChangeListener() {
            @Override
            public void onStopTrackingTouch(SeekBar seekBar) {
            }

            @Override
            public void onStartTrackingTouch(SeekBar seekBar) {
```

```
            }

            @Override
            public void onProgressChanged(SeekBar seekBar, int progress, boolean fromUser) {
                //将当前的拖动条的数组 progress 作为 LayoutParams 构造器参数
                 LinearLayout.LayoutParams layoutParams9 = new LinearLayout.LayoutParams
(progress, progress);
                //使图片居中显示
                layoutParams9.leftMargin = (width - progress) / 2;
                //为 ImageView 设置 LayoutParams,实现图片大小由 progress 控制
                imageView.setLayoutParams(layoutParams9);
            }
        });
    }
}
```

代码文件：codes\04\4.1\SeekBarDemo\cn\edu\hstc\activity\MainActivity.java

在该主程序代码中，为 SeekBar 设置了 OnSeekBarChangeListener 监听器并重写了 onProgressChanged 方法，实现将当前的 progress 值，也就是拖动滑块后拖动条的数值作为图片的 width 与 height 值，并且通过获取的当前手机屏幕宽度以及 progress 值控制使图片居中显示。

将应用部署在模拟器上，运行效果如图 4.6 所示。

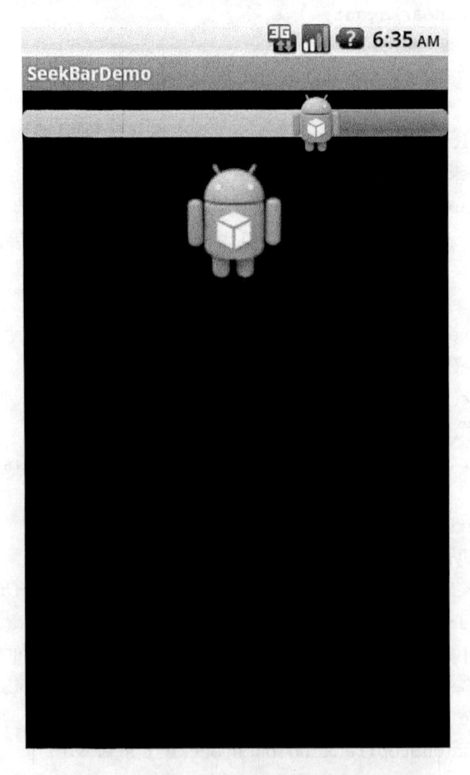

图 4.6　使用 SeekBar 控制图片大小

4.1.6 评分组件的介绍与应用

在很多网站中，我们经常可以看到用户打分的情况，比如天猫买家给商品评分，那么在Android应用中，该如何实现用户参与评分呢？这时候只需要使用RatingBar这个评分组件就能轻松解决问题了。

RatingBar作为SeekBar的一种扩展，在用法和功能上都与后者十分相似，它们都允许用户通过拖动来改变进度，只是评分组件RatingBar的默认效果为若干个星星，用于代表等级，当然，也可以通过设置RatingBar的style属性来自定义其显示效果。

RatingBar支持如表4.3所示的XML属性。

表4.3 RatingBar支持的XML属性

属性名称	描述
android:isIndicator	设置该RatingBar是否允许用户改变其值，true为不允许，此时将作为指示器使用
android:numStars	显示的星星数量，必须是一个整型值，如100
android:rating	设置默认的星级，必须是浮点类型，如1.2
android:stepSize	设置每次最少需要改变多少个星级，必须是浮点类型，如1.2

如果该RatingBar是允许用户改变其值的，那么为了响应用户的操作，可以为RatingBar绑定一个OnRatingBarChangeListener监听器。下面通过一个实例来演示星级评分控件的功能和用法。

实例：使用RatingBar让用户参与评分。界面代码如下：

```
<LinearLayout xmlns:android="http://schemas.android.com/apk/res/android"
    xmlns:tools="http://schemas.android.com/tools"
    android:layout_width="match_parent"
    android:layout_height="match_parent"
    android:background="@android:color/black"
    android:orientation="vertical" >

    <TextView
        android:id="@+id/txt_show"
        android:layout_width="wrap_content"
        android:layout_height="wrap_content"
        android:paddingLeft="5dp"
        android:paddingTop="10dp"
        android:textColor="@android:color/white"
        android:text="您的评分是: 5 分" />

    <RatingBar
        android:id="@+id/ratingBar"
        android:layout_width="wrap_content"
        android:layout_height="wrap_content"
        android:paddingTop="5dp"
        android:max="5"
        android:progress="5"
        android:numStars="5"
        android:stepSize="0.5" />

</LinearLayout>
```

代码文件：codes\04\4.1\RatingBarDemo\res\layout\activity_main.xml

```
package cn.edu.hstc.ratingbardemo.activity;

import android.app.Activity;
import android.os.Bundle;
import android.widget.RatingBar;
import android.widget.RatingBar.OnRatingBarChangeListener;
import android.widget.TextView;

public class MainActivity extends Activity {
    @Override
    protected void onCreate(Bundle savedInstanceState) {
        super.onCreate(savedInstanceState);
        setContentView(R.layout.activity_main);
        final TextView showView = (TextView) MainActivity.this.findViewById(R.id.txt_show);
        RatingBar ratingBar = (RatingBar) MainActivity.this.findViewById(R.id.ratingBar);

        //为 RatingBar 绑定 OnRatingBarChangeListener 监听器，
        //并在 onRatingChanged 方法中实现将当前的 RatingBar 数组作为分数显示出来
        ratingBar.setOnRatingBarChangeListener(new OnRatingBarChangeListener() {
            @Override
            public void onRatingChanged(RatingBar ratingBar, float rating, boolean fromUser) {
                showView.setText("您的评分是：" + String.valueOf(rating) + "分");
            }
        });
    }
}
```

代码文件：codes\04\4.1\RatingBarDemo\cn\edu\hstc\activity\MainActivity.java

将应用部署在模拟器上，运行效果如图 4.7 所示。

图 4.7　用户评分软件

4.1.7　选项卡

选项卡 TabHost 是一个已经过时的组件，现在已经不建议使用了，有其他组件替代了 TabHost，不过这里还是需要对其进行一个简单的介绍，以便在某些简单的场合使用它。

TabHost 是一个容器，而容器中装的就是一个一个的 Tab 标签页，每个标签页相当于获得了一个与外部容器相同大小的组件摆放区域，这样，每个 Tab 都对应了一个独立的布局，从而使一个 Activity 对应了多个功能布局。例如，许多手机系统都会在同一个窗口用多个标签页来显示包括“未接来电”“已接来电”“已拨电话”的通话记录。

TabHost 提供了如下两个方法来创建选项卡、添加选项卡：

- newTabSpec(String tag)——创建选项卡；
- addTab(TabHost. TabSpec tabSpec)——添加选项卡。

下面通过一个实例来演示如何 TabHost 组件，运行效果如图 4.8 所示。

图 4.8　使用 TabHost 示例

实例：使用 TabHost。主界面代码如下：

```
<TabHost xmlns:android = "http://schemas.android.com/apk/res/android"
    android:id = "@android:id/tabhost"
    android:layout_width = "match_parent"
    android:layout_height = "match_parent" >

    <RelativeLayout
        android:layout_width = "fill_parent"
        android:layout_height = "fill_parent" >

        <!-- TabWidget 组件 id 值不可变 -->
        <TabWidget
            android:id = "@android:id/tabs"
            android:layout_width = "fill_parent"
            android:layout_height = "wrap_content"
            android:layout_alignParentBottom = "true" >
        </TabWidget>

        <!-- FrameLayout 布局, id 值不可变 -->
        <FrameLayout
            android:id = "@android:id/tabcontent"
            android:layout_width = "fill_parent"
            android:layout_height = "fill_parent"
```

```
        android:layout_above = "@android:id/tabs" >
    </FrameLayout>

  </RelativeLayout>

</TabHost>
```

代码文件：codes\04\4.1\TabHostDemo\res\layout\activity_main.xml

本实例中还需要四个界面文件，用于存放每个标签页所对应的布局，由于比较简单，这里只给出其中一个布局界面，其他布局文件可参考本书配套资源中的 demo 完整代码。

```
<?xml version = "1.0" encoding = "utf - 8"?>
<RelativeLayout xmlns:android = "http://schemas.android.com/apk/res/android"
    android:layout_width = "match_parent"
    android:layout_height = "match_parent"
    android:orientation = "vertical" >

    <TextView
        android:layout_width = "wrap_content"
        android:layout_height = "wrap_content"
        android:textSize = "20sp"
        android:layout_centerInParent = "true"
        android:text = "Scan" />

</RelativeLayout>
```

代码文件：codes\04\4.1\TabHostDemo\res\layout\activity_scan.xml

主程序代码如下：

```
package cn.edu.hstc.tabhostdemo.activity;

import android.app.TabActivity;
import android.content.Intent;
import android.os.Bundle;
import android.widget.TabHost;

public class MainActivity extends TabActivity {
    private TabHost tabHost;
    private Intent scan, history, generator, setting;

    @Override
    protected void onCreate(Bundle savedInstanceState) {
        super.onCreate(savedInstanceState);
        setContentView(R.layout.activity_main);
        tabHost = getTabHost();
        initIntent();
        addSpec();
    }

    /**
     * 初始化各个 Tab 标签对应的 intent,即单击各个标签页跳转到的页面
```

```
     */
    private void initIntent() {
        scan = new Intent(this, ScanActivity.class);
        history = new Intent(this, HistoryActivity.class);
        generator = new Intent(this, GeneratorActivity.class);
        setting = new Intent(this, SettingActivity.class);
    }

    private void addSpec() {
        tabHost.addTab(this.buildTabSpec("Scan", "Scan", R.drawable.scan_selected_item,
scan));
        tabHost.addTab(this.buildTabSpec("History", "History", R.drawable.history_selected_
item, history));
        tabHost.addTab(this.buildTabSpec("Generator", "Generator", R.drawable.generator_
selected_item, generator));
        tabHost.addTab(this.buildTabSpec("Setting", "Setting", R.drawable.setting_selected_
item, setting));
    }

    /**
     * 创建标签项
     * @param tabName 标签标识
     * @param tabLable 标签文字
     * @param icon 标签图标
     * @param content 标签对应的内容
     * @return
     */
    private TabHost.TabSpec buildTabSpec(String tabName, String tabLable, int icon, Intent
content) {
        return tabHost
                .newTabSpec(tabName)
                .setIndicator(tabLable,
                        getResources().getDrawable(icon)).setContent(content);
    }
}
```

代码文件：codes\04\4.1\TabHostDemo\cn\edu\hstc\activity\MainActivity.java

通过以上主程序代码可以看出，使用 TabHost 的一般步骤如下：

① 在布局界面中定义 TabHost 组件，为该组件定义 TabWidget 标签项，并定义了一个 FrameLayout 用于存放各个标签页；

② Activity 继承 TabActivity；

③ 调用 TabActivity 的 getTabHost()方法获取 TabHost 对象；

④ 为 TabHost 创建选项卡、添加选项卡。

需要指出的是，上面的代码通过 setContent(Intent intent)方法来设置标签页的内容，也就是单击标签项所对应的布局。

4.1.8 滚动视图

当布局中的内容非常多，已经容纳不下时，这时将会用到垂直滚动视图 ScrollView 或水平滚动视图 HorizontalScrollView。ScrollView 和 HorizontalScrollView 的功能基本相似，作用都是为普通组件或布局添加滚动条，不同之处在于前者添加的是垂直滚动条，而后者添加的是水平滚动条。

上述两个组件都只能包含一个组件，但它们都不是传统意义上的容器，都只是为其他容器添加滚动条，并且，ScrollView 与 HorizontalScrollView 可以互相嵌套。

下面通过一个实例来进一步了解 ScrollView 跟 HorizontalScrollView。

实例：结合 ScrollView 和 HorizontalScrollView 实现横竖两方向都可滚动。

```
<ScrollView xmlns:android = "http://schemas.android.com/apk/res/android"
    xmlns:tools = "http://schemas.android.com/tools"
    android:layout_width = "match_parent"
    android:layout_height = "match_parent" >

    <HorizontalScrollView
        android:layout_width = "fill_parent"
        android:layout_height = "wrap_content" >

        <LinearLayout
            android:layout_width = "fill_parent"
            android:layout_height = "wrap_content"
            android:orientation = "vertical" >

            <LinearLayout
                android:layout_width = "fill_parent"
                android:layout_height = "wrap_content"
                android:orientation = "horizontal" >

                <ImageView
                    android:layout_width = "wrap_content"
                    android:layout_height = "wrap_content"
                    android:src = "@drawable/ic_launcher" />

                <ImageView
                    android:layout_width = "wrap_content"
                    android:layout_height = "wrap_content"
                    android:src = "@drawable/ic_launcher" />

                <ImageView
                    android:layout_width = "wrap_content"
                    android:layout_height = "wrap_content"
                    android:src = "@drawable/ic_launcher" />

                <ImageView
                    android:layout_width = "wrap_content"
                    android:layout_height = "wrap_content"
                    android:src = "@drawable/ic_launcher" />

                <ImageView
                    android:layout_width = "wrap_content"
```

```
        android:layout_height = "wrap_content"
        android:src = "@drawable/ic_launcher" />

    < ImageView
        android:layout_width = "wrap_content"
        android:layout_height = "wrap_content"
        android:src = "@drawable/ic_launcher" />

    < ImageView
        android:layout_width = "wrap_content"
        android:layout_height = "wrap_content"
        android:src = "@drawable/ic_launcher" />

    < ImageView
        android:layout_width = "wrap_content"
        android:layout_height = "wrap_content"
        android:src = "@drawable/ic_launcher" />

    < ImageView
        android:layout_width = "wrap_content"
        android:layout_height = "wrap_content"
        android:src = "@drawable/ic_launcher" />

    < ImageView
        android:layout_width = "wrap_content"
        android:layout_height = "wrap_content"
        android:src = "@drawable/ic_launcher" />
</LinearLayout >
< LinearLayout
    android:layout_width = "fill_parent"
    android:layout_height = "wrap_content"
    android:orientation = "horizontal" >

    < ImageView
        android:layout_width = "wrap_content"
        android:layout_height = "wrap_content"
        android:src = "@drawable/ic_launcher" />

    < ImageView
        android:layout_width = "wrap_content"
        android:layout_height = "wrap_content"
        android:src = "@drawable/ic_launcher" />

    < ImageView
        android:layout_width = "wrap_content"
        android:layout_height = "wrap_content"
        android:src = "@drawable/ic_launcher" />

    < ImageView
        android:layout_width = "wrap_content"
        android:layout_height = "wrap_content"
```

```
        android:src = "@drawable/ic_launcher" />

    < ImageView
        android:layout_width = "wrap_content"
        android:layout_height = "wrap_content"
        android:src = "@drawable/ic_launcher" />

    < ImageView
        android:layout_width = "wrap_content"
        android:layout_height = "wrap_content"
        android:src = "@drawable/ic_launcher" />

    < ImageView
        android:layout_width = "wrap_content"
        android:layout_height = "wrap_content"
        android:src = "@drawable/ic_launcher" />

    < ImageView
        android:layout_width = "wrap_content"
        android:layout_height = "wrap_content"
        android:src = "@drawable/ic_launcher" />

    < ImageView
        android:layout_width = "wrap_content"
        android:layout_height = "wrap_content"
        android:src = "@drawable/ic_launcher" />

    < ImageView
        android:layout_width = "wrap_content"
        android:layout_height = "wrap_content"
        android:src = "@drawable/ic_launcher" />
</LinearLayout >
< LinearLayout
    android:layout_width = "fill_parent"
    android:layout_height = "wrap_content"
    android:orientation = "horizontal" >

    < ImageView
        android:layout_width = "wrap_content"
        android:layout_height = "wrap_content"
        android:src = "@drawable/ic_launcher" />

    < ImageView
        android:layout_width = "wrap_content"
        android:layout_height = "wrap_content"
        android:src = "@drawable/ic_launcher" />

    < ImageView
        android:layout_width = "wrap_content"
        android:layout_height = "wrap_content"
        android:src = "@drawable/ic_launcher" />
```

```
    < ImageView
        android:layout_width = "wrap_content"
        android:layout_height = "wrap_content"
        android:src = "@drawable/ic_launcher" />

    < ImageView
        android:layout_width = "wrap_content"
        android:layout_height = "wrap_content"
        android:src = "@drawable/ic_launcher" />

    < ImageView
        android:layout_width = "wrap_content"
        android:layout_height = "wrap_content"
        android:src = "@drawable/ic_launcher" />

    < ImageView
        android:layout_width = "wrap_content"
        android:layout_height = "wrap_content"
        android:src = "@drawable/ic_launcher" />

    < ImageView
        android:layout_width = "wrap_content"
        android:layout_height = "wrap_content"
        android:src = "@drawable/ic_launcher" />

    < ImageView
        android:layout_width = "wrap_content"
        android:layout_height = "wrap_content"
        android:src = "@drawable/ic_launcher" />

    < ImageView
        android:layout_width = "wrap_content"
        android:layout_height = "wrap_content"
        android:src = "@drawable/ic_launcher" />
</LinearLayout >
< LinearLayout
    android:layout_width = "fill_parent"
    android:layout_height = "wrap_content"
    android:orientation = "horizontal" >

    < ImageView
        android:layout_width = "wrap_content"
        android:layout_height = "wrap_content"
        android:src = "@drawable/ic_launcher" />

    < ImageView
        android:layout_width = "wrap_content"
        android:layout_height = "wrap_content"
        android:src = "@drawable/ic_launcher" />
```

```
        <ImageView
            android:layout_width="wrap_content"
            android:layout_height="wrap_content"
            android:src="@drawable/ic_launcher" />

        <ImageView
            android:layout_width="wrap_content"
            android:layout_height="wrap_content"
            android:src="@drawable/ic_launcher" />

        <ImageView
            android:layout_width="wrap_content"
            android:layout_height="wrap_content"
            android:src="@drawable/ic_launcher" />

        <ImageView
            android:layout_width="wrap_content"
            android:layout_height="wrap_content"
            android:src="@drawable/ic_launcher" />

        <ImageView
            android:layout_width="wrap_content"
            android:layout_height="wrap_content"
            android:src="@drawable/ic_launcher" />

        <ImageView
            android:layout_width="wrap_content"
            android:layout_height="wrap_content"
            android:src="@drawable/ic_launcher" />

        <ImageView
            android:layout_width="wrap_content"
            android:layout_height="wrap_content"
            android:src="@drawable/ic_launcher" />

        <ImageView
            android:layout_width="wrap_content"
            android:layout_height="wrap_content"
            android:src="@drawable/ic_launcher" />
    </LinearLayout>
    <LinearLayout
        android:layout_width="fill_parent"
        android:layout_height="wrap_content"
        android:orientation="horizontal" >

        <ImageView
            android:layout_width="wrap_content"
            android:layout_height="wrap_content"
            android:src="@drawable/ic_launcher" />

        <ImageView
```

```
        android:layout_width = "wrap_content"
        android:layout_height = "wrap_content"
        android:src = "@drawable/ic_launcher" />

    < ImageView
        android:layout_width = "wrap_content"
        android:layout_height = "wrap_content"
        android:src = "@drawable/ic_launcher" />

    < ImageView
        android:layout_width = "wrap_content"
        android:layout_height = "wrap_content"
        android:src = "@drawable/ic_launcher" />

    < ImageView
        android:layout_width = "wrap_content"
        android:layout_height = "wrap_content"
        android:src = "@drawable/ic_launcher" />

    < ImageView
        android:layout_width = "wrap_content"
        android:layout_height = "wrap_content"
        android:src = "@drawable/ic_launcher" />

    < ImageView
        android:layout_width = "wrap_content"
        android:layout_height = "wrap_content"
        android:src = "@drawable/ic_launcher" />

    < ImageView
        android:layout_width = "wrap_content"
        android:layout_height = "wrap_content"
        android:src = "@drawable/ic_launcher" />

    < ImageView
        android:layout_width = "wrap_content"
        android:layout_height = "wrap_content"
        android:src = "@drawable/ic_launcher" />

    < ImageView
        android:layout_width = "wrap_content"
        android:layout_height = "wrap_content"
        android:src = "@drawable/ic_launcher" />
</LinearLayout >
< LinearLayout
    android:layout_width = "fill_parent"
    android:layout_height = "wrap_content"
    android:orientation = "horizontal" >

    < ImageView
        android:layout_width = "wrap_content"
```

```
            android:layout_height = "wrap_content"
            android:src = "@drawable/ic_launcher" />

        < ImageView
            android:layout_width = "wrap_content"
            android:layout_height = "wrap_content"
            android:src = "@drawable/ic_launcher" />

        < ImageView
            android:layout_width = "wrap_content"
            android:layout_height = "wrap_content"
            android:src = "@drawable/ic_launcher" />

        < ImageView
            android:layout_width = "wrap_content"
            android:layout_height = "wrap_content"
            android:src = "@drawable/ic_launcher" />

        < ImageView
            android:layout_width = "wrap_content"
            android:layout_height = "wrap_content"
            android:src = "@drawable/ic_launcher" />

        < ImageView
            android:layout_width = "wrap_content"
            android:layout_height = "wrap_content"
            android:src = "@drawable/ic_launcher" />

        < ImageView
            android:layout_width = "wrap_content"
            android:layout_height = "wrap_content"
            android:src = "@drawable/ic_launcher" />

        < ImageView
            android:layout_width = "wrap_content"
            android:layout_height = "wrap_content"
            android:src = "@drawable/ic_launcher" />

        < ImageView
            android:layout_width = "wrap_content"
            android:layout_height = "wrap_content"
            android:src = "@drawable/ic_launcher" />

        < ImageView
            android:layout_width = "wrap_content"
            android:layout_height = "wrap_content"
            android:src = "@drawable/ic_launcher" />
    </LinearLayout >
    < LinearLayout
        android:layout_width = "fill_parent"
        android:layout_height = "wrap_content"
```

```
android:orientation = "horizontal" >

< ImageView
    android:layout_width = "wrap_content"
    android:layout_height = "wrap_content"
    android:src = "@drawable/ic_launcher" />

< ImageView
    android:layout_width = "wrap_content"
    android:layout_height = "wrap_content"
    android:src = "@drawable/ic_launcher" />

< ImageView
    android:layout_width = "wrap_content"
    android:layout_height = "wrap_content"
    android:src = "@drawable/ic_launcher" />

< ImageView
    android:layout_width = "wrap_content"
    android:layout_height = "wrap_content"
    android:src = "@drawable/ic_launcher" />

< ImageView
    android:layout_width = "wrap_content"
    android:layout_height = "wrap_content"
    android:src = "@drawable/ic_launcher" />

< ImageView
    android:layout_width = "wrap_content"
    android:layout_height = "wrap_content"
    android:src = "@drawable/ic_launcher" />

< ImageView
    android:layout_width = "wrap_content"
    android:layout_height = "wrap_content"
    android:src = "@drawable/ic_launcher" />

< ImageView
    android:layout_width = "wrap_content"
    android:layout_height = "wrap_content"
    android:src = "@drawable/ic_launcher" />

< ImageView
    android:layout_width = "wrap_content"
    android:layout_height = "wrap_content"
    android:src = "@drawable/ic_launcher" />

< ImageView
    android:layout_width = "wrap_content"
    android:layout_height = "wrap_content"
    android:src = "@drawable/ic_launcher" />
```

```
</LinearLayout>
<LinearLayout
    android:layout_width="fill_parent"
    android:layout_height="wrap_content"
    android:orientation="horizontal">

    <ImageView
        android:layout_width="wrap_content"
        android:layout_height="wrap_content"
        android:src="@drawable/ic_launcher" />

    <ImageView
        android:layout_width="wrap_content"
        android:layout_height="wrap_content"
        android:src="@drawable/ic_launcher" />

    <ImageView
        android:layout_width="wrap_content"
        android:layout_height="wrap_content"
        android:src="@drawable/ic_launcher" />

    <ImageView
        android:layout_width="wrap_content"
        android:layout_height="wrap_content"
        android:src="@drawable/ic_launcher" />

    <ImageView
        android:layout_width="wrap_content"
        android:layout_height="wrap_content"
        android:src="@drawable/ic_launcher" />

    <ImageView
        android:layout_width="wrap_content"
        android:layout_height="wrap_content"
        android:src="@drawable/ic_launcher" />

    <ImageView
        android:layout_width="wrap_content"
        android:layout_height="wrap_content"
        android:src="@drawable/ic_launcher" />

    <ImageView
        android:layout_width="wrap_content"
        android:layout_height="wrap_content"
        android:src="@drawable/ic_launcher" />

    <ImageView
        android:layout_width="wrap_content"
        android:layout_height="wrap_content"
        android:src="@drawable/ic_launcher" />
```

```
    <ImageView
        android:layout_width="wrap_content"
        android:layout_height="wrap_content"
        android:src="@drawable/ic_launcher" />
</LinearLayout>
<LinearLayout
    android:layout_width="fill_parent"
    android:layout_height="wrap_content"
    android:orientation="horizontal" >

    <ImageView
        android:layout_width="wrap_content"
        android:layout_height="wrap_content"
        android:src="@drawable/ic_launcher" />

    <ImageView
        android:layout_width="wrap_content"
        android:layout_height="wrap_content"
        android:src="@drawable/ic_launcher" />

    <ImageView
        android:layout_width="wrap_content"
        android:layout_height="wrap_content"
        android:src="@drawable/ic_launcher" />

    <ImageView
        android:layout_width="wrap_content"
        android:layout_height="wrap_content"
        android:src="@drawable/ic_launcher" />

    <ImageView
        android:layout_width="wrap_content"
        android:layout_height="wrap_content"
        android:src="@drawable/ic_launcher" />

    <ImageView
        android:layout_width="wrap_content"
        android:layout_height="wrap_content"
        android:src="@drawable/ic_launcher" />

    <ImageView
        android:layout_width="wrap_content"
        android:layout_height="wrap_content"
        android:src="@drawable/ic_launcher" />

    <ImageView
        android:layout_width="wrap_content"
        android:layout_height="wrap_content"
        android:src="@drawable/ic_launcher" />

    <ImageView
```

```
        android:layout_width = "wrap_content"
        android:layout_height = "wrap_content"
        android:src = "@drawable/ic_launcher" />

    < ImageView
        android:layout_width = "wrap_content"
        android:layout_height = "wrap_content"
        android:src = "@drawable/ic_launcher" />
</LinearLayout >
< LinearLayout
    android:layout_width = "fill_parent"
    android:layout_height = "wrap_content"
    android:orientation = "horizontal" >

    < ImageView
        android:layout_width = "wrap_content"
        android:layout_height = "wrap_content"
        android:src = "@drawable/ic_launcher" />

    < ImageView
        android:layout_width = "wrap_content"
        android:layout_height = "wrap_content"
        android:src = "@drawable/ic_launcher" />

    < ImageView
        android:layout_width = "wrap_content"
        android:layout_height = "wrap_content"
        android:src = "@drawable/ic_launcher" />

    < ImageView
        android:layout_width = "wrap_content"
        android:layout_height = "wrap_content"
        android:src = "@drawable/ic_launcher" />

    < ImageView
        android:layout_width = "wrap_content"
        android:layout_height = "wrap_content"
        android:src = "@drawable/ic_launcher" />

    < ImageView
        android:layout_width = "wrap_content"
        android:layout_height = "wrap_content"
        android:src = "@drawable/ic_launcher" />

    < ImageView
        android:layout_width = "wrap_content"
        android:layout_height = "wrap_content"
        android:src = "@drawable/ic_launcher" />

    < ImageView
        android:layout_width = "wrap_content"
```

```
        android:layout_height = "wrap_content"
        android:src = "@drawable/ic_launcher" />

    < ImageView
        android:layout_width = "wrap_content"
        android:layout_height = "wrap_content"
        android:src = "@drawable/ic_launcher" />

    < ImageView
        android:layout_width = "wrap_content"
        android:layout_height = "wrap_content"
        android:src = "@drawable/ic_launcher" />
</LinearLayout >
< LinearLayout
    android:layout_width = "fill_parent"
    android:layout_height = "wrap_content"
    android:orientation = "horizontal" >

    < ImageView
        android:layout_width = "wrap_content"
        android:layout_height = "wrap_content"
        android:src = "@drawable/ic_launcher" />

    < ImageView
        android:layout_width = "wrap_content"
        android:layout_height = "wrap_content"
        android:src = "@drawable/ic_launcher" />

    < ImageView
        android:layout_width = "wrap_content"
        android:layout_height = "wrap_content"
        android:src = "@drawable/ic_launcher" />

    < ImageView
        android:layout_width = "wrap_content"
        android:layout_height = "wrap_content"
        android:src = "@drawable/ic_launcher" />

    < ImageView
        android:layout_width = "wrap_content"
        android:layout_height = "wrap_content"
        android:src = "@drawable/ic_launcher" />

    < ImageView
        android:layout_width = "wrap_content"
        android:layout_height = "wrap_content"
        android:src = "@drawable/ic_launcher" />

    < ImageView
        android:layout_width = "wrap_content"
        android:layout_height = "wrap_content"
```

```
            android:src = "@drawable/ic_launcher" />

        < ImageView
            android:layout_width = "wrap_content"
            android:layout_height = "wrap_content"
            android:src = "@drawable/ic_launcher" />

        < ImageView
            android:layout_width = "wrap_content"
            android:layout_height = "wrap_content"
            android:src = "@drawable/ic_launcher" />

        < ImageView
            android:layout_width = "wrap_content"
            android:layout_height = "wrap_content"
            android:src = "@drawable/ic_launcher" />
    </LinearLayout >
    < LinearLayout
        android:layout_width = "fill_parent"
        android:layout_height = "wrap_content"
        android:orientation = "horizontal" >

        < ImageView
            android:layout_width = "wrap_content"
            android:layout_height = "wrap_content"
            android:src = "@drawable/ic_launcher" />

        < ImageView
            android:layout_width = "wrap_content"
            android:layout_height = "wrap_content"
            android:src = "@drawable/ic_launcher" />

        < ImageView
            android:layout_width = "wrap_content"
            android:layout_height = "wrap_content"
            android:src = "@drawable/ic_launcher" />

        < ImageView
            android:layout_width = "wrap_content"
            android:layout_height = "wrap_content"
            android:src = "@drawable/ic_launcher" />

        < ImageView
            android:layout_width = "wrap_content"
            android:layout_height = "wrap_content"
            android:src = "@drawable/ic_launcher" />

        < ImageView
            android:layout_width = "wrap_content"
            android:layout_height = "wrap_content"
            android:src = "@drawable/ic_launcher" />
```

```
    <ImageView
        android:layout_width="wrap_content"
        android:layout_height="wrap_content"
        android:src="@drawable/ic_launcher" />

    <ImageView
        android:layout_width="wrap_content"
        android:layout_height="wrap_content"
        android:src="@drawable/ic_launcher" />

    <ImageView
        android:layout_width="wrap_content"
        android:layout_height="wrap_content"
        android:src="@drawable/ic_launcher" />

    <ImageView
        android:layout_width="wrap_content"
        android:layout_height="wrap_content"
        android:src="@drawable/ic_launcher" />
</LinearLayout>
<LinearLayout
    android:layout_width="fill_parent"
    android:layout_height="wrap_content"
    android:orientation="horizontal" >

    <ImageView
        android:layout_width="wrap_content"
        android:layout_height="wrap_content"
        android:src="@drawable/ic_launcher" />

    <ImageView
        android:layout_width="wrap_content"
        android:layout_height="wrap_content"
        android:src="@drawable/ic_launcher" />

    <ImageView
        android:layout_width="wrap_content"
        android:layout_height="wrap_content"
        android:src="@drawable/ic_launcher" />

    <ImageView
        android:layout_width="wrap_content"
        android:layout_height="wrap_content"
        android:src="@drawable/ic_launcher" />

    <ImageView
        android:layout_width="wrap_content"
        android:layout_height="wrap_content"
        android:src="@drawable/ic_launcher" />
```

```
        <ImageView
            android:layout_width="wrap_content"
            android:layout_height="wrap_content"
            android:src="@drawable/ic_launcher" />

        <ImageView
            android:layout_width="wrap_content"
            android:layout_height="wrap_content"
            android:src="@drawable/ic_launcher" />

        <ImageView
            android:layout_width="wrap_content"
            android:layout_height="wrap_content"
            android:src="@drawable/ic_launcher" />

        <ImageView
            android:layout_width="wrap_content"
            android:layout_height="wrap_content"
            android:src="@drawable/ic_launcher" />

        <ImageView
            android:layout_width="wrap_content"
            android:layout_height="wrap_content"
            android:src="@drawable/ic_launcher" />
    </LinearLayout>
    <LinearLayout
        android:layout_width="fill_parent"
        android:layout_height="wrap_content"
        android:orientation="horizontal" >

        <ImageView
            android:layout_width="wrap_content"
            android:layout_height="wrap_content"
            android:src="@drawable/ic_launcher" />

        <ImageView
            android:layout_width="wrap_content"
            android:layout_height="wrap_content"
            android:src="@drawable/ic_launcher" />

        <ImageView
            android:layout_width="wrap_content"
            android:layout_height="wrap_content"
            android:src="@drawable/ic_launcher" />

        <ImageView
            android:layout_width="wrap_content"
            android:layout_height="wrap_content"
            android:src="@drawable/ic_launcher" />

        <ImageView
```

```
            android:layout_width = "wrap_content"
            android:layout_height = "wrap_content"
            android:src = "@drawable/ic_launcher" />

        < ImageView
            android:layout_width = "wrap_content"
            android:layout_height = "wrap_content"
            android:src = "@drawable/ic_launcher" />

        < ImageView
            android:layout_width = "wrap_content"
            android:layout_height = "wrap_content"
            android:src = "@drawable/ic_launcher" />

        < ImageView
            android:layout_width = "wrap_content"
            android:layout_height = "wrap_content"
            android:src = "@drawable/ic_launcher" />

        < ImageView
            android:layout_width = "wrap_content"
            android:layout_height = "wrap_content"
            android:src = "@drawable/ic_launcher" />

        < ImageView
            android:layout_width = "wrap_content"
            android:layout_height = "wrap_content"
            android:src = "@drawable/ic_launcher" />
    </LinearLayout >
    < LinearLayout
        android:layout_width = "fill_parent"
        android:layout_height = "wrap_content"
        android:orientation = "horizontal" >

        < ImageView
            android:layout_width = "wrap_content"
            android:layout_height = "wrap_content"
            android:src = "@drawable/ic_launcher" />

        < ImageView
            android:layout_width = "wrap_content"
            android:layout_height = "wrap_content"
            android:src = "@drawable/ic_launcher" />

        < ImageView
            android:layout_width = "wrap_content"
            android:layout_height = "wrap_content"
            android:src = "@drawable/ic_launcher" />

        < ImageView
            android:layout_width = "wrap_content"
```

```
            android:layout_height = "wrap_content"
            android:src = "@drawable/ic_launcher" />

        <ImageView
            android:layout_width = "wrap_content"
            android:layout_height = "wrap_content"
            android:src = "@drawable/ic_launcher" />

        <ImageView
            android:layout_width = "wrap_content"
            android:layout_height = "wrap_content"
            android:src = "@drawable/ic_launcher" />

        <ImageView
            android:layout_width = "wrap_content"
            android:layout_height = "wrap_content"
            android:src = "@drawable/ic_launcher" />

        <ImageView
            android:layout_width = "wrap_content"
            android:layout_height = "wrap_content"
            android:src = "@drawable/ic_launcher" />

        <ImageView
            android:layout_width = "wrap_content"
            android:layout_height = "wrap_content"
            android:src = "@drawable/ic_launcher" />

        <ImageView
            android:layout_width = "wrap_content"
            android:layout_height = "wrap_content"
            android:src = "@drawable/ic_launcher" />
    </LinearLayout>
    <LinearLayout
        android:layout_width = "fill_parent"
        android:layout_height = "wrap_content"
        android:orientation = "horizontal" >

        <ImageView
            android:layout_width = "wrap_content"
            android:layout_height = "wrap_content"
            android:src = "@drawable/ic_launcher" />

        <ImageView
            android:layout_width = "wrap_content"
            android:layout_height = "wrap_content"
            android:src = "@drawable/ic_launcher" />

        <ImageView
            android:layout_width = "wrap_content"
            android:layout_height = "wrap_content"
```

```
        android:src = "@drawable/ic_launcher" />

    < ImageView
        android:layout_width = "wrap_content"
        android:layout_height = "wrap_content"
        android:src = "@drawable/ic_launcher" />

    < ImageView
        android:layout_width = "wrap_content"
        android:layout_height = "wrap_content"
        android:src = "@drawable/ic_launcher" />

    < ImageView
        android:layout_width = "wrap_content"
        android:layout_height = "wrap_content"
        android:src = "@drawable/ic_launcher" />

    < ImageView
        android:layout_width = "wrap_content"
        android:layout_height = "wrap_content"
        android:src = "@drawable/ic_launcher" />

    < ImageView
        android:layout_width = "wrap_content"
        android:layout_height = "wrap_content"
        android:src = "@drawable/ic_launcher" />

    < ImageView
        android:layout_width = "wrap_content"
        android:layout_height = "wrap_content"
        android:src = "@drawable/ic_launcher" />

    < ImageView
        android:layout_width = "wrap_content"
        android:layout_height = "wrap_content"
        android:src = "@drawable/ic_launcher" />
</LinearLayout >
< LinearLayout
    android:layout_width = "fill_parent"
    android:layout_height = "wrap_content"
    android:orientation = "horizontal" >

    < ImageView
        android:layout_width = "wrap_content"
        android:layout_height = "wrap_content"
        android:src = "@drawable/ic_launcher" />

    < ImageView
        android:layout_width = "wrap_content"
        android:layout_height = "wrap_content"
        android:src = "@drawable/ic_launcher" />
```

```
    <ImageView
        android:layout_width="wrap_content"
        android:layout_height="wrap_content"
        android:src="@drawable/ic_launcher" />

    <ImageView
        android:layout_width="wrap_content"
        android:layout_height="wrap_content"
        android:src="@drawable/ic_launcher" />

    <ImageView
        android:layout_width="wrap_content"
        android:layout_height="wrap_content"
        android:src="@drawable/ic_launcher" />

    <ImageView
        android:layout_width="wrap_content"
        android:layout_height="wrap_content"
        android:src="@drawable/ic_launcher" />

    <ImageView
        android:layout_width="wrap_content"
        android:layout_height="wrap_content"
        android:src="@drawable/ic_launcher" />

    <ImageView
        android:layout_width="wrap_content"
        android:layout_height="wrap_content"
        android:src="@drawable/ic_launcher" />

    <ImageView
        android:layout_width="wrap_content"
        android:layout_height="wrap_content"
        android:src="@drawable/ic_launcher" />

    <ImageView
        android:layout_width="wrap_content"
        android:layout_height="wrap_content"
        android:src="@drawable/ic_launcher" />
</LinearLayout>
<LinearLayout
    android:layout_width="fill_parent"
    android:layout_height="wrap_content"
    android:orientation="horizontal" >

    <ImageView
        android:layout_width="wrap_content"
        android:layout_height="wrap_content"
        android:src="@drawable/ic_launcher" />
```

```
    < ImageView
        android:layout_width = "wrap_content"
        android:layout_height = "wrap_content"
        android:src = "@drawable/ic_launcher" />

    < ImageView
        android:layout_width = "wrap_content"
        android:layout_height = "wrap_content"
        android:src = "@drawable/ic_launcher" />

    < ImageView
        android:layout_width = "wrap_content"
        android:layout_height = "wrap_content"
        android:src = "@drawable/ic_launcher" />

    < ImageView
        android:layout_width = "wrap_content"
        android:layout_height = "wrap_content"
        android:src = "@drawable/ic_launcher" />

    < ImageView
        android:layout_width = "wrap_content"
        android:layout_height = "wrap_content"
        android:src = "@drawable/ic_launcher" />

    < ImageView
        android:layout_width = "wrap_content"
        android:layout_height = "wrap_content"
        android:src = "@drawable/ic_launcher" />

    < ImageView
        android:layout_width = "wrap_content"
        android:layout_height = "wrap_content"
        android:src = "@drawable/ic_launcher" />

    < ImageView
        android:layout_width = "wrap_content"
        android:layout_height = "wrap_content"
        android:src = "@drawable/ic_launcher" />

    < ImageView
        android:layout_width = "wrap_content"
        android:layout_height = "wrap_content"
        android:src = "@drawable/ic_launcher" />
</LinearLayout >< LinearLayout
    android:layout_width = "fill_parent"
    android:layout_height = "wrap_content"
    android:orientation = "horizontal" >

    < ImageView
        android:layout_width = "wrap_content"
```

```
            android:layout_height = "wrap_content"
            android:src = "@drawable/ic_launcher" />

        < ImageView
            android:layout_width = "wrap_content"
            android:layout_height = "wrap_content"
            android:src = "@drawable/ic_launcher" />

        < ImageView
            android:layout_width = "wrap_content"
            android:layout_height = "wrap_content"
            android:src = "@drawable/ic_launcher" />

        < ImageView
            android:layout_width = "wrap_content"
            android:layout_height = "wrap_content"
            android:src = "@drawable/ic_launcher" />

        < ImageView
            android:layout_width = "wrap_content"
            android:layout_height = "wrap_content"
            android:src = "@drawable/ic_launcher" />

        < ImageView
            android:layout_width = "wrap_content"
            android:layout_height = "wrap_content"
            android:src = "@drawable/ic_launcher" />

        < ImageView
            android:layout_width = "wrap_content"
            android:layout_height = "wrap_content"
            android:src = "@drawable/ic_launcher" />

        < ImageView
            android:layout_width = "wrap_content"
            android:layout_height = "wrap_content"
            android:src = "@drawable/ic_launcher" />

        < ImageView
            android:layout_width = "wrap_content"
            android:layout_height = "wrap_content"
            android:src = "@drawable/ic_launcher" />

        < ImageView
            android:layout_width = "wrap_content"
            android:layout_height = "wrap_content"
            android:src = "@drawable/ic_launcher" />
    </LinearLayout >
    < LinearLayout
        android:layout_width = "fill_parent"
        android:layout_height = "wrap_content"
```

```
android:orientation = "horizontal" >

< ImageView
    android:layout_width = "wrap_content"
    android:layout_height = "wrap_content"
    android:src = "@drawable/ic_launcher" />

< ImageView
    android:layout_width = "wrap_content"
    android:layout_height = "wrap_content"
    android:src = "@drawable/ic_launcher" />

< ImageView
    android:layout_width = "wrap_content"
    android:layout_height = "wrap_content"
    android:src = "@drawable/ic_launcher" />

< ImageView
    android:layout_width = "wrap_content"
    android:layout_height = "wrap_content"
    android:src = "@drawable/ic_launcher" />

< ImageView
    android:layout_width = "wrap_content"
    android:layout_height = "wrap_content"
    android:src = "@drawable/ic_launcher" />

< ImageView
    android:layout_width = "wrap_content"
    android:layout_height = "wrap_content"
    android:src = "@drawable/ic_launcher" />

< ImageView
    android:layout_width = "wrap_content"
    android:layout_height = "wrap_content"
    android:src = "@drawable/ic_launcher" />

< ImageView
    android:layout_width = "wrap_content"
    android:layout_height = "wrap_content"
    android:src = "@drawable/ic_launcher" />

< ImageView
    android:layout_width = "wrap_content"
    android:layout_height = "wrap_content"
    android:src = "@drawable/ic_launcher" />

< ImageView
    android:layout_width = "wrap_content"
    android:layout_height = "wrap_content"
    android:src = "@drawable/ic_launcher" />
```

```
</LinearLayout>
<LinearLayout
    android:layout_width="fill_parent"
    android:layout_height="wrap_content"
    android:orientation="horizontal">

    <ImageView
        android:layout_width="wrap_content"
        android:layout_height="wrap_content"
        android:src="@drawable/ic_launcher" />

    <ImageView
        android:layout_width="wrap_content"
        android:layout_height="wrap_content"
        android:src="@drawable/ic_launcher" />

    <ImageView
        android:layout_width="wrap_content"
        android:layout_height="wrap_content"
        android:src="@drawable/ic_launcher" />

    <ImageView
        android:layout_width="wrap_content"
        android:layout_height="wrap_content"
        android:src="@drawable/ic_launcher" />

    <ImageView
        android:layout_width="wrap_content"
        android:layout_height="wrap_content"
        android:src="@drawable/ic_launcher" />

    <ImageView
        android:layout_width="wrap_content"
        android:layout_height="wrap_content"
        android:src="@drawable/ic_launcher" />

    <ImageView
        android:layout_width="wrap_content"
        android:layout_height="wrap_content"
        android:src="@drawable/ic_launcher" />

    <ImageView
        android:layout_width="wrap_content"
        android:layout_height="wrap_content"
        android:src="@drawable/ic_launcher" />

    <ImageView
        android:layout_width="wrap_content"
        android:layout_height="wrap_content"
        android:src="@drawable/ic_launcher" />
```

```
                <ImageView
                    android:layout_width="wrap_content"
                    android:layout_height="wrap_content"
                    android:src="@drawable/ic_launcher" />
            </LinearLayout>
        </LinearLayout>
    </HorizontalScrollView>

</ScrollView>
```

代码文件：codes\04\4.1\ScrollViewDemo\res\layout\activity_main.xml

主程序代码比较简单，只是通过 setContentView(R.layout.activity_main)方法中将界面文件布局设置为该应用启动界面。

将应用部署在模拟器上，运行效果如图 4.9 所示。

当在水平方向上左右滑动界面时，将会看到有一条水平滚动条出现，并且界面可以左右移动，看到更多的水平方向的小机器人；当在垂直方向上下滑动界面时，将会看到一条垂直滚动条，并且界面可以上下移动，也可以看到更多的垂直方向的小机器人。这就是滚动视图给应用带来的效果。

图 4.9　滚动视图的使用

4.1.9　列表视图

列表视图 ListView 是一个非常好用的组件，在 Android 应用中也非常常见，比如通讯录就是一个列表视图，腾讯 QQ 的好友列表也是一个列表视图。可见，ListView 在手机系统中使用非常广泛。

ListView 以垂直列表的形式显示所有列表项，并且能够根据数据的长度进行自适应显示，也就是说，通常因为数据多，ListView 是自动加了滚动条的，创建 ListView 有两种方法：

➢ 加载界面文件中的 ListView 对象并给 ListView 设置适配器 Adapter，即数据源；

➢ 让 Activity 继承 ListActivity。

ListView 提供了如表 4.4 所示的 XML 属性。

表 4.4　ListView 所支持的 XML 属性

XML 属性	说　　明
android:choiceMode	设置 ListView 的选择行为
android:divider	设置列表项的分隔条(既可用颜色分隔，也可用 Drawable 分隔)
android:dividerHeight	设置分隔条的高度
android:entries	指定一个数组资源作为生成 ListView 的数据源

下面通过一个实例来介绍如何使用ListView，在该实例中，将分别介绍在界面文件中直接为ListView绑定数据源，在主程序中通过ArrayAdapter、SimpleAdapter、SimpleCursorAdapter，通过Activity继承ListActivity这几种方式来创建ListView。

实例：ListView的各种创建方式。应用的启动界面代码如下：

```
<?xml version = "1.0" encoding = "utf - 8"?>
< LinearLayout xmlns:android = "http://schemas.android.com/apk/res/android"
    android:layout_width = "fill_parent"
    android:layout_height = "fill_parent"
    android:orientation = "vertical" >

    < Button
        android:id = "@ + id/btn_entries"
        android:layout_width = "fill_parent"
        android:layout_height = "wrap_content"
        android:text = "在界面代码中直接为 ListView 指定填充内容的数组"
        android:layout_marginTop = "5dp" />

    < Button
        android:id = "@ + id/btn_ArrayAdapter"
        android:layout_width = "fill_parent"
        android:layout_height = "wrap_content"
        android:text = "ArrayAdapter 作为 ListView 的适配器" />

    < Button
        android:id = "@ + id/btn_SimpleAdapter"
        android:layout_width = "fill_parent"
        android:layout_height = "wrap_content"
        android:text = "SimpleAdapter 作为 ListView 的适配器" />

    < Button
        android:id = "@ + id/btn_SimpleCursorAdapter"
        android:layout_width = "fill_parent"
        android:layout_height = "wrap_content"
        android:text = "SimpleCursorAdapter 作为 ListView 的适配器" />

    < Button
        android:id = "@ + id/btn_listActivity"
        android:layout_width = "fill_parent"
        android:layout_height = "wrap_content"
        android:text = "基于 ListActivity 实现列表" />

</LinearLayout >
```

代码文件：codes\04\4.1\ListViewDemo\res\layout\activity_main.xml

在该界面代码文件中，在一个垂直方向的线性布局中从上而下放置了五个按钮，各个按钮代表了不同的创建ListView的方式，接下来在主程序中为各个按钮添加时间监听器，代码如下：

```
package cn.edu.hstc.listviewdemo.activtiy;
```

```
import android.app.Activity;
import android.content.Intent;
import android.os.Bundle;
import android.view.View;
import android.widget.Button;
import cn.edu.hstc.listviewdemo.activity.R;

public class MainActivity extends Activity {
    private Button entriesBtn;
    private Button arrayAdapterBtn;
    private Button simpleAdapterBtn;
    private Button simpleCursorAdapterBtn;
    private Button testListActivityBtn;

    @Override
    public void onCreate(Bundle savedInstanceState) {
        super.onCreate(savedInstanceState);
        setContentView(R.layout.activity_main);
        initButton();
    }

    private void initButton() {
        //加载界面代码中各个 Button 按钮
        entriesBtn = (Button) findViewById(R.id.btn_entries);
        arrayAdapterBtn = (Button) findViewById(R.id.btn_ArrayAdapter);
        simpleAdapterBtn = (Button) findViewById(R.id.btn_SimpleAdapter);
        simpleCursorAdapterBtn = (Button) findViewById(R.id.btn_SimpleCursorAdapter);
        testListActivityBtn = (Button) findViewById(R.id.btn_listActivity);

        //为各个按钮添加时间监听器
        entriesBtn.setOnClickListener(btnListener);
        arrayAdapterBtn.setOnClickListener(btnListener);
        simpleAdapterBtn.setOnClickListener(btnListener);
        simpleCursorAdapterBtn.setOnClickListener(btnListener);
        testListActivityBtn.setOnClickListener(btnListener);
    }

    //创建一个时间监听器类,在该类中根据按钮的 Id 为按钮添加不同的动作,
    //即跳转到不同的二级界面中,展示不同的创建 ListView 方式
    private View.OnClickListener btnListener = new View.OnClickListener() {
        @Override
        public void onClick(View v) {
            Intent intent = null;
            switch (v.getId()) {
            case R.id.btn_entries:
                intent = new Intent(MainActivity.this, EntriesActivity.class);
                MainActivity.this.startActivity(intent);
                break;
            case R.id.btn_ArrayAdapter:
                intent = new Intent(MainActivity.this, ArrayAdapterActivity.class);
```

```
                MainActivity.this.startActivity(intent);
                break;
            case R.id.btn_SimpleAdapter:
                intent = new Intent(MainActivity.this, SimpleAdapterActivity.class);
                MainActivity.this.startActivity(intent);
                break;
            case R.id.btn_SimpleCursorAdapter:
                intent = new Intent(MainActivity.this, SimpleCursorAdapterActivity.class);
                MainActivity.this.startActivity(intent);
                break;
            case R.id.btn_listActivity:
                intent = new Intent(MainActivity.this, TestListActivity.class);
                MainActivity.this.startActivity(intent);
                break;
            default:
                break;
            }
        }
    };
}
```

代码文件：codes\04\4.1\ListViewDemo\cn\edu\hstc\activity\MainActivity.java

将程序部署在模拟器上，可以看到如图 4.10 所示的运行界面。

图 4.10 ListViewDemo 运行启动界面

单击启动界面中的第一个 Button 按钮，应用将会跳转到 EntriesActivity 界面，该界面的布局如下：

```
<?xml version = "1.0" encoding = "utf - 8"?>
< LinearLayout xmlns:android = "http://schemas.android.com/apk/res/android"
    android:layout_width = "fill_parent"
    android:layout_height = "fill_parent"
    android:orientation = "vertical" >

    <! -- 直接使用数组资源给出列表项 -->
    <! -- 为 ListView 指定颜色分隔线,并制定分隔线高度 -->
    < ListView
        android:layout_width = "fill_parent"
        android:layout_height = "wrap_content"
        android:entries = "@array/sports"
        android:divider = "@color/yellow"
        android:dividerHeight = "0.5dp" />

</LinearLayout >
```

代码文件：codes\04\4.1\ListViewDemo\res\layout\activity_entries.xml

在该界面代码中，通过 android:entries="@array/sports"直接使用数组作为 ListView 的列表项，这里用到了数组资源 sports，所以需要新建一个 value 文件，代码如下：

```
<?xml version = "1.0" encoding = "utf - 8"?>
< resources >

    < string - array name = "sports">
        < item >篮球</item >
        < item >羽毛球</item >
        < item >乒乓球</item >
        < item >排球</item >
        < item >足球</item >
        < item >毽球</item >
    </string - array >

</resources >
```

代码文件：codes\04\4.1\ListViewDemo\res\values\sports.xml

EntriesActivity.java 的代码非常简单，只是通过 setContentView()方法将 activity_listview_entries.xml 作为其界面布局，这里就不展示了。

二级界面 EntriesActivity 的运行效果如图 4.11 所示。

图 4.11　EntriesActivity 运行效果

单击启动界面的第二个按钮，将跳转到 ArrayAdapterActivity 界面中。在该界面中，使用 ArrayAdapter 作为 ListView 的适配器，界面布局代码与 activity_listview_entries.xml 类似，只是没有为

ListView 指定数组资源，这里就不进行展示了。下面给出 ArrayAdapterActivity.java 的代码内容，如下所示：

```
package cn.edu.hstc.listviewdemo.activtiy;

import java.util.ArrayList;
import java.util.List;

import android.app.Activity;
import android.os.Bundle;
import android.widget.ArrayAdapter;
import android.widget.ListView;
import cn.edu.hstc.listviewdemo.activity.R;

public class ArrayAdapterActivity extends Activity {
    @Override
    protected void onCreate(Bundle savedInstanceState) {
        super.onCreate(savedInstanceState);
        setContentView(R.layout.activity_listview_arrayadapter);
        //加载界面布局中的 ListView 对象
        ListView arrayAdapterListView = (ListView) ArrayAdapterActivity.this.findViewById
(R.id.listview_arrayadapter);
        //将 getData()方法返回的数组包装在一个 ArrayAdapter 中
        ArrayAdapter<String> arrayAdapter = new ArrayAdapter<String>(this, android.R.
layout.simple_list_item_1, getData());
        //为 ListView 设置 Adapter
        arrayAdapterListView.setAdapter(arrayAdapter);
    }

    private List<String> getData() {
        List<String> data = new ArrayList<String>();
        data.add("三国演义");
        data.add("水浒传");
        data.add("西游记");
        data.add("红楼梦");
        return data;
    }
}
```

上面的程序创建了一个 ArrayAdapter，创建 ArrayAdapter 时必须指定一个 textViewResourceId，该参数决定每个列表项的外观样式，如下所示：

- simple_list_item_1——每个列表项都是一个普通的 TextView；
- simple_list_item_2——每个列表项都是一个普通的 TextView，但字体略大；
- simple_list_item_checked——每个列表项都是一个已勾选的列表项；
- simple_list_item_multiple_choice——每个列表项都是带多选框的文本；
- simple_list_item_single_choice——每个列表项都是带单选按钮的文本。

二级界面 ArrayAdapterActivity 的运行效果如图 4.12 所示。

单击启动界面中的第三个按钮，将跳转到使用 SimpleAdapter 作为 ListView 适配器的二级界面 SimpleAdapterActivity，其界面布局与 ArrayAdapterActivity 的界面布局类似，这里也不进行展示。这里关注的是 SimpleAdapterActivity 的实现，代码如下所示：

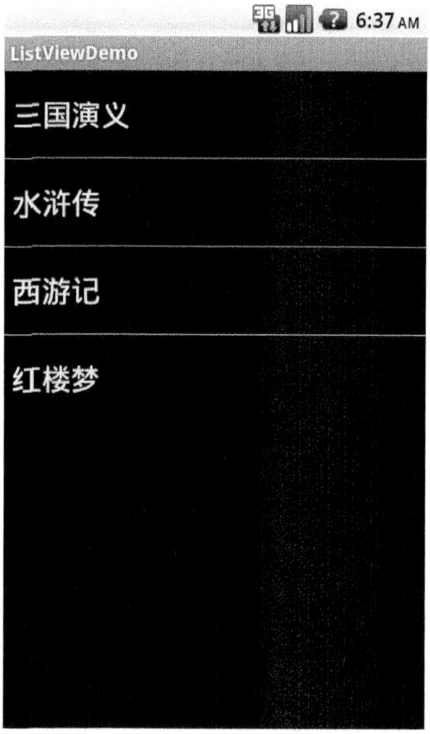

图 4.12　ArrayAdapterActivity 运行效果

```
package cn.edu.hstc.listviewdemo.activtiy;

import java.util.ArrayList;
import java.util.HashMap;
import java.util.List;
import java.util.Map;

import android.app.Activity;
import android.os.Bundle;
import android.widget.ListView;
import android.widget.SimpleAdapter;
import cn.edu.hstc.listviewdemo.activity.R;

public class SimpleAdapterActivity extends Activity {
    @Override
    protected void onCreate(Bundle savedInstanceState) {
        super.onCreate(savedInstanceState);
        setContentView(R.layout.activity_listview_simpleadapter);
        //加载界面中的 ListView 对象
         ListView listView = (ListView) SimpleAdapterActivity.this.findViewById(R.id.
listview_simpleadapter);

        //这里 R.layout.item_listview_simpleadapter 自定义了 ListView 的列表项
        //其中 new String[]{"image", "title", "info"}与 getData 方法中的 map 对象的 key 一一对应
        SimpleAdapter adapter = new SimpleAdapter(this, getData(), R.layout.item_listview_
simpleadapter,
                new String[]{"image", "title", "info"}, new int[]{R.id.image_news, R.id.txt_
title, R.id.txt_info});
        listView.setAdapter(adapter);
    }

    private List<Map<String, Object>> getData() {
        List<Map<String, Object>> list = new ArrayList<Map<String,Object>>();
        Map<String, Object> map = new HashMap<String, Object>();

        map.put("image", R.drawable.image001);
        map.put("title", "欧冠-巴萨 3-1 尤文加冕 3 冠王");
        map.put("info", "苏牙与内少破门,梅西策动进球,巴萨 10 年 4 夺欧冠.");
        list.add(map);
```

```
        map = new HashMap<String, Object>();
        map.put("image", R.drawable.image002);
        map.put("title", "梁咏琪混血女儿萌化网友");
        map.put("info", "宝宝带着小兔帽子,甜美出境,网友: 这以后得多美.");
        list.add(map);

        map = new HashMap<String, Object>();
        map.put("image", R.drawable.image003);
        map.put("title", "高考作文结束 作文题目引吐槽");
        map.put("info", "2015年全国高考大幕拉开,还记得当年高考时的情景么?");
        list.add(map);
        return list;
    }
}
```

以上程序展示了使用 SimpleAdapter 创建 ListView 的方式,在代码中,可以看到,程序使用 item_listview_simpleadapter.xml 自定义了列表项的展示,该文件代码如下:

```
<?xml version="1.0" encoding="utf-8"?>
<LinearLayout xmlns:android="http://schemas.android.com/apk/res/android"
    android:layout_width="fill_parent"
    android:layout_height="fill_parent"
    android:orientation="horizontal" >

    <ImageView
        android:id="@+id/image_news"
        android:layout_width="80dp"
        android:layout_height="50dp"
        android:scaleType="fitXY"
        android:layout_margin="5px" />

    <LinearLayout
        android:layout_width="fill_parent"
        android:layout_height="wrap_content"
        android:layout_gravity="center_vertical"
        android:orientation="vertical" >

        <TextView
            android:id="@+id/txt_title"
            android:layout_width="fill_parent"
            android:layout_height="wrap_content"
            android:textColor="#FFFFFFFF"
            android:textSize="22px" />

        <TextView
            android:id="@+id/txt_info"
            android:layout_width="fill_parent"
            android:layout_height="wrap_content"
            android:paddingTop="3dp"
            android:textColor="#FFFFFFFF"
            android:textSize="16px" />
```

```
    </LinearLayout>

</LinearLayout>
```

通过以上代码可以看出,SimpleAdapterActivity 的 ListView 的列表项包含了一个 ImageView 作为新闻图片,显示在界面最左边,在图片的右边,从上而下,显示了一条新闻的标题和简介。SimpleAdapterActivity 的运行效果如图 4.13 所示。

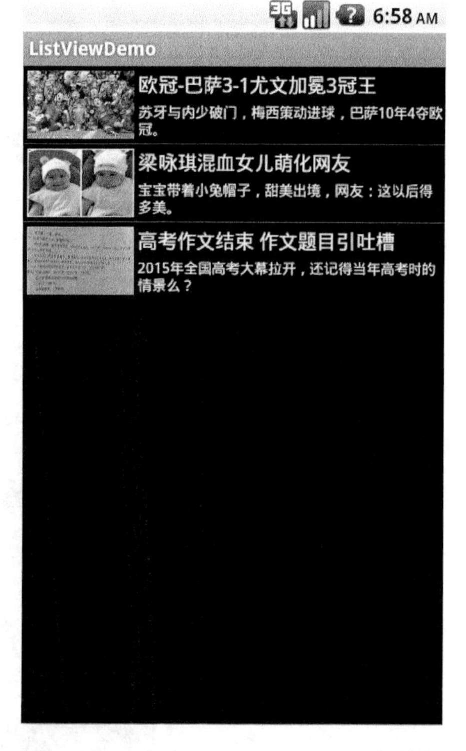

图 4.13 SimpleAdapterActivity 运行效果

单击启动界面的第四个按钮,将跳转到 SimpleCursorAdapterActivity 界面中,其布局与 SimpleAdapterActivity 的布局类似,这里也不展示。我们主要关注的是 SimpleCursorAdapterActivity 的实现代码,实现如下:

```
package cn.edu.hstc.listviewdemo.activtiy;

import android.app.Activity;
import android.database.Cursor;
import android.os.Bundle;
import android.provider.Contacts.People;
import android.widget.ListAdapter;
import android.widget.ListView;
import android.widget.SimpleCursorAdapter;
import cn.edu.hstc.listviewdemo.activity.R;

@SuppressWarnings("deprecation")
public class SimpleCursorAdapterActivity extends
Activity {
    private ListView listView;

    @Override
    protected void onCreate(Bundle savedInstanceState) {
        super.onCreate(savedInstanceState);
        setContentView(R.layout.activity_listview_simplecursoradapter);
        //加载界面文件中的 ListView 对象
        listView = (ListView) SimpleCursorAdapterActivity.this.findViewById(R.id.listview_
simplecursoradapter);

        //查询手机通讯录
        Cursor cursor = getContentResolver().query(People.CONTENT_URI, null, null, null, null);

        //将查询结果 cursor 作为 SimpleCursorAdapter 的填充数据源
        ListAdapter listAdapter = new SimpleCursorAdapter(this, android.R.layout.simple_
expandable_list_item_1, cursor,
                new String[]{People.NAME}, new int[]{android.R.id.text1});

        //为 ListView 对象设置适配器
        listView.setAdapter(listAdapter);
    }
}
```

由于程序需要查询手机通讯录，所以必须在 AndroidManifest. xml 中为程序添加权限：

```
<uses-permission android:name="android.permission.READ_CONTACTS">
```

SimpleCursorAdapterActivity 的运行效果如图 4.14 所示。

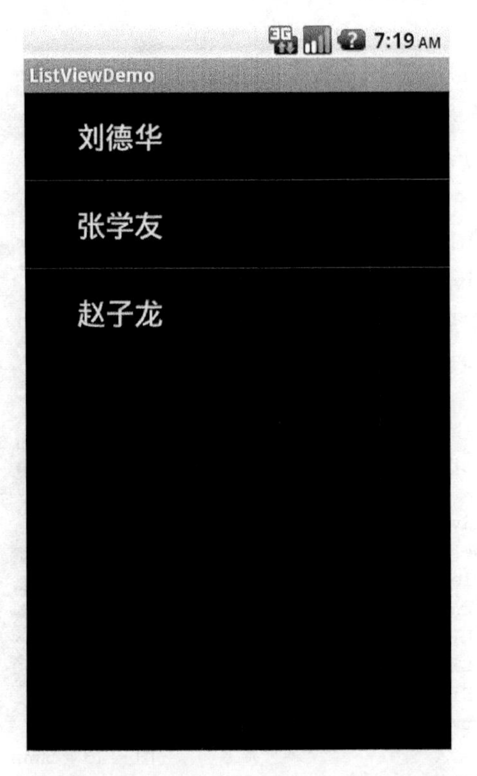

图 4.14 SimpleCursorAdapterActivity 运行效果

单击启动界面的最后一个按钮，跳转到 TestListActivity 界面，该界面不需要指定布局文件。TestListActivity. java 文件代码如下：

```
package cn.edu.hstc.listviewdemo.activtiy;

import android.app.ListActivity;
import android.os.Bundle;
import android.widget.ArrayAdapter;

public class TestListActivity extends ListActivity {
    @Override
    protected void onCreate(Bundle savedInstanceState) {
        super.onCreate(savedInstanceState);
        String[] arr = {"Java", "Android","HTML5"};
        ArrayAdapter<String> adapter = new ArrayAdapter<String>(this, android.R.layout.
simple_list_item_multiple_choice, arr);
        //ListActivity类所带方法,设置该窗体显示列表
        setListAdapter(adapter);
    }
}
```

从以上代码可以看出，程序并没有为 TestListActivity 设置一个布局，而是通过 setListAdapter 方法直接让该窗口显示一个列表，这是因为 ListActivity 有它自己默认的布局，即列表，而 TestListActivity 继承自 ListActivity，所以并不一定要为其指定布局文件。

实际上，我们依旧可以通过 setContentView 方法为继承了 ListActivity 的 Activity 指定一个自定义的布局界面，但这个自定义的布局中需要包含一个 id 为“@＋id/android:list”的 ListView，这是一种固定写法，开发者只需按照约定来编写代码就可以了。例如，以下代码文件：

```
<?xml version = "1.0" encoding = "utf - 8"?>
< LinearLayout xmlns:android = "http://schemas.android.com/apk/res/android"
    android:layout_width = "fill_parent"
    android:layout_height = "fill_parent"
    android:orientation = "vertical" >

    < ListView
        android:id = "@ + id/android:list"
        android:layout_width = "fill_parent"
        android:layout_height = "wrap_content"
        android:divider = "@color/white"
        android:dividerHeight = "0.5dp" />

</LinearLayout >
```

代码文件：codes\04\4.1\ListViewDemo\res\layout\activity_listactivity.xml

TestListActivity 的运行界面如图 4.15 所示。

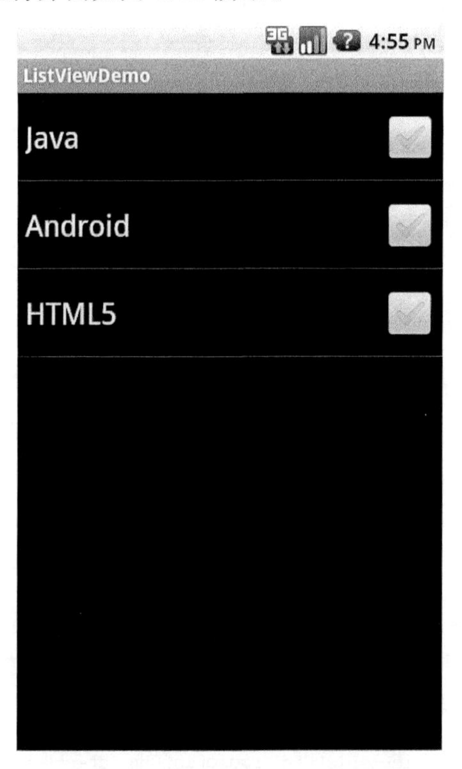

图 4.15　TestListActivity 运行效果

4.2 使用对话框

对话框虽然在程序中不是必备的，但是用好对话框能使应用增色不少，因为采用对话框可以大大地增强应用的友好性。在这个用户体验至上的时代，好的界面就是成功的一半。所以，要学好对话框的使用，并在适当的地方灵活运用。比较常用的场景是：用户登录、网络正在下载、下载成功或失败的提示。当然还有很多场景可以用到，比如短信来了、电池没电了等，一般都需要用到弹出对话框。

以下用一个示例来演示对话框的使用，运行应用后，在主界面会显示几个 Button，单击不同的 Button 会弹出不同的对话框。

在主界面中，从上至下放置了六个 Buuton 按钮，分别对应不同风格的对话框，界面布局文件比较简单，此处不再列出代码，详见代码文件：codes\04\4.2\DialogDemo\res\layout\activity_main.xml。

将程序部署于模拟器上，运行效果如图 4.16 所示。

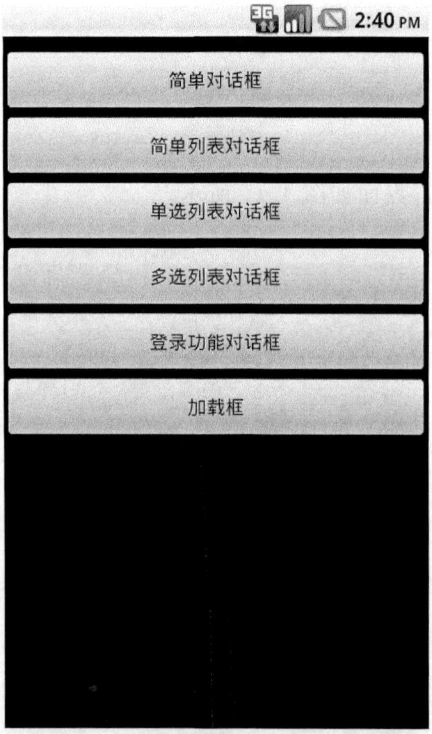

图 4.16 对话框应用主界面

接着，我们来看看主界面对应的 Java 代码，其中，应首先在 onCreate 方法中初始化各个控件并设置监听器，此处 MainActivity 类实现 OnClickListener 接口，其对象将被设置为各个 Button 控件的事件监听器，关键代码如下：

```
private Button simpleBtn, simpleListBtn, radioListBtn, checkBoxListBtn,
            inputBtn, loadBtn;
```

```
@Override
protected void onCreate(Bundle savedInstanceState) {
    super.onCreate(savedInstanceState);
    setContentView(R.layout.activity_main);
    initButton();
}

private void initButton() {
    simpleBtn = (Button) findViewById(R.id.btn_simple);
    simpleBtn.setOnClickListener(MainActivity.this);

    simpleListBtn = (Button) findViewById(R.id.btn_simple_list);
    simpleListBtn.setOnClickListener(MainActivity.this);

    radioListBtn = (Button) findViewById(R.id.btn_radio_list);
    radioListBtn.setOnClickListener(MainActivity.this);

    checkBoxListBtn = (Button) findViewById(R.id.btn_checkbox_list);
    checkBoxListBtn.setOnClickListener(MainActivity.this);

    inputBtn = (Button) findViewById(R.id.btn_input);
    inputBtn.setOnClickListener(MainActivity.this);

    loadBtn = (Button) findViewById(R.id.btn_load);
    loadBtn.setOnClickListener(MainActivity.this);
}

@Override
public void onClick(View v) {
    switch (v.getId()) {
    case R.id.btn_simple:
        createSimpleDialog();
        break;
    case R.id.btn_simple_list:
        createSimpleListDialog();
        break;
    case R.id.btn_radio_list:
        createRadioListDialog();
        break;
    case R.id.btn_checkbox_list:
        createCheckBoxListDialog();
        break;
    case R.id.btn_input:
        createInputDialog();
        break;
    case R.id.btn_load:
        createLoadDialog();
        break;
    default:
        break;
```

```
        }
    }
```

单击界面中的第一个按钮，将会看到弹出如图 4.17 所示的对话框，该对话框也是 Android 最原始最简单的对话框。

图 4.17　简单对话框

第一个按钮“简单对话框”的功能实现的关键代码如下：

```
/**
     * 创建简单对话框
     */
    protected void createSimpleDialog() {
        AlertDialog.Builder builder = new Builder(MainActivity.this);
        builder.setIcon(R.drawable.simple_dialog_icon);
        builder.setTitle("提示");
        builder.setMessage("确认退出吗?");
        builder.setPositiveButton("确认", new DialogInterface.OnClickListener() {
            @Override
            public void onClick(DialogInterface dialog, int which) {
                dialog.dismiss();
                MainActivity.this.finish();
            }
        });
        builder.setNegativeButton("取消", new DialogInterface.OnClickListener() {
            @Override
            public void onClick(DialogInterface dialog, int which) {
```

```
                dialog.dismiss();
            }
        });
        builder.create().show();
    }
```

单击弹出框的“确定”按钮，弹出框消失，接着退出主界面；单击弹出框的“取消”按钮，弹出框消失。

从创建简单对话框的方法中，我们知道，要想创建一个简单的对话框，可以执行如下步骤：

- 通过 AlertDialog.Builder builder = new Builder(context)语句首先生成 AlertDialog.Builder 的对象，这样就可以开始构造 AlertDialog；
- builder.setIcon(R.drawable.simple_dialog_icon)语句给 AlertDialog 预设值一个图标，这里预设的是 simple_dialog_icon.png 这张图片；
- builder.setTitle("提示")语句为 AlertDialog 预设一个标题；
- builder.setMessage("确认退出吗?")语句为 AlertDialog 预设了提示消息；
- builder.setPositiveButton()方法设置弹出框确定按钮的一些属性，第一个参数为按钮上显示出来的内容，第二个参数为 DialogInterface.onClickListener()监听器对象，这个监听器与 2.1 节中介绍的 UI 控件第一种响应方法类似，该监听器实现 onClick()回调方法，当单击 Dialog 的按钮时系统回调该方法，一般将对话框的处理逻辑写到回调方法中；
- builder.setNegativeButton()方法和 builder.setPositiveButton()方法对应，用于设置取消按钮的一些属性；
- 预设好所有关于 Dialog 的属性后，执行 builder.create().show()后将生成一个配置好的 Dialog。

AlertDialog 是 Dialog 的一个直接子类，AlertDialog 也是 Android 系统中最常用的对话框之一。一个 AlertDialog 可以有两个或三个 Button。不能直接通过 AlertDialog 的构造方法来生成一个 AlertDialog，一般是使用它的一个内部静态类 AlertDialog.Builder 来构造的。

单击主界面中的第二个按钮，弹出如图 4.18 所示对话框。

图 4.18　简单列表对话框

生成简单列表对话框的关键代码如下：

```
/**
     * 创建简单列表对话框
     */
    protected void createSimpleListDialog() {
        final Builder b = new AlertDialog.
Builder(MainActivity.this);
        b.setTitle("简单列表对话框");
```

```
        b.setItems(new String[]{"绿色","黄色","黑色"}, new DialogInterface.
OnClickListener() {
            @Override
            public void onClick(DialogInterface dialog, int which) { //该方法的 which 参数代
表用户单击了哪个列表项
                switch (which) {
                case 0:
                    simpleListBtn.setTextColor(R.color.green);
                    break;
                case 1:
                    simpleListBtn.setTextColor(R.color.yellow);
                    break;
                case 2:
                    simpleListBtn.setTextColor(R.color.black);
                    break;
                default:
                    break;
                }
            }
        });
        b.create().show();
    }
```

上面的程序中调用 AlertDialog.Builder 的 setItems 方法为对话框设置了多个列表项，此处生成的只是三个普通列表项。单击任意一个列表项，将为主界面的“简单列表对话框”按钮设置对应的字体颜色。

单击主界面中的第三个按钮“单选列表对话框”，将弹出如图 4.19 所示对话框。

图 4.19　单选列表对话框

生成单选列表对话框的关键代码如下：

```
/**
    * 创建单选列表对话框
    */
   protected void createRadioListDialog() {
       final Builder b = new AlertDialog.Builder(MainActivity.this);
       b.setTitle("单选列表对话框");
        b.setSingleChoiceItems(new String[]{"绿色", "黄色", "黑色"}, 1, new
DialogInterface.OnClickListener() {
           @Override
           public void onClick(DialogInterface dialog, int which) {
               switch (which) {
               case 0:
                   radioListBtn.setTextColor(R.color.green);
                   break;
               case 1:
                   radioListBtn.setTextColor(R.color.yellow);
                   break;
               case 2:
                   radioListBtn.setTextColor(R.color.black);
                   break;
               default:
                   break;
               }
           }
       });
       b.setPositiveButton("确定", null);
       b.create().show();
   }
```

上面的程序中调用 AlertDialog.Builder 的 setSingleChoiceItems 方法为对话框设置了多个单选列表项。单击任意一个列表项，将为主界面的“单选列表对话框”按钮设置对应的字体颜色。

单击主界面中的“多选列表对话框”按钮，弹出如图 4.20 所示对话框。

创建多选列表对话框的关键代码如下：

```
/**
    * 创建多选列表对话框
    */
   protected void createCheckBoxListDialog() {
       final boolean[] checkStatus = new boolean[] {true, false, true};
       final String[] sports = new String[] {"篮球", "羽毛球", "毽球"};
       final Builder b = new AlertDialog.Builder(MainActivity.this);
       b.setTitle("多选列表对话框");
       //下面的 setMultiChoiceItems 方法中的第二个参数 checkStatus 设置了默认勾选的列表项
       b.setMultiChoiceItems(sports, checkStatus, new DialogInterface.OnMultiChoiceClickListener() {
           @Override
           public void onClick(DialogInterface dialog, int which, boolean isChecked) {
               String result = "您喜欢的运动为: ";
```

```
                for (int i = 0; i < checkStatus.length; i++) {
                    if (checkStatus[i]) {
                        result += sports[i] + ",";
                    }
                }
                Toast.makeText(MainActivity.this, result.substring(0, result.length() - 
1), 3000).show();
            }
        });
        b.setPositiveButton("确定", null);
        b.create().show();
    }
```

上面的程序中调用 AlertDialog.Builder 的 setMultiChoiceItems 方法为对话框设置了多个多选列表项。单击任意一个列表项，将弹出类似“您喜欢的运动为篮球、羽毛球”这样的提示。

单击主界面上的登录功能对话框，弹出如图 4.21 所示对话框。

图 4.20　多选列表对话框

图 4.21　登录对话框

创建登录对话框的关键代码如下：

```
/**
     * 创建登录对话框
     */
    protected void createInputDialog() {
        LayoutInflater inflater = LayoutInflater.from(MainActivity.this);
```

```
        final View textEntryView = inflater.inflate(R.layout.input_dialog_layout, null);
        final Builder b = new AlertDialog.Builder(MainActivity.this);
        b.setTitle("用户登录");
        b.setView(textEntryView);
        b.setPositiveButton("确定", new DialogInterface.OnClickListener() {
            @Override
            public void onClick(DialogInterface arg0, int arg1) {
                EditText nameEdt = (EditText) textEntryView.findViewById(R.id.edt_name);
                EditText passEdt = (EditText) textEntryView.findViewById(R.id.edt_pass);
                if (nameEdt.getText().toString().trim().equals("") || passEdt.getText().
toString().trim().equals("")) {
                     Toast.makeText(MainActivity.this, "用户名或密码不能为空!", 3000).
show();
                } else {
                    Toast.makeText(MainActivity.this, "您按了确定键", 3000).show();
                }
            }
        });
        b.setNegativeButton("取消", new DialogInterface.OnClickListener() {
            @Override
            public void onClick(DialogInterface arg0, int arg1) {
                Toast.makeText(MainActivity.this, "您按了取消键", 3000).show();
            }
        });
        b.create().show();
    }
```

通过以上代码可以看出，创建该登录对话框的方法与前面创建其他对话框的方法类似，但是这个方法中没有设置 Dialog 的 message，而是直接给 Dialog 设置了一个定制化的 View 实例。

- 通过 LayoutInflater 类的 inflate 方法，可以将一个 XML 的布局变成一个 View 实例；
- 通过 b.setView(textEntryView)语句将自定义的 View 放置到 Dialog 中去，这是 Dialog 的精髓所在，这里的 textEntryView 和 input_dialog_layout.xml 这个文件定义的布局相关联。input_dialog_layout.xml 的布局文件比较简单，此处不再列出。读者可以参考 codes\04\4.2\DialogDemo\res\layout\ input_dialog_layout.xml 文件。

如果用户名或密码两个或其中一个为空的时候，单击“确定”按钮，就会提示“用户名或密码不能为空!”，否则将提示“您按了确定键”；而当用户单击“取消”按钮，就会提示“您按了取消键”。

单击主界面中的加载框按钮，弹出如图 4.22 所示对话框。

创建加载框的关键代码如下：

```
/**
     * 创建加载框
     */
    protected void createLoadDialog() {
```

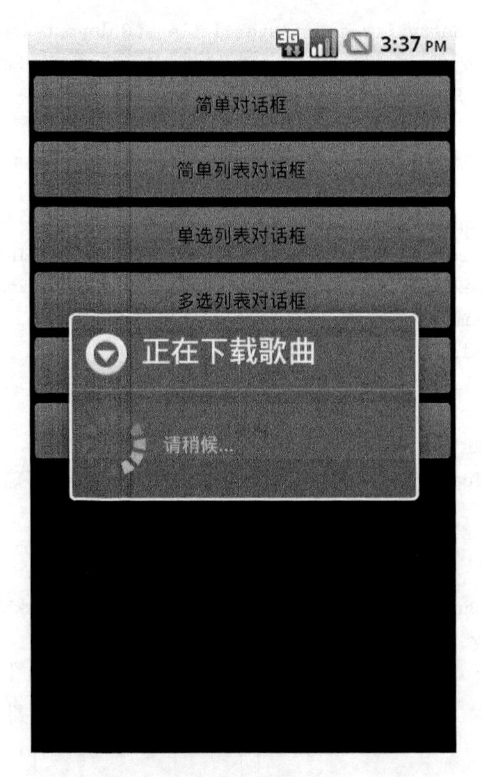

图 4.22 加载框

```
        ProgressDialog dialog = new ProgressDialog(MainActivity.this);
        dialog.setTitle("正在下载歌曲...");
        dialog.setMessage("请稍候...");
        dialog.show();
    }
```

通过以上程序可以看出，创建加载框和创建其他弹出框不一样，不需要通过AlertDialog.Builder这个内部静态类进行构造，而是直接使用ProgressDialog类的构造方法进行构造。ProgressDialog是AlertDialog的一个子类，同样通过setTitle()方法设置标题以及通过setMessage()设置内容。

4.3 Toast和Notification的应用

前面介绍的Dialog对话框其实已经起到向用户提示消息的作用，但其更多用于当程序有大量消息、图片需要向用户提示时，Android系统还提供了一套友好的、更轻量级的对话框供程序只有少量信息要向用户呈现时使用，这种机制不会打断用户的当前操作，可谓非常巧妙，这就是Android的消息提示。

下面通过一个例子来演示使用Toast和Notification两种方式提示用户。该实例在主界面自上而下放置三个Button控件，用来供用户单击进入不同的功能。该主界面代码如下：

```
<LinearLayout xmlns:android = "http://schemas.android.com/apk/res/android"
    android:layout_width = "match_parent"
    android:layout_height = "match_parent"
    android:orientation = "vertical" >

    <Button
        android:id = "@ + id/btn_simple_toast"
        android:layout_width = "fill_parent"
        android:layout_height = "wrap_content"
        android:layout_marginTop = "5dip"
        android:text = "普通的 Toast 消息提示" />

    <Button
        android:id = "@ + id/btn_image_toast"
        android:layout_width = "fill_parent"
        android:layout_height = "wrap_content"
        android:text = "带图片的 Toast 消息提示" />

    <Button
        android:id = "@ + id/btn_create_notification"
        android:layout_width = "fill_parent"
        android:layout_height = "wrap_content"
        android:text = "创建 Notification 消息提示" />

    <Button
        android:id = "@ + id/btn_delete_notification"
        android:layout_width = "fill_parent"
        android:layout_height = "wrap_content"
        android:text = "删除 Notification 消息提示" />

</LinearLayout>
```

该程序部署于模拟器上，运行效果如图 4.23 所示。

该实例主程序代码中关于初始化界面中的控件以及实现各个控件的业务代码块如下所示：

```
private Button simpleToast, imageToast, createNotification, deleteNotification;

@Override
protected void onCreate(Bundle savedInstanceState) {
    super.onCreate(savedInstanceState);
    setContentView(R.layout.activity_main);

    simpleToast = (Button) findViewById(R.id.btn_simple_toast);
    simpleToast.setOnClickListener(MainActivity.this);

    imageToast = (Button) findViewById(R.id.btn_image_toast);
    imageToast.setOnClickListener(MainActivity.this);

    createNotification = (Button) findViewById(R.id.btn_create_notification);
```

```
        createNotification.setOnClickListener(MainActivity.this);

        deleteNotification = (Button) findViewById(R.id.btn_delete_notification);
        deleteNotification.setOnClickListener(MainActivity.this);
    }

    @Override
    public void onClick(View v) {
        switch (v.getId()) {
        case R.id.btn_simple_toast:
            createSimpleToast();
            break;
        case R.id.btn_image_toast:
            createImageToast();
            break;
        case R.id.btn_create_notification:
            createNotification();
            break;
        case R.id.btn_delete_notification:
            deleteNotification();
            break;
        default:
            break;
        }
    }
```

单击界面中的第一个按钮,将弹出如图 4.24 所示的提示。

图 4.23　Android 消息提示实例主界面

图 4.24　普通 Toast 提示消息

实现弹出该普通 Toast 提示消息的关键代码如下：

```
/**
    * 创建普通的 Toast 消息提示
    */
   private void createSimpleToast() {
       //makeText 方法中的第三个参数的作用是设置该 Toast 提示消息的持续时间
       //Toast.LENGTH_SHORT 表示短时间显示,Toast.LENGTH_LONG 表示长时间显示
       //也可以自定义秒数,比如将参数设置为 3000,表示该 Toast 提示消息持续时间为 3 秒
       //最后一点要说的是,不要忘了最后的.show(),这样才能将该 Toast 显示出来
   Toast.makeText(MainActivity.this, "这是一个简单的 Toast 提示", Toast.LENGTH_LONG).show();
   }
```

上面程序中关于实现弹出普通 Toast 提示消息的代码比较简单，只需要通过 Toast 的一个静态方法即可构造该 Toast。下面直接进入带图片的 Toast 提示消息的学习，单击主界面中的第二个 Button 按钮，弹出如图 4.25 所示的提示。

图 4.25　带图片的 Toast 提示消息

实现带图片的 Toast 提示消息的关键代码如下：

```
/**
 * 创建带图片的 Toast 消息提示
 */
private void createImageToast() {
 Toast toast = Toast.makeText(MainActivity.this, "这是一个带图片的 Toast 提示", Toast.
LENGTH_LONG);
 toast.setGravity(Gravity.CENTER, 0, 0);
```

```
    //获取 Toast 提示里原有的 View
    View toastView = toast.getView();
    //创建一个 ImageView
    ImageView image = new ImageView(MainActivity.this);
    image.setImageResource(R.drawable.toast);
    //创建一个 LinearLayout 容器
    LinearLayout layout = new LinearLayout(MainActivity.this);
    //向 layout 中添加图片、原有的 View
    layout.addView(image);
    layout.addView(toastView);
    //将该 layout 作为该 Toast 消息提示所展示的 View
    toast.setView(layout);
    toast.show();
}
```

上面代码已经有相关注释，其中最核心的代码就是获取 Toast 自带的 View，并重新为该 Toast 设定一个新的 View。

从以上两种 Toast 提示消息的学习可以看出，Toast 是一种非常方便的提示消息框，它具有如下两个特点：

- Toast 提示消息不会获得焦点；
- Toast 提示消息过一段时间会自动消失。

从以上程序也可了解到，开发 Toast 提示消息的步骤也是非常简单的，具体步骤如下：

➢ 调用 Toast 的构造器或 makeText()静态方法创建一个 Toast 对象；
➢ 调用 Toast 的方法来设置该消息提示的对齐方式、页边距等；
➢ 调用 Toast 的 show()方法将它显示出来；
➢ 如果需要开发带图片的 Toast，则需要调用 setView()方法设置该 Toast 显示的 View 组件，该方法允许开发者自定义自己的 Toast 显示内容，但一般建议需要显示图片的复杂提示则使用 Dialog 对话框来完成。

单击主界面上的第四个按钮，启动 Notification，将会在手机状态栏显示"启动其他 Activity 的通知"，过一会儿标题会消失，通知图标继续显示在状态栏上，如图 4.26 所示。

实现创建 Notification 通知的关键代码如下：

```
/**
 * 创建 Notification
 */
private void createNotification() {
 //创建一个启动其他 Activity 的 Intent
 Intent intent = new Intent(MainActivity.this, NotificationActivity.class);
 PendingIntent pi = PendingIntent.getActivity(MainActivity.this, 0, intent, 0);
 //创建一个 Notification
 Notification notify = new Notification();
 //为 Notification 设置图标,该图标显示在状态栏
 notify.icon = R.drawable.ic_launcher;
 //为 Notification 设置文本内容,该图标显示在状态栏
 notify.tickerText = "启动其他 Activity 的通知";
 //为 Notification 设置发送时间
 notify.when = System.currentTimeMillis();
```

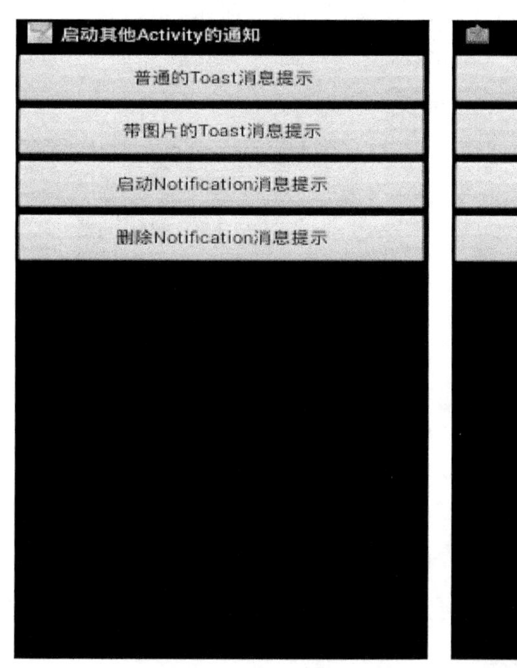

(a) 状态栏显示通知图标与标题

(b) 状态栏只显示通知图标

图 4.26　状态栏的两种情况

```
    //为 Notification 设置声音
    notify.defaults = Notification.DEFAULT_SOUND;
    //为 Notification 设置震动、默认闪光灯
    notify.defaults = Notification.DEFAULT_ALL;
    //设置事件消息
    notify.setLatestEventInfo(MainActivity.this, "普通通知", "点击查看", pi);
    //获取系统的 NotificationManager 服务
    NotificationManager notificationManager = (NotificationManager) getSystemService(NOTIFICATION_
SERVICE);
    //发送通知
    notificationManager.notify(0x1123, notify);
}
```

从以上代码可以看出，创建一个 Notification 并不难，只需要通过以下步骤即可：

- 通过构造器创建一个 Notification 对象；
- 为该 Notification 对象设置各个属性，比如图标、文本标题等等；
- 调用 getSystemService(NOTIFICATION_SERVICE)方法获取系统的 NotificationManager 服务；
- 通过 NotificationManager 发送 Notification。

代码中有一行代码用于设置 Notification 的事件信息，该行代码为 notify.setLatestEventInfo(MainActivity.this, "普通通知", "点击查看", pi)，第四个参数为一个跳转到其他 Activity 的 Intent 对象，表示当用户单击该 Notification 时会启动该 Intent 对应的程序。

将状态栏向下拖动将看到 Notification 的详情，如图 4.27 所示，单击该详情，将会跳到 NotificationMessageActivity 页面，如图 4.28 所示。该页面所对应的界面布局以及主程序比较简单，此处不再赘述。

图 4.27　Notification 通知详情

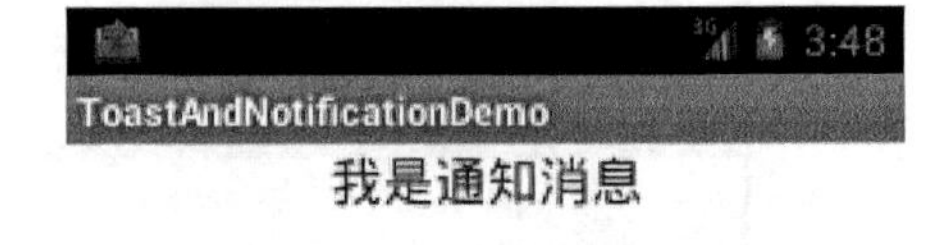

我是通知消息

图 4.28　跳转至其他页面

单击程序主界面的第四个按钮，删除 Notification，将会直接删除该条 Notification 通知，此时状态栏中对应的 Notification 图标也会消失。关键代码如下：

```
/**
 * 删除 Notification
 */
private void deleteNotification() {
 NotificationManager notificationManager = (NotificationManager) getSystemService(NOTIFICATION_
SERVICE);
 notificationManager.cancel(0x1123);
}
```

以上代码通过获取系统的 NotificationManager 服务，然后通过 cancel()方法即可取消该条 Notification，cancel()方法需要传入一个参数，该参数值与启动 Notification 时 notificationManager.notify(0x1123, notify)语句的第一个参数值对应，此处都是 0x1123，通过该标识的传入来通知 Android 取消对应的 Notification 通知。

4.4　使用菜单

受到手机屏幕大小的制约，Android 应用的菜单并不会像桌面应用那样直接展示出来，Android 程序中的菜单默认是看不见的，只有当用户单击手机上的 MENU 键时，系统才会展示该应用所带的菜单。菜单在 Android 系统中的应用还是比较常见的。

Android 平台提供了三种菜单的实现方式，即选项菜单(OptionMenu)、子菜单(SubMenu)、上下文菜单(ContextMenu)。

下面通过一个简单的应用来介绍如何开发菜单程序，程序主界面布局文件如下：

```
<RelativeLayout xmlns:android = "http://schemas.android.com/apk/res/android"
    android:layout_width = "match_parent"
    android:layout_height = "match_parent"
    android:padding = "5dip" >

    <Button
        android:id = "@ + id/btn_pay"
        android:layout_width = "wrap_content"
        android:layout_height = "wrap_content"
        android:layout_alignParentRight = "true"
        android:text = "长按选择..." />

    <EditText
        android:id = "@ + id/edt_pay"
        android:layout_width = "fill_parent"
        android:layout_height = "wrap_content"
        android:layout_toLeftOf = "@id/btn_pay"
        android:editable = "false"
        android:text = "付款方式" />

</RelativeLayout>
```

代码文件：codes\04\4.4\MenuDemo\res\layout\activity_main.xml

将程序部署于 Android 模拟器上，看到如图 4.29 所示运行结果。

图 4.29　菜单应用

从运行效果上看，该界面上没有显示任何控件，这个与上面的布局界面有些不一样，这是因为在主程序中将界面上的 EditText 控件以及 Button 控件给隐藏了。代码片段如下：

```
//声明主界面上的文本输入控件以及 Button 控件
private EditText payEdt;
private Button payBtn;

//为每个上下文菜单定义一个标识
final int MENU1 = 0X111;
```

```
final int MENU2 = 0X112;
final int MENU3 = 0X113;

//定义显示或隐藏控件的子菜单项标识
final int EVISIBIE = 0x114;
final int EINVISIBLE = 0x115;

//定义普通菜单项标识
final int BSUBMIT = 0x116;

@Override
protected void onCreate(Bundle savedInstanceState) {
    super.onCreate(savedInstanceState);
    setContentView(R.layout.activity_main);
    //通过资源 ID 加载界面控件
    payEdt = (EditText) findViewById(R.id.edt_pay);
    payBtn = (Button) findViewById(R.id.btn_pay);

    //隐藏界面上的 EditText 控件以及 Button 按钮
    payEdt.setVisibility(View.INVISIBLE);
    payBtn.setVisibility(View.INVISIBLE);

    //为 Button 按钮注册上下文菜单
    registerForContextMenu(payBtn);
}
```

通过红色框中的两行代码将布局上的两个控件一并隐藏掉，接下来，程序将实现一个带有两个子菜单的选项菜单以及一个不带子菜单的普通选项菜单，代码片段如下：

```
/**
 * 创建选项菜单和子菜单
 */
@Override
public boolean onCreateOptionsMenu(Menu menu) {
    //向 menu 中添加显示或隐藏支付方式的子菜单
SubMenu vis = menu.addSubMenu("显示或隐藏支付方式");
//设置菜单头的标题
vis.setHeaderTitle("选择是否显示支付方式");
vis.add(0, EVISIBIE, 0, "显示");
vis.add(0, EINVISIBLE, 0, "隐藏");
//向 menu 中添加普通菜单项
menu.add(0, BSUBMIT, 0, "提交");
    return super.onCreateOptionsMenu(menu);
}
```

以上代码片段重写了 Activity 类中的 onCreateOptionsMenu(Menu menu)方法，该方法实现用户按下手机上的 menu 按键，将会在手机屏幕正下方出现两个选项菜单，这个也是 menu 菜单的默认显示位置，单击“显示或隐藏支付方式”这个菜单项，将会弹出一个带着两个选项的子菜单，供用户选择“显示”或“隐藏”；另外一个 menu 菜单项是“提交”，该菜单项不带任何子菜单。效果如图 4.30 和图 4.31 所示。

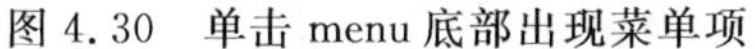

图 4.30 单击 menu 底部出现菜单项

图 4.31 单击第一个菜单项出现了对应的子菜单

此时，我们还并未真正实现“显示”“隐藏”以及“提交”这三个按钮的业务功能，因此需要重写 Activity 类中 onOptionsItemSelected(MenuItem item)回调方法，方法根据各个菜单项的标识，实现不同的操作，代码片段如下：

```
/**
 * 选项菜单和子菜单被单击后的回调方法
 */
@Override
public boolean onOptionsItemSelected(MenuItem item) {
switch (item.getItemId()) {
    case EVISIBIE:
        payEdt.setVisibility(View.VISIBLE);
        payBtn.setVisibility(View.VISIBLE);
        break;
    case EINVISIBLE:
        payEdt.setVisibility(View.INVISIBLE);
        payBtn.setVisibility(View.INVISIBLE);
        break;
    case BSUBMIT:
        Toast.makeText(MainActivity.this, "您单击了普通菜单项,正在提交...", Toast.LENGTH_
LONG).show();
        break;
    default:
        break;
    }
```

```
return true;
}
```

以上程序实现单击子菜单中的“显示”按钮，将会实现显示界面上的 EditText 以及 Button 控件，单击子菜单中的“隐藏”，将会重新隐藏掉这两个控件；用户单击 menu 菜单项“提交”，将会弹出一个 Toast 消息，提示用户正在提交。运行效果如图 4.32 和图 4.33 所示。

图 4.32　显示 EditText 及 Button

图 4.33　单击提交时的 Toast 提示

以上主要介绍了如何开发选项菜单以及子菜单。从上面的应用的开发过程可以看出，开发选项菜单以及子菜单其实非常简单，步骤如下：

- 重写 onCreateOptionsMenu(Menu menu)方法，在该方法中为 menu 添加选项菜单以及子菜单，即决定用户按下手机 menu 按键后在屏幕下方所显示的菜单有哪些，该方法最后返回一个布尔类型值，如果返回 false，则不会显示菜单；
- 重写 onOptionsItemSelected(MenuItem item)方法，在该方法中定义各个菜单项被选中后的动作事件，该方法会在用户选中菜单项时被自动调用。

在此需要向读者说明的是，按下 menu 后的默认样式是在屏幕底部弹出一个菜单，这个菜单就叫做选项菜单 OptionsMenu，一般情况下，选项菜单最多显示两排，每排显示三个菜单项，这些菜单项可以同时包含文字和图标，也被称作 Icon Menus，如果多于六项，从第六项开始会被隐藏，在第六项会出现一个 More 菜单项，单击 More 才出现第六项及以后的菜单项，这些菜单项也被称为扩展菜单(Expanded Menus)。

除了上面介绍的选项菜单以及子菜单，还有另外一种叫做上下文菜单(ContextMenu)的菜单。接下来将继续在该实例上扩展，为其开发上下文菜单。首先，来看一下效果，长按

图 4.33 中的“长按选择”按钮，将会弹出一个带着三个菜单项的菜单，该菜单就是上下文菜单，如图 4.34 所示。

图 4.34　长按弹出上下文菜单

实现上下文菜单的关键代码片段如下：

```
/**
 * 每次创建上下文菜单时都会触发该方法
 */
@Override
public void onCreateContextMenu(ContextMenu menu, View v, ContextMenuInfo menuInfo) {
 menu.add(0, MENU1, 0, "支付宝");
 menu.add(0, MENU2, 0, "财付通");
 menu.add(0, MENU3, 0, "网上银行");
 //将这三个菜单项设置为单选菜单项
 menu.setGroupCheckable(0, true, true);
 //设置上下文菜单的标题和图标
 menu.setHeaderTitle("选择支付方式");
 menu.setHeaderIcon(R.drawable.pay);
}
```

重写 onCreateContextMenu(ContextMenu menu, View v, ContextMenuInfo menuInfo)方法，该方法自定义上下文菜单内容，在该实例中，我们为上下文菜单添加了三个菜单项并设置为单选菜单项，通过 ContextMenu 的 setHeaderTitle()和 setHeaderIcon()方法分别设置了菜单的标题和图标。此时还并未实现各个菜单项的单击事件，可以通过重写 onContextItemSelected(MenuItem item)方法实现，代码片段如下：

```
/**
 * 重写单击菜单项时触发的方法
 */
@Override
public boolean onContextItemSelected(MenuItem item) {
switch (item.getItemId()) {
    case MENU1:
        item.setChecked(true);
        payEdt.setText("支付宝");
        break;
    case MENU2:
        item.setChecked(true);
        payEdt.setText("财付通");
        break;
    case MENU3:
        item.setChecked(true);
        payEdt.setText("网上银行");
        break;
    default:
        break;
    }
return true;
}
```

上面程序根据各个菜单项的唯一标识，为各个菜单项设置了选中时的动作，即设置EditText文本框内容为相应的支付方式。此时单击图4.34中的第一个选项“支付宝”，EditText里的文字将变为“支付宝”，效果如图4.35所示。

图4.35 设置上下文菜单选项选中事件

通过开发该实例的上下文菜单，可以看出，其开发步骤与开发选项菜单类似，此处不再赘述。至此，我们已学习完了前面所说的三种类型的菜单的使用。

4.5 本章小结

对于一个手机应用，用户最直接的感受是软件的界面，与第 3 章一样，本章重点介绍的也是能让软件与用户实现良好的交互、能让软件有一个友好的界面的一些控件，包括自动完成文本框、下拉列表、日期时间选择器、进度条与拖动条、评分组件、选项卡、滚动视图与列表视图。除此之外，掌握使用对话框与菜单，也能大大增加软件的可交互性，而用于显示简单的提示消息，并且一段时间后会自动隐藏的 Toast，使用起来十分方便，也是非常常用的一种消息提示方式。

第5章 使用资源文件

查看 Android 项目文件夹，可以看到在第二级目录下有一个叫 res 的文件夹，该文件夹中存放的是 Android 应用中使用到的资源，包括字符串资源、颜色资源、数组资源、菜单资源等，在应用程序中可以直接对这些资源定义进行应用。

除了 res 目录下可以存放资源以外，assets 目录也可用于存放资源。一般情况下，应用无法直接访问的原生态资源将会被放到 assets 目录下，程序需要通过 AssetManager 以二进制流的形式来读取资源。而 res 目录下的资源，Android SDK 会在编译该应用时，自动在 R.java 文件中为这些资源创建索引，程序可以直接通过 res 清单类进行访问。

前面介绍的实例中大多都是直接将字符串值直接写在布局文件中或者 Java 程序代码中，其实这是一个不好的编程习惯，现在通过本章关于 Android 资源文件的学习后，应该在编程时将 Android 应用会使用到的资源放在 res 文件目录下并通过资源文件来管理，然后在布局文件或 Java 代码中采用 res 清单类进行访问。

5.1 资源的类型和存储方式

Android 资源包含了保存在 assets 目录下无法直接访问的原生资源以及保存在 res 目录下通过 R 清单类来访问的这两种类型的资源。

在 Android 项目文件夹的 res 目录下，不同的子目录存放着不同类型的应用资源，表 5.1 显示了 Android 不同资源在 res 目录下的存储方式。

表 5.1 res 目录下各资源存储方式

目　　录	存放的资源
/res/anim/	存放定义补间动画的 XML 文件
/res/color/	存放定义不同状态下颜色列表的 XML 文件
/res/drawable/	该目录下存放各种位图文件(如 *.png、*.9.png、*.jpg、*.gif)等，除此之外还可编译成如下各种 Drawable 对象的 XML 文件： BitmapDrawable NinePatchDrawable 对象 StateListDrawable 对象 ShapeDrawable 对象 AnimationDrawable 对象 Drawable 的其他各种子类的对象

续表

目　　录	存放的资源
/res/layout/	存放各种用户界面的布局文件
/res/menu/	存放为应用程序定义各种菜单的资源，包括选项菜单、子菜单、上下文菜单资源
/res/raw/	该目录下存放任意类型的原生资源。在 Java 代码中通过调用 Resource 对象的 openRawResource(int id)方法获取该资源的二进制输入流。 实际上，如果应用程序使用原生资源，推荐把这些原生资源保存到/assrts 目录下，然后在应用程序中使用 AssetManager 来访问这些资源。
/res/values/	存放各种简单的 XML 文件。这些简单值包括字符串值、整数值、颜色值、数组等。 字符串、整数值、颜色值、数组等各种值都是存放在该目录下，而且这些资源文件的根目录都是< resources.../>元素，若为该< resource.../>元素添加不同的子元素，则代表不同的资源，例如： string/integer/bool 子元素——代表添加一个字符串值/整数值/boolean 值 color 子元素——代表添加一个颜色值 array 子元素或 string-array、int-array 子元素——代表添加一个数组 style 子元素——代表添加一个样式 dimen——代表添加一个尺寸 由于各种简单值都可以定义在/res/values/目录下的资源文件中，如果在同一份资源文件中定义各种值，势必增加程序维护的难度。为此，Android 建议使用不同的文件来存放不同类型的值： arrays. xml——定义数组资源 colors. xml——定义颜色值资源 dimens. xml——定义尺寸值资源 strings. xml——定义字符串资源 styles. xml——定义样式资源
/res/xml/	任意的原生 XML 文件。这些 XML 文件可在 Java 代码中使用 Resources. getXML()访问

将应用程序中的各种资源分别保存在 Android 项目文件夹下的 res 目录下后，就可以在 Java 代码中使用这些资源，也可以在 XML 文件中使用这些资源。

5.2　通过字体设置功能使用字符串、颜色、尺寸资源

本节所使用的实例需要用到的字符串资源、颜色资源、尺寸资源所对应的 XML 文件都将被保存在/res/values 目录下，它们默认的文件名以及在 R 类中对应的内部类如表 5.2 所示。

表 5.2　资源文件及其在 R 类中对应的内部类

资 源 类 型	资源文件默认文件名	对应 R 类中的内部类名称
字符串资源	/res/values/strings. xml	R. string
颜色资源	/res/values/colors. xml	R. color
尺寸资源	/res/values/dimens. xml	R. dimen

本节通过一个具有字体设置功能的实例来向读者介绍如何使用字符串、颜色、尺寸资源。首先，字符串资源文件 strings.xml 内容如下：

```
<?xml version = "1.0" encoding = "utf - 8"?>
<resources>
    <string name = "app_name"> ResourceDemo </string>
    <string name = "text">我是文字</string>
    <string name = "text_change">我是改变后的文字</string>
    <string name = "btn_change_text">改变文本显示文字的内容</string>
    <string name = "btn_change_color">改变文本显示文字的颜色</string>
    <string name = "btn_change_size">改变文本显示文字的大小</string>
</resources>
```

从上面的字符串资源文件的 XML 代码可以看出，字符串资源文件的根元素是<resources>，该元素中每个<string>子元素定义一个字符串常量，其中<string>标签的 name 属性指定该常量的名称，<string>元素的开始标签和结束标签之间的内容定义了字符串值。如上面代码中的<string name="app_name"> ResourceDemo </string>语句。

接着，我们一起来看看颜色资源文件 colors.xml 中的内容，如下：

```
<?xml version = "1.0" encoding = "utf - 8"?>
<resources>
    <color name = "black">#000000 </color><! -- 黑色 -->
    <color name = "red">#FF0000 </color> <! -- 红色 -->
    <color name = "white">#FFFFFF </color> <! -- 白色 -->
    <color name = "yellow">#FFFF00 </color> <! -- 黄色 -->
</resources>
```

从上面的颜色资源文件的 XML 代码可以看出，颜色资源文件的根元素是<resources>，该元素中的每个<color>子元素定义了一个颜色值常量，其中<color>标签的 name 属性指定该颜色的名称，<color>元素的开始标签与结束标签之间的内容定义了颜色值。如上面代码中的<color name="red">#FF0000 </color>语句。

下面再来看看尺寸资源文件 dimens.xml 中的内容，如下：

```
<?xml version = "1.0" encoding = "utf - 8"?>
<resources>
    <dimen name = "text_size_16"> 16.0sp </dimen>
    <dimen name = "text_size_20"> 16.0sp </dimen>
</resources>
```

从上面的尺寸资源文件的 XML 代码可以看出，尺寸资源文件的根元素是<resources>，该元素中的每个<dimen>子元素定义了一个颜色值常量，其中<dimen>标签的 name 属性指定该颜色的名称，<dimen>元素的开始标签与结束标签之间的内容定义了颜色值。如上面代码中的<dimen name=" text_size_16"> 16.0sp </dimen>语句。

至此，我们已将实例中会使用到三种资源定义好了，包括字符串资源、颜色资源、尺寸资源。现在开始编写实例的布局的 XML 代码，如下：

```
<?xml version = "1.0" encoding = "utf - 8"?>
<LinearLayout xmlns:android = "http://schemas.android.com/apk/res/android"
```

```
    android:layout_width = "match_parent"
    android:layout_height = "match_parent"
    android:gravity = "center"
    android:orientation = "vertical" >

    <TextView
        android:id = "@ + id/txt_show"
        android:layout_width = "wrap_content"
        android:layout_height = "wrap_content"
        android:paddingBottom = "5dp"
        android:text = "@string/text"
        android:textColor = "@color/white"
        android:textSize = "@dimen/text_size_16" />

    <Button
        android:id = "@ + id/btn_change_text"
        android:layout_width = "wrap_content"
        android:layout_height = "wrap_content"
        android:text = "@string/btn_change_text" />

    <Button
        android:id = "@ + id/btn_change_color"
        android:layout_width = "wrap_content"
        android:layout_height = "wrap_content"
        android:text = "@string/btn_change_color" />

    <Button
        android:id = "@ + id/btn_change_size"
        android:layout_width = "wrap_content"
        android:layout_height = "wrap_content"
        android:text = "@string/btn_change_size" />

</LinearLayout>
```

代码文件：codes\05\5.2\ResourceDemo\res\layout\activity_main.xml

以上布局文件比较简单，从上至下放置了一个 TextView，用于显示文字，三个 Button 按钮分别代表不用的功能，按下第一个 Button 按钮将会改变 TextView 所显示的文字，按下第二个 Button 按钮将会改变 TextView 显示文字的颜色，按下第三个 Button 按钮将会使 TextView 所显示的文字字体变大。

从以上布局文件中可以看出，在 XML 文件中使用布局文件遵循的语法格式为：@< resource_type >/< resource_name >，例如，使用字符串资源时用到@string/text，说明如下：

- < resource_type >——R 类中代表不同资源类型的内部类；
- < resource_name >——指定资源的名称，在这里是 XML 资源元素中由 android:name 属性所指定的名称，其实该资源名称也可以是无后缀的文件名，如放在 drawable-hdpi、drawable-ldpi、drawable-mdpi 这三个文件夹下的图片资源。

其实上面的语法格式并不是完整的语法格式，完整的语法格式为：@[< package_name >:]< resource_type >/< resource_name >，package_name 指定了资源类所在应用的包。如果所引用的资源和当前资源位于同一个包下，则< package_name >可以省略。

在学习了如何在 XML 文件里使用资源文件后，接下来介绍如何在 Java 代码中使用字符串资源、颜色资源、尺寸资源。本实例的 MainActivity 的 Java 代码如下：

```
package cn.edu.hstc.resourcedemo.activity;

import android.app.Activity;
import android.os.Bundle;
import android.view.View;
import android.view.View.OnClickListener;
import android.widget.TextView;

public class MainActivity extends Activity {
    private TextView show;

    @Override
    protected void onCreate(Bundle savedInstanceState) {
        super.onCreate(savedInstanceState);
        setContentView(R.layout.activity_main);
        show = (TextView) findViewById(R.id.txt_show);
        initButton();
    }

    private void initButton() {
        findViewById(R.id.btn_change_text).setOnClickListener(new OnClickListener() {
            @Override
            public void onClick(View v) {
                //改变文本显示内容,将字符串资源文件对应的字符串赋予文本显示控件
                show.setText(R.string.text_change);
            }
        });

        findViewById(R.id.btn_change_color).setOnClickListener(new OnClickListener() {
            @Override
            public void onClick(View v) {
                //使用颜色资源,改变文本字体颜色
                show.setTextColor(getResources().getColor(R.color.red));
            }
        });

        findViewById(R.id.btn_change_size).setOnClickListener(new OnClickListener() {
            @Override
            public void onClick(View v) {
                //使用尺寸资源,改变文本字体大小
                show.setTextSize(getResources().getDimensionPixelOffset(R.dimen.text_size_20));
            }
        });
    }
}
```

代码文件：codes\05\5.2\ResourceDemo\cn\edu\hstc\activity\MainActivity.java

从上面的 Java 代码可以看出，在 Java 代码中按如下语法格式使用资源：[< package_name >].R.< resource_type >.< resource_name >，与在 XML 中使用资源的语法格式类似，如果所使用的资源和当前资源位于同一个包下，则< package_name >可以省略。

将上面的程序部署在 Android 模拟器上，运行效果如图 5.1 所示。

单击第一个按钮，可以看到如图 5.2 的运行效果。

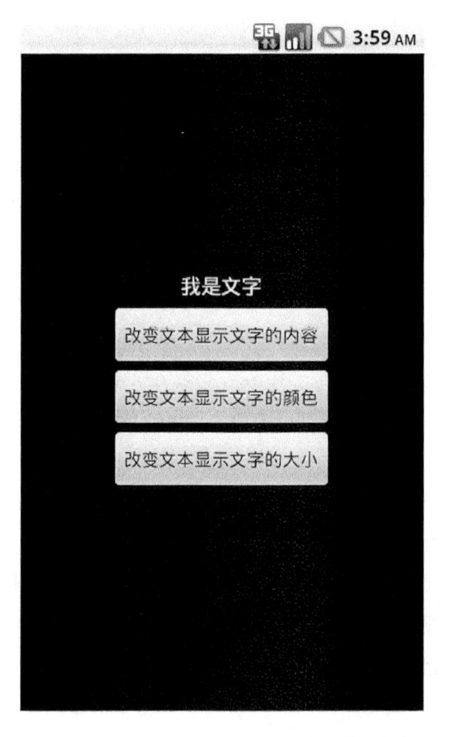

图 5.1 在 XML 布局文件中使用各种资源

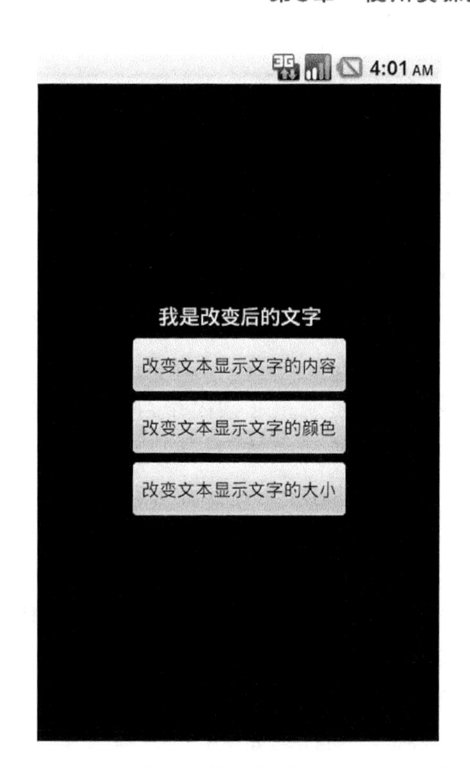

图 5.2 在 Java 代码中使用字符串资源改变文字内容

单击第二个按钮,可以看到如图 5.3 的运行效果。

单击第三个按钮,可以看到如图 5.4 的运行效果。

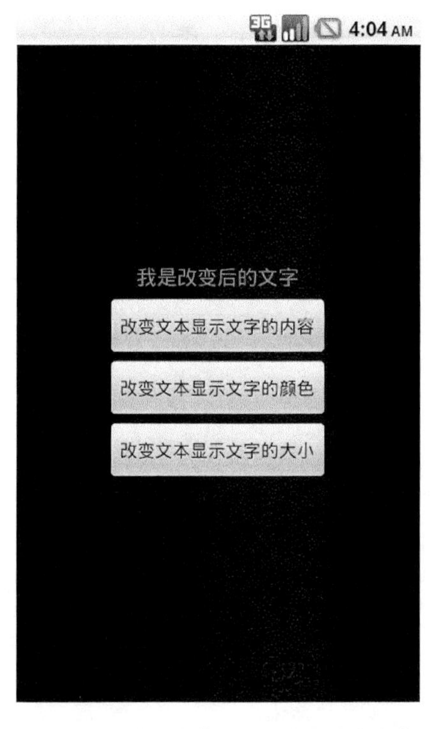

图 5.3 在 Java 代码中使用颜色资源改变文字颜色

图 5.4 在 Java 代码中使用尺寸资源改变文字大小

5.3 使用图片资源

图片资源是最简单的一种 Drawable 资源，只需要把 *.png、*.jpg、*.gif 等格式的图片放入/res/drawable-xxx 目录下，Android SDK 就会在编译应用时自动加载该图片，并在 R 资源清单类中生成该资源的索引。

这里需要注意的是，Android 不允许图片资源的文件名中出现大写字母，且不能以数字开头，否则 Android SDK 无法为该图片在 R 类中生成资源索引。

使用图片资源也非常简单，在 Java 代码中可以使用[< package >.]R. drawable. < file_name >的语法格式来访问该图片资源，而在 XML 代码中可以使用@[< package_name >:] drawable/file_name 的语法格式来访问图片资源。

为了在程序中获得实际的 Drawable 对象，Resources 提供了 Drawable getDrawable (int id)方法，该方法可根据 Drawable 资源在 R 清单类中的 ID 来获取实际的 Drawable 对象。

本节通过一个实例来演示如何在 XML 代码中使用图片资源以及在 Java 代码中使用图片资源。实例的布局文件比较简单，只是在界面底部放置一个 Button 按钮，然后在该按钮的上方放置一个图片显示控件，该 Button 按钮的功能是将图片容器里的图片替换成另一张图片。界面布局代码如下：

```
<?xml version = "1.0" encoding = "utf - 8"?>
< RelativeLayout xmlns:android = "http://schemas.android.com/apk/res/android"
    android:layout_width = "match_parent"
    android:layout_height = "match_parent" >

    < Button
        android:id = "@ + id/button"
        android:layout_width = "fill_parent"
        android:layout_height = "wrap_content"
        android:text = "换一张图片"
        android:layout_alignParentBottom = "true" />

    < ImageView
        android:id = "@ + id/image"
        android:layout_width = "fill_parent"
        android:layout_height = "fill_parent"
        android:layout_above = "@id/button"
        android:background = "@drawable/a"
        android:scaleType = "fitXY" />

</RelativeLayout >
```

代码文件：codes\05\5.3\ImageResourceDemo\res\layout\activity_main.xml

上面的程序演示了在 XML 代码中如何使用图片资源。

接下来看看在 Java 代码中如何使用图片资源，代码如下：

```
package cn.edu.hstc.imageresourcedemo.activity;

import com.example.imageresourcedemo.R;
```

```
import android.app.Activity;
import android.os.Bundle;
import android.view.View;
import android.widget.ImageView;

public class MainActivity extends Activity {
    ImageView image = null;
    @Override
    protected void onCreate(Bundle savedInstanceState) {
        super.onCreate(savedInstanceState);
        setContentView(R.layout.activity_main);
        image = (ImageView) findViewById(R.id.image);
        findViewById(R.id.button).setOnClickListener(new View.OnClickListener() {
            @Override
            public void onClick(View v) {
                //在 Java 代码中使用图片资源
                image.setImageDrawable(getResources().getDrawable(R.drawable.b));
            }
        });
    }
}
```

代码文件: codes\05\5.3\ImageResourceDemo\cn\edu\hstc\activity\MainActivity.java

上面的程序演示了如何在 Java 代码中使用图片资源。

将程序部署在模拟器上,运行效果如图 5.5 所示。

单击布局最下面的"换一张图片"按钮,会看到界面上的图片被替换成另外一张,运行效果如图 5.6 所示。

图 5.5 在 XML 界面布局中直接使用图片资源

图 5.6 在 Java 代码中使用图片资源

5.4 通过声音播放功能使用样式资源、主题资源和原始资源

假如经常需要对某个类型的组件设置大致相似的属性和对应的属性值，也就是说，该类型的组件在该应用中保持一样的长相，例如字体、颜色、背景色等。如果每次都对该 View 组件进行重复的属性的指定，这明显是不科学的，也不利于后期项目的维护。

类似于 Web 开发中定义一个一个的 css 文件，文件中的一块一块的 css 代码，Android 提供了样式资源的概念：一个样式等于一组格式的集合。为一个组件设置了某个样式后，该样式所包含的全部格式将会应用于该组件。这就是样式资源。

主题资源与样式资源非常类似，主要区别在于：

- 主题资源不能作用于单个 View 组件，主题应用对整个应用中的所有的 Activity 起作用或对指定的 Activity 起作用。
- 主题资源定义的格式应该是改变窗口外观的格式，例如，窗口标题、窗口边框等。

原始资源，指的是声音资源等 Android 应用会用到的大量的原生态的资源。类似声音文件以及其他各种类型的文件，只要 Android 没有为其提供专门的支持，这种资源都被称为原始资源。Android 的原始资源被存放在/res/raw 目录下以及/assets 目录下。只是 Android SDK 会在 R 清单类中为/res/raw 目录下的资源生成一个索引，而/assets 目录下的资源会是更彻底的原始资源，Android 应用需要通过 AssetManager 来管理该目录下的原始资源。

本节通过一个简单的声音播放功能来介绍如何使用样式资源、主题资源和原始资源。实例需要用到的样式资源以及主题资源文件 my_style.xml 代码如下：

```
<?xml version = "1.0" encoding = "utf - 8"?>
< resources xmlns:android = "http://schemas.android.com/apk/res/android">
    <! -- 定义一个样式,指定字号大小以及字体颜色 -->
    < style name = "style1">
        < item name = "android:textSize"> 20sp </item >
        < item name = "android:textColor"> # FF0000 </item >
    </style >

    <! -- 定义一个样式,继承前一个样式 -->
    < style name = "style2" parent = "@style/style1">
        < item name = "android:background"> # FFFF00 </item >
        < item name = "android:padding"> 8dp </item >
        <! -- 覆盖父样式中指定的属性 -->
        < item name = "android:textColor"> # 000000 </item >
    </style >

    <! -- 定义主题资源 -->
    < style name = "mytheme">
        < item name = "android:windowNoTitle"> true </item >
        < item name = "android:windowFullscreen"> true </item >
        < item name = "android:windowBackground">@drawable/ic_launcher </item >
    </style >
```

```
</resources>
```

代码文件：codes\05\5.4\AudioDemo\res\values\my_style.xml

从上面的代码可以看出，样式以及主题资源文件的根元素是<resources>，该元素内可包含多个<style>子元素，每个<style>元素定义一个样式或主题，<style>元素指定如下两个属性：

- name——指定样式或主题的名称；
- parent——指定该样式或主题所继承的父样式或父主题。当继承某个父样式或主题，该样式或主题将会获得父样式或主题中定义的全部资源，当然，当前样式或主题也可以对父样式或主题的属性进行重新设置。

例如，在上面的代码中，定义了两个样式，第二个样式继承了第一个样式并对第一个样式的 android:textColor 属性进行重新设定。

接下来，看一下在布局 XML 文件中如何使用上面所定义的样式和主题，实例的布局文件代码如下：

```
<?xml version = "1.0" encoding = "utf - 8"?>
<LinearLayout xmlns:android = "http://schemas.android.com/apk/res/android"
    android:layout_width = "match_parent"
    android:layout_height = "match_parent"
    android:orientation = "vertical"
    android:padding = "5dp" >

    <LinearLayout
        android:layout_width = "fill_parent"
        android:layout_height = "wrap_content"
        android:orientation = "horizontal" >

        <TextView
            android:layout_width = "wrap_content"
            android:layout_height = "wrap_content"
            android:text = "声源 1: " />

        <EditText
            android:layout_width = "200dp"
            android:layout_height = "wrap_content"
            android:text = "test1.mp3"
            style = "@style/style1" />

        <Button
            android:id = "@ + id/button1"
            android:layout_width = "wrap_content"
            android:layout_height = "wrap_content"
            android:text = "播放" />
    </LinearLayout>

    <LinearLayout
        android:layout_width = "fill_parent"
        android:layout_height = "wrap_content"
```

```
        android:orientation="horizontal">

        <TextView
            android:layout_width="wrap_content"
            android:layout_height="wrap_content"
            android:text="声源1: " />

        <EditText
            android:layout_width="200dp"
            android:layout_height="wrap_content"
            android:text="test2.mp3"
            style="@style/style2" />

        <Button
            android:id="@+id/button2"
            android:layout_width="wrap_content"
            android:layout_height="wrap_content"
            android:text="播放" />
    </LinearLayout>

</LinearLayout>
```

代码文件：codes\05\5.4\AudioDemo\res\layout\activity_main.xml

从上面的代码可以看出，在XML中使用样式资源的语法格式为@[<package_name>:]style/file_name。接着再看看如何使用主题资源，一般使用主题资源是通过Android全局配置文件AndroidManifest.xml来指定Activity或整个应用所使用到的主题。代码如下：

```
<manifest xmlns:android="http://schemas.android.com/apk/res/android"
    package="com.example.audiodemo"
    android:versionCode="1"
    android:versionName="1.0">

    <uses-sdk
        android:minSdkVersion="8"
        android:targetSdkVersion="8" />

    <!-- 在全局配置文件中使用主题资源 -->
    <application
        android:allowBackup="true"
        android:icon="@drawable/ic_launcher"
        android:label="@string/app_name"
        android:theme="@style/mytheme">
        <activity
            android:name="cn.edu.hstc.audiodemo.activity.MainActivity"
            android:label="@string/app_name">
            <intent-filter>
                <action android:name="android.intent.action.MAIN" />

                <category android:name="android.intent.category.LAUNCHER" />
            </intent-filter>
        </activity>
```

```
    </application>

</manifest>
```

代码文件：codes\05\5.4\AudioDemo\ AndroidManifest.xml

上面的代码在<application>标签指定了 android:theme 的属性值为 my_style.xml 中对应的主题资源，即可让该主题资源作用于整个应用的 Activity，下面的每个<Activity>标签也可以单独指定该属性，覆盖<application>标签中指定的主题资源。从代码可以看出，在 XML 文件中使用主题资源的语法格式与使用样式资源一样。

接下来，看一下布局文件所对应的 Activity Java 代码，在该 Activity 文件中，将对一开始放在/res/raw 目录下的 test1.mp3 文件以及/assets 目录下的 test2.mp3 文件这两个原生资源进行访问。主程序代码如下：

```
package cn.edu.hstc.audiodemo.activity;

import com.example.audiodemo.R;

import android.app.Activity;
import android.content.res.AssetFileDescriptor;
import android.content.res.AssetManager;
import android.media.MediaPlayer;
import android.os.Bundle;
import android.view.View;

public class MainActivity extends Activity {
    MediaPlayer mediaPlayer1 = null;
    MediaPlayer mediaPlayer2 = null;
    @Override
    protected void onCreate(Bundle savedInstanceState) {
        super.onCreate(savedInstanceState);
        setContentView(R.layout.activity_main);
        //直接根据声音文件的 ID 来创建 MediaPlayer
        mediaPlayer1 = MediaPlayer.create(this, R.raw.test1);
        //获取应用的 AssetManager
        AssetManager am = getAssets();
        try {
            //获取指定文件对应的 AssetFileDescriptor
            AssetFileDescriptor afd = am.openFd("test2.mp3");
            mediaPlayer2 = new MediaPlayer();
            //使用 MediaPlayer 加载指定的声音文件
            mediaPlayer2.setDataSource(afd.getFileDescriptor());
            mediaPlayer2.prepare();
        } catch (Exception e) {
            e.printStackTrace();
        }

        //获取第一个按钮并为其绑定事件监听器
        findViewById(R.id.button1).setOnClickListener(new View.OnClickListener() {
            @Override
```

```
            public void onClick(View v) {
                //播放声音
                mediaPlayer1.start();
            }
        });
        //获取第二个按钮并为其绑定事件监听器
        findViewById(R.id.button2).setOnClickListener(new View.OnClickListener() {
            @Override
            public void onClick(View v) {
                //播放声音
                mediaPlayer2.start();
            }
        });
    }
}
```

代码文件：codes\05\5.4\AudioDemo\cn\edu\hstc\activity\MainActivity.java

上面的代码分别获取/res/raw 和/assets 目录下的资源作为两个 MediaPlayer 的声源文件，并赋予两个 Button 控件不同的事件监听器来播放声音。从上面的 Java 代码可以看出，使用/res/raw 目录下的原始资源的语法格式为[< package >.]R.raw.< file_name >，当然，由以往的经验可以知道，在 XML 代码中访问/res/raw 目录下的原始资源的语法格式为：@[< package_name >:]raw/file_name。

而获取/assets 目录下的原始资源则需要通过一个专门管理/assets 目录下的原始资源的管理器类 AssetManager，本示例通过该类的 AssetFileDescriptor openFd(String fileName)方法，根据文件名来获取原始资源对应的 AssetFileDescriptor，再通过 AssetFileDescriptor 的 getFileDescriptor()方法来获取对应的原始资源。

事实上，还有另外的一种获取/assets 目录下的资源的方法，就是通过 InputStream open (String fileName)方法，根据文件名来获取原始资源对应的输入流。

至此，对如何使用样式资源、主题资源以及原始资源已经全部介绍完毕，接下来，把应用部署在 Android 模拟器上，将会看到如图 5.7 所示的运行效果。

分别单击界面上的两个 Button 按钮，可以听到不同的音乐。从运行效果也可以看出，两个文本框控件的样式受到了 my_style.xml 文件中定义的样式资源 sytle1 以及 style2 的影响，分别呈现出不同的效果，而整个窗体则受到了 my_style.xml 文件中的定义的主题资源的影响，呈现了没有状态栏、全屏、背景为 Android Robot 的效果。

图 5.7 使用样式、主题、原始资源

5.5 本章小结

本章主要介绍了 Android 应用资源的使用，通过使用各种资源文件，Android 应用可以把各种字符串、图片、颜色、界面布局等交给 XML 文件配置管理，形成了一种高度解耦的设计模式。各种 Android 资源的存储方式以及如何使用是本章的学习重点。当然，除了本章所介绍的几种资源以外，还有很多其他类型的资源，在日后的学习工作中，读者可以慢慢进行了解。

第6章 通过商品发布器详细介绍Activity

本章将通过自身设计的一个小应用——商品发布器来向读者详细介绍 Android 的一个重要的组件 Activity。6.1 节在代码的角度介绍商品发布器的实现，6.2 节通过剖析该商品发布器介绍 Activity 的建立、配置、启动和关闭，并从中理解 Activity 的回调机制和生命周期。

6.1 实现商品发布器

本节的主要内容是介绍如何实现商品发布器。首先，看一下该应用的启动界面的布局代码，也就是 activity_main.xml 文件的内容，如下：

```
<?xml version = "1.0" encoding = "utf - 8"?>
< TableLayout xmlns:android = "http://schemas.android.com/apk/res/android"
    android:layout_width = "match_parent"
    android:layout_height = "match_parent"
    android:padding = "10dp"
    android:stretchColumns = "1" >

    < TableRow
        android:layout_width = "fill_parent"
        android:layout_height = "wrap_content" >

        < TextView
            android:layout_width = "wrap_content"
            android:layout_height = "wrap_content"
            android:text = "商品名称：" />

        < EditText
            android:id = "@ + id/edt_name"
            android:layout_width = "wrap_content"
            android:layout_height = "wrap_content"
            android:hint = "请输入商品名称" />
    </TableRow >

    < TableRow
        android:layout_width = "fill_parent"
        android:layout_height = "wrap_content" >
```

```
        <TextView
            android:layout_width="wrap_content"
            android:layout_height="wrap_content"
            android:text="商品种类:" />

        <Spinner
            android:id="@+id/sp_type"
            android:layout_width="wrap_content"
            android:layout_height="wrap_content"
            android:prompt="@string/hint"
            android:entries="@array/type" />
    </TableRow>

    <TableRow
        android:layout_width="fill_parent"
        android:layout_height="wrap_content" >

        <TextView
            android:layout_width="wrap_content"
            android:layout_height="wrap_content"
            android:text="商品价格:" />

        <EditText
            android:id="@+id/edt_price"
            android:layout_width="wrap_content"
            android:layout_height="wrap_content"
            android:hint="请输入商品价格(RMB)" />
    </TableRow>

    <TableRow
        android:layout_width="fill_parent"
        android:layout_height="wrap_content" >

        <Button
            android:id="@+id/btn_release"
            android:layout_width="wrap_content"
            android:layout_height="wrap_content"
            android:text="发布" />

        <Button
            android:id="@+id/btn_show"
            android:layout_width="wrap_content"
            android:layout_height="wrap_content"
            android:text="查看已有商品" />
    </TableRow>

</TableLayout>
```

代码实现在界面上从上而下分为四行,第一行第一列和第二列分别显示“商品名称:”这四个字以及一个供用户输入商品名称的文本输入框,第二行第一列和第二列分别显示“商品种类："这四个字以及一个供用户选择商品种类的下拉列表控件，第三行第一列和第二列分别显示"商品价格："这四个字以及一个供用户输入商品价格的文本输入框，第四行第一列和第二列分别显示"发布"按钮和"查看已有商品"按钮。

布局最外层采用表格布局 TableLayout 方式，这种布局方式已经在前面的章节有所介绍，此处不再赘述。代码中使用到了一个下拉控件 Spinner，该 Spinner 的填充数据源来自一个数组 type，如代码中 android:entries="@array/type"这一行所示，于是应用需要访问一个数组资源文件，该资源文件位于/res/values/下，文件内容如下：

```
<?xml version = "1.0" encoding = "utf - 8"?>
< resources >
    < string - array name = "type">
        < item >食品</item >
        < item >生活用品</item >
        < item >电脑配件</item >
        < item >电子元件</item >
    </string - array >
</resources >
```

在这里，我们需要讲一下需求，应用在启动界面中编辑要发布的商品信息，然后单击"发布"按钮，将该商品上传至服务器端，服务器端获取该商品信息并保存在服务端数据库表中。用户单击界面中的"查看已有商品"按钮将会跳转到另一个页面，在该页面中去访问服务端并 Get 到数据库表中保存的所有商品，然后显示在该页面上的 ListView 控件。因此需要在/res/layout/目录下新建一个新的界面布局文件 activity_show. xml，该文件内容如下：

```
<?xml version = "1.0" encoding = "utf - 8"?>
< LinearLayout xmlns:android = "http://schemas.android.com/apk/res/android"
    android:layout_width = "match_parent"
    android:layout_height = "match_parent"
    android:background = "@drawable/segmented_bg"
    android:orientation = "vertical" >

    < RelativeLayout
        android:layout_width = "fill_parent"
        android:layout_height = "40dp"
        android:background = "@drawable/topback" >

        < ImageView
            android:id = "@ + id/img_back"
            android:layout_width = "wrap_content"
            android:layout_height = "wrap_content"
            android:paddingLeft = "5dp"
            android:layout_alignParentLeft = "true"
            android:layout_centerVertical = "true"
            android:src = "@drawable/backbtn" />

        < TextView
            android:layout_width = "wrap_content"
            android:layout_height = "wrap_content"
```

```
            android:layout_centerInParent = "true"
            android:text = "查看已有商品"
            android:textColor = "@color/white"
            android:textSize = "22sp" />
    </RelativeLayout>

    <ListView
        android:id = "@ + id/listView"
        android:layout_width = "fill_parent"
        android:layout_height = "wrap_content"
        android:divider = "@color/black"
        android:background = "@drawable/segmented_bg"
        android:dividerHeight = "1px" />

</LinearLayout>
```

在该程序中，需要用到一个实体类，用来代表商品，在 src 目录下新建包 cn. edu. hstc. commodityrelease. entity 并新建类 Commodity，该类内容如下：

```
package cn.edu.hstc.commodityrelease.entity;

public class Commodity {
    public Integer theId;
    public String name;
    public String type;
    public String price;

    public Integer getTheId() {
        return theId;
    }

    public void setTheId(Integer theId) {
        this.theId = theId;
    }

    public String getName() {
        return name;
    }

    public void setName(String name) {
        this.name = name;
    }

    public String getType() {
        return type;
    }

    public void setType(String type) {
        this.type = type;
    }
```

```
        public String getPrice() {
            return price;
        }

        public void setPrice(String price) {
            this.price = price;
        }

        public Commodity(Integer theId, String name, String type, String price) {
            this.theId = theId;
            this.name = name;
            this.type = type;
            this.price = price;
        }
    }
```

由于第二个界面用到了 ListView 这个控件，该控件需要设置适配器，这里为该显示商品的 ListView 定制了一个适配器，新建 cn.edu.hstc.commodityrelease.util 包并新建 MyListViewAdapter 类，该类程序代码如下：

```
package cn.edu.hstc.commodityrelease.util;

import java.util.ArrayList;
import java.util.List;

import android.content.Context;
import android.view.LayoutInflater;
import android.view.View;
import android.view.ViewGroup;
import android.widget.BaseAdapter;
import android.widget.TextView;
import cn.edu.hstc.commodityrelease.entity.Commodity;

import com.example.commodityrelease.R;

public class MyListViewAdapter extends BaseAdapter {
    //定义一个存放商品信息的 List 集合
    private List<Commodity> commodityList = new ArrayList<Commodity>();
    //声明 LayoutInflater 类,用以后期载入 XML 界面
    private LayoutInflater inflater;

    /**
     * 在构造器中初始化 List 集合以及初始化 LayoutInflater
     */
    public MyListViewAdapter(Context context, List<Commodity> commodityList) {
        this.commodityList = commodityList;
         inflater = (LayoutInflater) context.getSystemService(Context.LAYOUT_INFLATER_
SERVICE);
    }

    @Override
```

```
    public int getCount() {
        return commodityList.size();
    }

    @Override
    public Object getItem(int position) {
        return commodityList.get(position);
    }

    @Override
    public long getItemId(int position) {
        return position;
    }

    @Override
    public View getView(int position, View convertView, ViewGroup parent) {
        ViewHolder holder = null;
        if (convertView == null) {
          //载入 ListView 的 Item 项的布局文件 listview_item
            convertView = inflater.inflate(R.layout.listview_item, null);
            holder = new ViewHolder();
            holder.name = (TextView) convertView.findViewById(R.id.list_txt_name);
            holder.type = (TextView) convertView.findViewById(R.id.list_txt_type);
            convertView.setTag(holder);
        } else {
            holder = (ViewHolder)convertView.getTag();
        }
        holder.name.setText(commodityList.get(position).name);
        holder.type.setText("[" + commodityList.get(position).type + "]");
        return convertView;
    }

    public void addItem(final Commodity commodity) {
        commodityList.add(commodity);
        notifyDataSetChanged();
    }

    public static class ViewHolder {
        public TextView name;
        public TextView type;
    }
}
```

上述代码中用到了一个 xml 文件 listview_item.xml，该文件描述了 ListView 的每个 item 的外观，该文件内容如下：

```
<?xml version="1.0" encoding="utf-8"?>
<RelativeLayout xmlns:android="http://schemas.android.com/apk/res/android"
    android:layout_width="match_parent"
    android:layout_height="match_parent"
    android:descendantFocusability="blocksDescendants"
```

```
    android:paddingLeft = "10dp"
    android:paddingRight = "10dp" >

    <TextView
        android:id = "@ + id/list_txt_type"
        android:layout_width = "wrap_content"
        android:layout_height = "wrap_content"
        android:layout_centerVertical = "true"
        android:textColor = "@color/white"
        android:textSize = "20sp" />

    <TextView
        android:id = "@ + id/list_txt_name"
        android:layout_width = "wrap_content"
        android:layout_height = "wrap_content"
        android:layout_centerVertical = "true"
        android:layout_toRightOf = "@id/list_txt_type"
        android:paddingLeft = "5dp"
        android:textColor = "@color/white"
        android:textSize = "20sp" />

    <ImageView
        android:layout_width = "wrap_content"
        android:layout_height = "wrap_content"
        android:layout_alignParentRight = "true"
        android:layout_centerVertical = "true"
        android:layout_marginTop = "6dp"
        android:layout_marginBottom = "6dp"
        android:background = "@drawable/to" />

</RelativeLayout>
```

接下来，看看实现第一个 Activity 的主程序代码 MainActivity.java，代码如下：

```
package cn.edu.hstc.commodityrelease.activity;

import java.util.HashMap;
import java.util.Map;

import org.json.JSONObject;

import android.app.Activity;
import android.app.AlertDialog;
import android.content.DialogInterface;
import android.content.DialogInterface.OnCancelListener;
import android.content.Intent;
import android.os.AsyncTask;
import android.os.Bundle;
import android.util.Log;
import android.view.View;
import android.view.View.OnClickListener;
```

```
import android.widget.Button;
import android.widget.EditText;
import android.widget.Spinner;
import android.widget.TextView;
import android.widget.Toast;
import cn.edu.hstc.commodityrelease.util.HttpUtil;

import com.example.commodityrelease.R;

public class MainActivity extends Activity implements OnClickListener {
    //声明界面布局中的各个组件
    private EditText name, price;
    private Spinner type;
    private Button release, show;

    private AlertDialog prompt;
    //声明异步类 AsyncTask
    private AsyncTask<String, Integer, String> access;
    private Toast toast;

    @Override
    protected void onCreate(Bundle savedInstanceState) {
        super.onCreate(savedInstanceState);
        //加载布局文件
        setContentView(R.layout.activity_main);
        init();
    }

    /**
     * 获得界面组件并为其设置监听器,这里监听器为 MainActivity 本身
     */
    private void init() {
        name = (EditText) findViewById(R.id.edt_name);
        type = (Spinner) findViewById(R.id.sp_type);
        price = (EditText) findViewById(R.id.edt_price);
        release = (Button) findViewById(R.id.btn_release);
        show = (Button) findViewById(R.id.btn_show);
        //为发布按钮设置监听器
        release.setOnClickListener(this);
        //为查看已有商品按钮设置监听器
        show.setOnClickListener(this);
    }

    @Override
    public void onClick(View v) {
        switch (v.getId()) {
        case R.id.btn_release:
            release();
            break;
        case R.id.btn_show:
            show();
```

```
                break;
            default:
                break;
        }
    }

    /**
     * 发布商品
     */
    private void release() {
        Map<String, String> map = new HashMap<String, String>();
        //以下将商品各个属性存放在 map 中
        if (name != null) {
            map.put("name", name.getText().toString());
        }
        if (type != null) {
            map.put("type", type.getSelectedItem().toString());
        }
        if (price != null) {
            map.put("price", price.getText().toString());
        }
        //触发该方法,请求 CommodityServlet 将商品属性作为参数提交到后台中
        acquire(map, "正在提交中...", true, "CommodityServlet");
    }

    /**
     * 查看已有商品
     */
    private void show() {
        Intent intent = new Intent(MainActivity.this, ShowActivity.class);
        //跳转到另一个页面 ShowActivity 中
        MainActivity.this.startActivity(intent);
    }

    /**
     * 请求方法,启动任务,请求后台路径
     * @param map
     * @param promptStr
     * @param dialog
     * @param servletStr
     */
    private void acquire(final Map<String, String> map, final String promptStr, final boolean
dialog, final String servletStr) {
        access = new AsyncTask<String, Integer, String>() {
            @Override
            protected void onPreExecute() {
                super.onPreExecute();
                if (dialog) {
                    prompt = HttpUtil.prompt(MainActivity.this, promptStr);
                    prompt.setOnCancelListener(new OnCancelListener() {
                        @Override
```

```
                    public void onCancel(DialogInterface dialog) {
                        access.cancel(true);
                    }
                });
            }
        }

        @Override
        protected String doInBackground(String... params) {
            String result = null;
            try {
                result = HttpUtil.request(HttpUtil.BASE_URL + servletStr, map);
            } catch (Exception e) {
                e.printStackTrace();
            }
            return result;
        }

        @Override
        protected void onPostExecute(String result) {
            Log.i("", "MainActivity result: " + result);
            prompt.dismiss();
            try {
                if (result != null) {
                    JSONObject object = new JSONObject(result);
                    if (object.getString("result").equals("插入成功!")) {
                        setToast("发布成功!");
                    } else if (object.getString("result").equals("插入失败!")) {
                        setToast("发布失败!");
                    } else if (object.getString("result").equals("后台出错!")) {
                        setToast("后台出错!");
                    }
                }
            } catch (Exception e) {
                e.printStackTrace();
            }
        }

    };
    access.execute();
}

private void setToast(String message) {
    if (toast!= null) {
        ((TextView) toast.getView()).setText(message);
    } else {
        toast = HttpUtil.createToast(MainActivity.this, message);
    }
    toast.show();
}
}
```

该类主要实现用户单击“发布”按钮时将数据提交到后台数据库中以及单击“查看已有商品”按钮时跳转到另一个页面的功能，类中调用了一个工具类 HttpUtil，该工具类实现具体的与后台进行数据交互的操作，在此不赘述，读者可以查看本书附带的源码。最后来看一下当用户单击“查看已有商品”按钮时跳转到的页面 ShowActivity 的具体实现，代码如下：

```
package cn.edu.hstc.commodityrelease.activity;

import java.util.ArrayList;
import java.util.HashMap;
import java.util.List;
import java.util.Map;

import org.json.JSONArray;
import org.json.JSONObject;

import android.app.Activity;
import android.app.AlertDialog;
import android.content.DialogInterface;
import android.content.DialogInterface.OnCancelListener;
import android.content.Intent;
import android.os.AsyncTask;
import android.os.Bundle;
import android.util.Log;
import android.view.View;
import android.widget.AdapterView;
import android.widget.AdapterView.OnItemClickListener;
import android.widget.ImageView;
import android.widget.ListView;
import android.widget.TextView;
import android.widget.Toast;
import cn.edu.hstc.commodityrelease.entity.Commodity;
import cn.edu.hstc.commodityrelease.util.HttpUtil;
import cn.edu.hstc.commodityrelease.util.MyListViewAdapter;

import com.example.commodityrelease.R;

public class ShowActivity extends Activity {
    //加载页面组件
    private ImageView backImg;
    //定义一个 List 集合，用于动态存储从后台获取到的商品列表
    private List<Commodity> commodityList = new ArrayList<Commodity>();
    //声明布局上的 ListView
    private ListView listView;
    //声明一个 ListView 的适配器 Adapter
    private MyListViewAdapter adapter;

    private AlertDialog prompt;
    private AsyncTask<String, Integer, String> access;
    private Toast toast;
```

```
    @Override
    protected void onCreate(Bundle savedInstanceState) {
        super.onCreate(savedInstanceState);
        //加载界面布局文件
        setContentView(R.layout.activity_show);
        //获取返回按钮
        backImg = (ImageView) findViewById(R.id.img_back);
        //为返回按钮添加事件监听器
        backImg.setOnClickListener(new View.OnClickListener() {
            @Override
            public void onClick(View v) {
                ShowActivity.this.finish();
            }
        });
        Map<String, String> map = new HashMap<String, String>();
        map.put("name", "");
        //请求后台 GetCommoditys,加载数据并完成页面更新操作
        acquire(map, "数据加载中...", true, "GetCommoditys");
    }

    /**
     * 请求后台,加载数据并完成页面更新操作
     * @param map
     * @param promptStr
     * @param dialog
     * @param servletStr
     */
    private void acquire(final Map<String, String> map, final String promptStr, final boolean
dialog, final String servletStr) {
        access = new AsyncTask<String, Integer, String>() {
            @Override
            protected void onPreExecute() {
                super.onPreExecute();
                if (dialog) {
                    prompt = HttpUtil.prompt(ShowActivity.this, promptStr);
                    prompt.setOnCancelListener(new OnCancelListener() {
                        @Override
                        public void onCancel(DialogInterface dialog) {
                            access.cancel(true);
                        }
                    });
                }
            }

            @Override
            protected String doInBackground(String... params) {
                String result = null;
                try {
                    result = HttpUtil.request(HttpUtil.BASE_URL + servletStr, map);
                } catch (Exception e) {
                    e.printStackTrace();
```

```
            }
            return result;
        }

        @Override
        protected void onPostExecute(String result) {
            prompt.dismiss();
            try {
                if (result != null) {
                    JSONObject object = new JSONObject(result);
                    if (object.getString("result").equals("后台出错!")) {
                        setToast("后台出错!");
                    } else if (object.getString("result").equals("[]")) {
                        setToast("暂无数据!");
                    } else {
                        JSONArray array = object.optJSONArray("result");
                        for (int i = 0; i < array.length(); i++) {
                            Commodity commodity = new Commodity(array.optJSONObject
(i).getInt("theId"), array.optJSONObject(i).optString("name"), array.optJSONObject(i).
optString("type"), array.optJSONObject(i).optString("price"));
                            commodityList.add(commodity);
                            Log.v("ShowActivity", commodity.getName());
                        }
                            adapter = new MyListViewAdapter(ShowActivity.this,
commodityList);
                        listView = (ListView) findViewById(R.id.listView);
                        listView.setAdapter(adapter);
                        listView.setOnItemClickListener(new OnItemClickListener() {
                            @Override
                            public void onItemClick(AdapterView<?> arg0, View arg1,
int arg2, long arg3) {
                                Commodity commodity = commodityList.get(arg2);
                                Bundle bundle = new Bundle();
                                bundle.putInt("theId", commodity.getTheId());
                                bundle.putString("name", commodity.getName());
                                bundle.putString("type", commodity.getType());
                                bundle.putString("price", commodity.getPrice());
                                    Intent intent = new Intent(ShowActivity.this,
UpdateActivity.class);
                                intent.putExtras(bundle);
                                startActivity(intent);
                                ShowActivity.this.finish();
                            }
                        });
                    }
                }
            } catch (Exception e) {
                e.printStackTrace();
            }
        }

    };
    access.execute();
```

```
    }

    private void setToast(String message) {
        if (toast!= null) {
            ((TextView) toast.getView()).setText(message);
        } else {
            toast = HttpUtil.createToast(ShowActivity.this, message);
        }
        toast.show();
    }
}
```

该类主要实现请求后台数据并将加载到的数据作为数据源填充在页面中的ListView组件的功能，类中所用到的与后台数据交互的技术与MainActivity中用到的技术类似，读者可以慢慢体会。

单击ShowActivity页面中的ListView的每一条数据，将会跳转到该条商品信息的修改页面，修改该条商品信息后单击保存修改按钮把数据提交到后台，由后台修改对应的数据库表数据。该Activity的布局与MainActivity类似，实现主程序也与MainActivity类似，此处不再赘述，读者可以在本书所附带的源码中查看。

而代码中所请求的后台实现，也在本书所附带的源码中，读者可直接发布后使用。至此，实现商品发布器的主要的客户端代码已介绍完毕。

将程序部署在Android模拟器中，启动应用，可以看到如图6.1所示启动界面。

单击“查看已有商品”按钮，跳转到如图6.2所示界面。

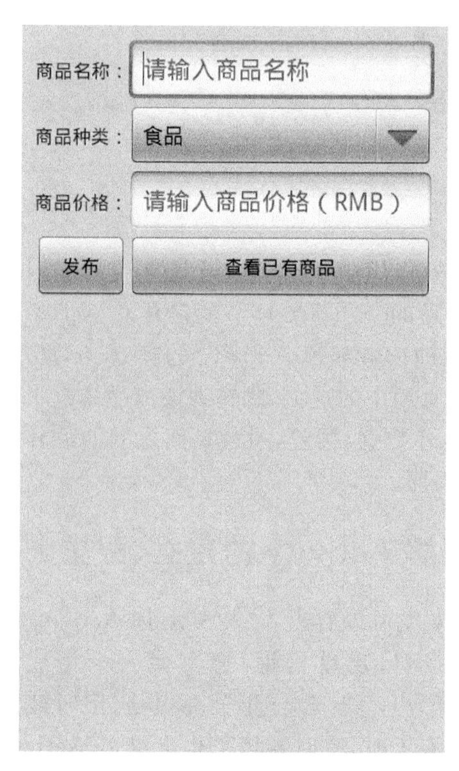

图6.1　商品发布器商品编辑界面

图6.2　查看已有商品

单击第一条数据，跳转到商品修改界面，如图6.3所示。

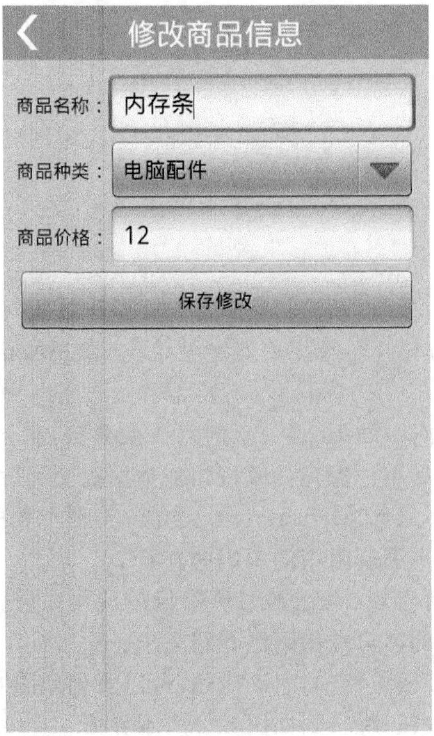

图6.3　商品修改页面

接下来对该商品发布器进行剖析，并详细介绍Activity。

6.2　剖析商品发布器

Activity作为Android四大组件之一，是Android应用中最重要以及最常用的应用组件（此处组件并非指界面控件），所以Android应用开发的一个重要环节就是开发Activity，本节主要是通过剖析6.1节所实现的商品发布器，帮助读者理解Activity的建立、配置、启动与关闭过程，然后再通过在实例中单击“已有商品”页面中的选项，跳转到该商品修改页面这个功能来介绍如何使用Bundle在Activity之间传递数据，接着，通过页面之间的跳转时的日志输出信息来理解Activity的回调机制和生命周期。

6.2.1　从商品发布器的启动界面理解Activity的建立、配置

首先通过查看商品发布器源码中的MainActivity类可以看出，该类继承自Activity基类并实现了android.view.View.OnClickListener接口，也就是说，想要使一个类成为Activity类，那么我们需要做的是让这个类继承Activity类，这与开发Web应用时建立Servlet需要继承HttpServlet类似。当然，在有特殊需求时，有时候该类也会继承Activity的子类，例如，继承ListActivity显示列表页或继承TabActivity实现标签页效果。

接下来我们还看到 MainActivity 中有下面这么一段代码：

```
@Override
protected void onCreate(Bundle savedInstanceState) {
    super.onCreate(savedInstanceState);
    //加载布局文件
    setContentView(R.layout.activity_main);
    init();
}
```

在方法上加上了@Override 注解，表示该方法是父类中的方法，也就是说，此处实现的 onCreate (Bundle savedInstanceState) 是属于重写 Activity 类中的该方法，这是实现 Activity 时需要实现的最常见的方法，该方法将会在 Activity 启动时被回调，这与开发 Servlet 时重写 HttpServlet 的 doGet(…) 和 doPost(…) 方法类似。在 onCreate (Bundle savedInstanceState)方法体中可以看到 setContentView(R. layout. activity_main)这行代码，该行代码将布局文件 activity_main. xml 中所展示的 View 显示在该 Activity 中，然后再通过自定义 init()方法去获取界面中各个组件以及修改各个组件的属性和方法即可。需要特别指出的是，创建 Activity 有时并不只是需要重写 onCreate 方法而已，而是需要实现多个 Activity 类中的回调方法。

在本应用实例中，通过开发 MainActivity，实现在界面单击“发布”按钮从而将界面中的 EditText 控件以及 Spinner 控件所对应的值提交到后台程序中，而后台将会把所接收到的值作为商品实体的属性存放到数据库中的商品表中；单击界面中的“查看已有商品”按钮跳转到商品列表页面。具体可以查看本书所附带源码中该实例关于 MainActivity 的实现源码，此处不再赘述，因为这些源码已经与具体的业务需求挂钩了，所以已不是建立 Activity 所需要做的通用操作了。

接下来，看一下商品发布器中是如何配置 MainActivity 的。查看项目源码全局配置文件 AndroidManifest. xml，代码如下：

```
<?xml version = "1.0" encoding = "utf - 8"?>
< manifest xmlns:android = "http://schemas.android.com/apk/res/android"
    package = "cn.edu.hstc.commodityrelease.activity"
    android:versionCode = "1"
    android:versionName = "1.0" >

    < uses - sdk
        android:minSdkVersion = "18"
        android:targetSdkVersion = "18" />

    < application
        android:allowBackup = "true"
        android:icon = "@drawable/ic_launcher"
        android:label = "@string/app_name"
        android:theme = "@style/AppTheme" >
        < activity
            android:name = "cn.edu.hstc.commodityrelease.activity.MainActivity"
            android:label = "@string/app_name" >
```

```
            <!-- 指定该 Activity 是程序的入口,即启动界面 -->
            <intent-filter>
                <action android:name="android.intent.action.MAIN" />

                <category android:name="android.intent.category.LAUNCHER" />
            </intent-filter>
        </activity>
        <!-- 定义应用中所出现的各个 Activity,必须指定其 name 属性 -->
        <activity android:name="cn.edu.hstc.commodityrelease.activity.ShowActivity" />
        <activity android:name="cn.edu.hstc.commodityrelease.activity.UpdateActivity" />
    </application>
    <!-- 定义应用中所会使用到的权限 -->
    <uses-permission android:name="android.permission.ACCESS_NETWORK_STATE"/>
    <uses-permission android:name="android.permission.INTERNET"/>

</manifest>
```

从代码中可以看到,开发了 Activity 之后,需要在全局配置文件中定义该 Activity,具体方法是为< application/>元素添加< activity/>子元素。应用中建立了多少个 Activity,则需要定义多少个< Activity/>元素。每个< Activity/>元素都需要指定 name 属性,该属性的属性值为该 Activity 类在项目中的全称,例如,本示例中的 cn.edu.hstc.commodityrelease.activity.ShowActivity。当然,还可以指定其他属性,如该 Activity 对应的图标 icon、该 Activity 的标签 label 等。如下配置片段:

```
<activity
    android:name="cn.edu.hstc.commodityrelease.activity.MainActivity"
    android:icon="@drawable/icon"
    android:label="@string/app_name" >
</activity>
```

以上的代码在配置 MainActivity 时还为其指定了一个< intent-filter/>元素,该元素用于指定该 Activity 相应的 Intent。例如代码中的以下片段:

```
<intent-filter>
    <action android:name="android.intent.action.MAIN" />
    <category android:name="android.intent.category.LAUNCHER" />
</intent-filter>
```

该代码片段指定了 MainActivity 为程序的入口。

至此,如何建立和配置 Activity 已全部介绍完毕。

6.2.2 使用 Bundle 将信息传递到商品修改页面

在本例中,单击"查看已有商品"页面中的某一条数据,程序将会获取 ListView 中该条商品信息,并将该条商品的 ID、名称、分类、价格都传递到商品修改页面,也就是说,此时的页面跳转,是带着数据信息跳转的。本例中跳转到商品修改页面的代码片段如下:

```
listView.setOnItemClickListener(new OnItemClickListener() {
    @Override
```

```
        public void onItemClick(AdapterView<?> arg0, View arg1, int arg2, long arg3) {
            Commodity commodity = commodityList.get(arg2);
            Bundle bundle = new Bundle();
            bundle.putInt("theId", commodity.getTheId());
            bundle.putString("name", commodity.getName());
            bundle.putString("type", commodity.getType());
            bundle.putString("price", commodity.getPrice());
            Intent intent = new Intent(ShowActivity.this, UpdateActivity.class);
            intent.putExtras(bundle);
            startActivity(intent);
            ShowActivity.this.finish();
        }
    });
```

从代码中可以看到，商品信息是存放到一个 Bundle 对象中的，Bundle 提供了多个方法来存入数据，例如代码中的 bundle.putInt("theId", commodity.getTheId())这一行代码以及 bundle.putString("name", commodity.getName())这一行代码。其实，Bundle 还提供了很多其他的类似存入 Integer 类型和 String 类型的方法来存入其他类型的数据，语法格式为：putXxx(String key, Xxx data)。

事实上，Bundle 还提供了一个 putSerializable(String key, Serialiizable data)的方法来存入一个可序列化的对象。

接着，Activity 之间的信使——Intent，通过方法 putExtras(Bundle bundle)将存入了数据的 Bundle 对象引入 Intent 对象中，这样，跳转时便将相应的数据传输到另一个 Activity 中。如上面代码片段中的 intent.putExtras(bundle)所示。

为了在另一个 Activity 中能够取出 Bundle 所携带的数据，Bundle 提供了以下方法：

- getXxx(String key)——从 Bundle 中取出 Integer、String 等各种类型的数据；
- getSerializable(String key, Serialiizable data)——从 Bundle 中取出一个可序列化的对象。

在商品修改页面中取出传递过来的商品信息的代码片段如下，读者可以从中体会上面两个取数据方法中的第一个。

```
Intent intent = getIntent();
Bundle bundle = intent.getExtras();
final Integer theId = bundle.getInt("theId");
String name = bundle.getString("name");
String type = bundle.getString("type");
String price = bundle.getString("price");
```

至此，使用 Bundle 在 Activity 间传递信息的介绍就完毕。

6.2.3 理解 Activity 的回调机制以及生命周期

当你开发了一个 Web 应用下的 Servlet 并运行于 Web 服务器中时，该 Servlet 何时创建实例，何时回调 Servlet 的方法，开发者是无法控制的，这种回调由服务器自行决定。

而 Android 应用中的 Activity 开发类似于 Web 开发中的 Servlet 开发，Activity 被开发出来后，开发者只需在 AndroidManifest.xml 文件中配置该 Activity 即可。至于该 Activity

何时被实例化，它的方法何时被调用，对开发者来说是透明的。

在不同的场景和时间点，Android 将会调用 Activity 的不同的特定的方法，这种调用就称为回调。当然，想要实现这些回调，Activity 开发必须继承 Activity 基类或 Activity 的子类，继承之后，这些特定的方法就已经被定义了，如 6.1 节所实现的商品发布器中的 ShowActivity 类中的 onCreate 方法。开发者可以根据业务需求选择性地重写 Activity 中的方法，Activity 将会回调这些特定的方法来完成具体的业务处理。这就是 Activity 的回调机制。

当商品发布器被部署在 Android 系统之后，随着商品发布器的运行，Activity 会在不同的状态之间不断切换，该 Activity 中特定的方法就会被回调，这些不同的状态就构成了 Activity 的生命周期。

当一个 Activity 处于 Android 应用中运行时，它的活动状态由 Android 以 Activity 栈的形式管理。当前活动的 Activity 位于栈顶。随着应用的运行阶段的不同，每个 Activity 都会在活动状态以及非活动状态之间不同切换。总结起来 Activity 大致有以下四个状态：

- 活动状态——当前 Activity 位于栈顶，用户可见，并可以获得焦点。
- 暂停状态——其他 Activity 位于栈顶，但该 Activity 依旧可见，只是不能获得焦点。
- 停止状态——该 Activity 不可见并且失去焦点。
- 销毁状态——该 Activity 结束，或该 Activity 所在的 Dalvik 进程被结束。

下面通过 Android 官方文档中的一张经典的生命周期流程图来显示 Activity 生命周期及相关的回调方法，如图 6.4 所示。

下面通过改造以下商品发布器查看已有商品页面所对应的 Activity—ShowActivity 类的源码来向读者展示 Activity 的生命周期。首先，在 ShowActivity 类中的 onCreate (Bundle savedInstanceState) 方法中加入一行代码，如下面代码片段中的方圈里的内容所示：

```
@Override
protected void onCreate(Bundle savedInstanceState) {
    super.onCreate(savedInstanceState);
    Log.i("ShowActivity", "------ onCreate ------");
    //加载界面布局文件
    setContentView(R.layout.activity_show);
    //获取返回按钮
    backImg = (ImageView) findViewById(R.id.img_back);
    //为返回按钮添加事件监听器
    backImg.setOnClickListener(new View.OnClickListener() {
        @Override
        public void onClick(View v) {
            ShowActivity.this.finish();
        }
    });
    Map<String, String> map = new HashMap<String, String>();
    map.put("name", "");
    //请求后台 GetCommoditys,加载数据并完成页面更新操作
    acquire(map, "数据加载中...", true, "GetCommoditys");
}
```

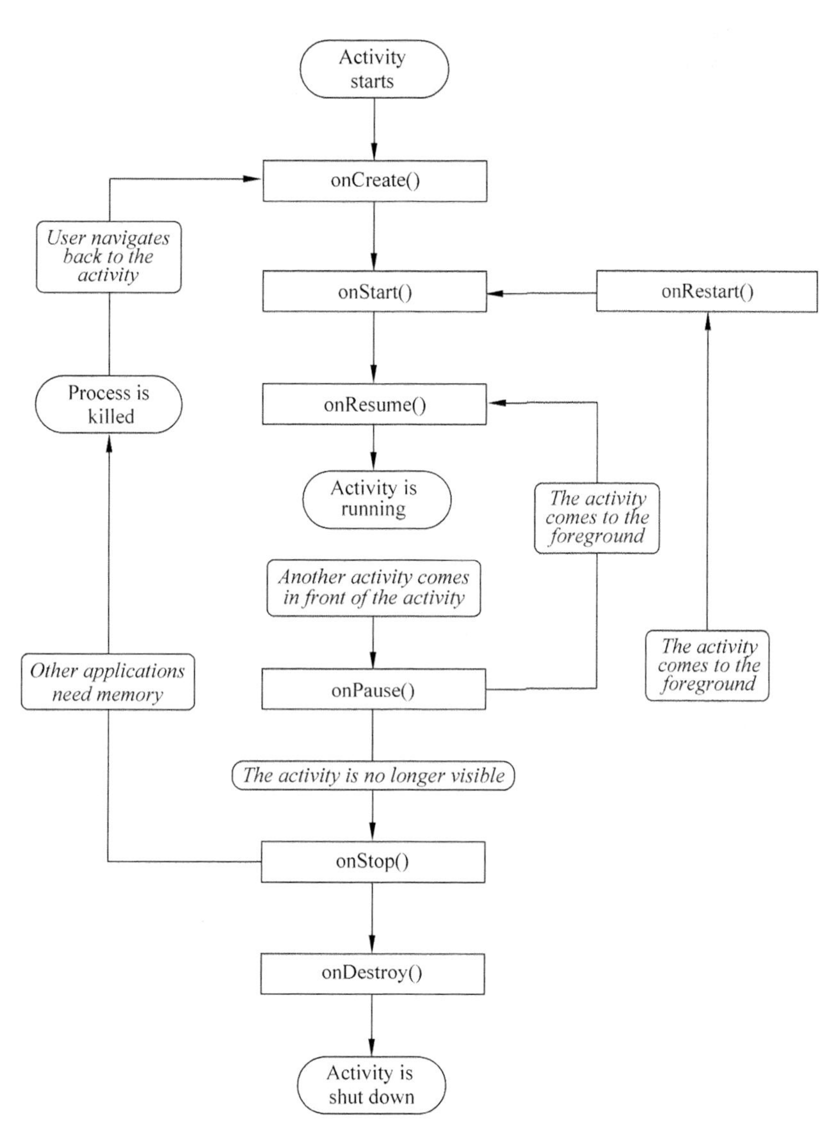

图 6.4 Activity 的生命周期以及回调方法

接着，在 ShowActivity 类中重写 Activity 基类的其他回调方法，即加入如下代码片段：

```
@Override
protected void onStart() {
    super.onStart();
    Log.i("ShowActivity", "------ onStart ------");
}

@Override
protected void onRestart() {
    super.onRestart();
    Log.i("ShowActivity", "------ onRestart ------");
}

@Override
```

```
protected void onResume() {
    super.onResume();
    Log.i("ShowActivity", "------ onResume ------");
}

@Override
protected void onPause() {
    super.onPause();
    Log.i("ShowActivity", "------ onPause ------");
}

@Override
protected void onStop() {
    super.onStop();
    Log.i("ShowActivity", "------ onStop ------");
}

@Override
protected void onDestroy() {
    super.onDestroy();
    Log.i("ShowActivity", "------ onDestroy ------");
}
```

将商品发布器重新部署在 Android 模拟器上，运行效果如图 6.1 所示。此时单击“查看已有商品”按钮，应用将会跳转到 ShowActivity 页，如图 6.2 所示，此时，查看 Eclipse 的 DDMS 面板，在 LogCat 窗口过滤出 info 信息，将会看到如图 6.5 所示的输出。

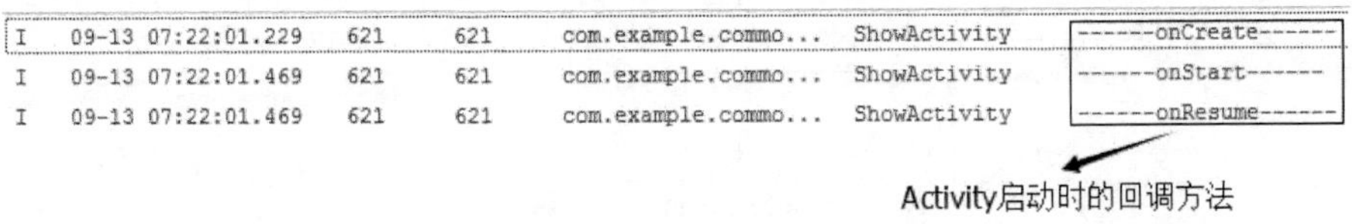

I	09-13 07:22:01.229	621	621	com.example.commo...	ShowActivity	------onCreate------
I	09-13 07:22:01.469	621	621	com.example.commo...	ShowActivity	------onStart------
I	09-13 07:22:01.469	621	621	com.example.commo...	ShowActivity	------onResume------

Activity启动时的回调方法

图 6.5　启动 Activity 时回调的方法

此时，按下 Android 模拟器上的 Home 键，返回系统桌面，当前 Activity，即“查看已有商品”页面将会失去焦点且不可见，但该 Activity 并未被销毁，只是进入暂停状态，此时看到 LogCat 窗口中有如图 6.6 所示的输出。

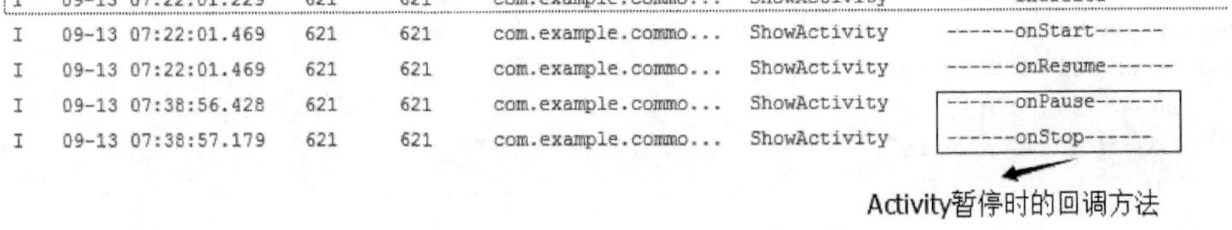

I	09-13 07:22:01.229	621	621	com.example.commo...	ShowActivity	------onCreate------
I	09-13 07:22:01.469	621	621	com.example.commo...	ShowActivity	------onStart------
I	09-13 07:22:01.469	621	621	com.example.commo...	ShowActivity	------onResume------
I	09-13 07:38:56.428	621	621	com.example.commo...	ShowActivity	------onPause------
I	09-13 07:38:57.179	621	621	com.example.commo...	ShowActivity	------onStop------

Activity暂停时的回调方法

图 6.6　暂停 Activity 时回调的方法

在模拟器程序列表中单击商品发布器图标，此时重新回到按下 Home 键之前的页面，也就是“查看已有商品”页面 ShowActivity，该 Activity 重新变为活动状态，可以看到 LogCat 窗口中有如图 6.7 所示输出。

I	09-13 07:22:01.229	621	621	com.example.commo...	ShowActivity	------onCreate------
I	09-13 07:22:01.469	621	621	com.example.commo...	ShowActivity	------onStart------
I	09-13 07:22:01.469	621	621	com.example.commo...	ShowActivity	------onResume------
I	09-13 07:38:56.428	621	621	com.example.commo...	ShowActivity	------onPause------
I	09-13 07:38:57.179	621	621	com.example.commo...	ShowActivity	------onStop------
I	09-13 07:43:48.460	621	621	com.example.commo...	ShowActivity	------onRestart------
I	09-13 07:43:48.460	621	621	com.example.commo...	ShowActivity	------onStart------
I	09-13 07:43:48.469	621	621	com.example.commo...	ShowActivity	------onResume------

Activity重新启动时的回调方法

图 6.7 重启 Activity 时回调的方法

单击"查看已有商品"页面中的商品列表中的任意一条数据，跳转到商品修改页面，依据程序设定，此时将会结束掉"查看已有商品"页面，可以看到 LogCat 窗口中有如图 6.8 所示输出。

I	09-13 07:22:01.229	621	621	com.example.commo...	ShowActivity	------onCreate------
I	09-13 07:22:01.469	621	621	com.example.commo...	ShowActivity	------onStart------
I	09-13 07:22:01.469	621	621	com.example.commo...	ShowActivity	------onResume------
I	09-13 07:38:56.428	621	621	com.example.commo...	ShowActivity	------onPause------
I	09-13 07:38:57.179	621	621	com.example.commo...	ShowActivity	------onStop------
I	09-13 07:43:48.460	621	621	com.example.commo...	ShowActivity	------onRestart------
I	09-13 07:43:48.460	621	621	com.example.commo...	ShowActivity	------onStart------
I	09-13 07:43:48.469	621	621	com.example.commo...	ShowActivity	------onResume------
I	09-13 07:48:31.379	621	621	com.example.commo...	ShowActivity	------onPause------
I	09-13 07:48:31.900	621	621	com.example.commo...	ShowActivity	------onStop------
I	09-13 07:48:31.900	621	621	com.example.commo...	ShowActivity	------onDestroy------

结束Activity时的回调方法

图 6.8 关闭 Activity 时回调的方法

通过上面的演示，相信读者对 Activity 的生命周期状态及在不同状态之间切换时所回调的方法有了清晰的认识。归纳如下：

- onCreate(Bundle savedInstanceState)——创建 Activity 时被回调。
- onStart()——启动 Activity 时被回调。
- onRestart()——重新启动 Activity 时被回调。
- onResume()——恢复 Activity 时被回调。
- onPause()——暂停 Activity 时被回调。
- onStop()——停止 Activity 时被回调。
- onDestroy()——销毁 Activity 时被回调。

6.3 本章小结

本章详细介绍了 Android 四大组件之一的 Activity。Activity 开发是 Android 开发中最常见的开发。本章介绍商品发布器的开发，重点掌握如何开发 Activity 以及如何在 AndroidManifest.xml 全局配置文件中配置 Activity。由于 Android 应用通常会由多个 Activity 组成，因此读者必须掌握如何利用 Bundle 在不同 Activity 之间传递数据。Activity 具有自身的生命周期，读者需要对此有清晰的了解。

第7章 通过计时器详细介绍Service及BroadcastReceiver

本章通过计时器这个小应用的实现来向读者介绍 Android 四大组件中的 Service 以及 BroadcastReceiver。7.1 节介绍实现开发计时器的流程以及代码，7.2 节通过剖析计时器介绍 Service 和 BroadcastReceiver 的创建、配置、启动等知识点。本章亦会涉及 Service 的更多知识点以及系统级的广播消息等。

7.1 实现计时器

本节的主要内容是介绍计时器的实现流程及代码。首先，该计时器的需求是当用户单击程序列表中的计时器应用图标时，将会以 Android 对话框的形式弹出一个小窗体，在该窗体中放置一个显示已过时间的文本输入框，该输入不可编辑，在该文本框正下方放置一个按钮，单击即可开始计时，并将已过时间实时显示在文本输入框中，时间精确到秒，用户在此单击该按钮可停止计时。

在 Eclipse 中新建一个 Android 应用工程，填写好应用名称、工程名称、包名等相关信息，选择 SDK 版本，完成其他相关操作，单击 Finish 按钮，项目工程便初始化完毕。所谓计时器，就是为了向用户展示已用时间，因而显示界面必不可少，新建一个 Layout 布局文件，完成计时器与用户交互的布局界面的代码编辑，界面代码如下：

```
<LinearLayout xmlns:android = "http://schemas.android.com/apk/res/android"
    android:layout_width = "fill_parent"
    android:layout_height = "fill_parent"
    android:orientation = "vertical"
    android:padding = "10dp" >

    <LinearLayout
        android:layout_width = "wrap_content"
        android:layout_height = "wrap_content"
        android:orientation = "horizontal" >

        <TextView
            android:layout_width = "wrap_content"
            android:layout_height = "wrap_content"
            android:text = "已经过: " />
```

```
        <EditText
            android:id="@+id/edit_time"
            android:layout_width="131.5dp"
            android:layout_height="wrap_content"
            android:background="@drawable/contact_edit_edittext_normal"
            android:cursorVisible="false"
            android:paddingLeft="2.5dp"
            android:editable="false" />
    </LinearLayout>

    <Button
        android:id="@+id/button"
        android:layout_width="fill_parent"
        android:layout_height="wrap_content"
        android:layout_marginTop="6dp"
        android:background="@drawable/style_button"
        android:text="开始计时" />

</LinearLayout>
```

代码文件：codes\07\7.1\CountDemo\res\layout\activity_main.xml

以上代码为界面中的文本输入框设置了一个 background 属性，该属性值是一个 drawable 样式资源 contact_edit_edittext_normal.xml，在该资源文件中自定义了文本框的样式，读者可以在本书附带源码中查看该资源文件源码。在代码中亦将该文本输入框设置为光标隐藏并且不可编辑，即用户单击该文本输入框，并不会弹出 Android 键盘供用户输入。界面中所放置的 Button 控件的样式通过样式资源文件 style_button.xml 实现自定义，读者也可通过学习本书附带案例源码获得该文件代码。

本计时器有一个特殊要求，就是当用户按下 Android 硬键盘上的 Home 键时，即使计时器界面此时已消失，不再与用户交互，但计时功能仍在后台继续进行，当用户重新单击程序列表中的计时器图标时，将会把界面重新调到前台，输入框将显示之前总共经过的时间并继续计时。要实现该需求，就必须使用到 Android 四大组件之一：Service 组件，将计时的核心程序放置在一个自定义的 Service 类中。该 Service 类的实现代码如下：

```
package cn.edu.hstc.countdemo.service;

import java.text.SimpleDateFormat;
import java.util.Date;
import java.util.Timer;
import java.util.TimerTask;

import android.app.Service;
import android.content.ComponentName;
import android.content.Intent;
import android.os.Bundle;
import android.os.IBinder;
import android.util.Log;
```

```
public class TimeService extends Service {
    private static String TIME_CHANGED_ACTION = "cn.edu.hstc.service.action.TIME_CHANGED_
ACTION";
    private Timer timer = null;
    private Intent timeIntent = null;
    private Bundle bundle = null;
    private SimpleDateFormat sdf = null;
    private boolean flag = true;

    @Override
    public void onCreate() {
        super.onCreate();
        Log.i("TimeService", "------onCreate------");
        this.init();                                        //初始化
        timer.schedule(new TimerTask() {                    //定时器发送广播
            @Override
            public void run() {
                if (flag) {
                    sendTimeChangedBroadcast();             //发送广播
                } else {
                    timer.cancel();
                    timer.purge();
                    timer = null;
                }
            }
        }, 1000, 1000);
    }

    @Override
    public void onStart(Intent intent, int startId) {
        super.onStart(intent, startId);
        Log.i("TimeService", "------onStart------");
    }

    @Override
    public int onStartCommand(Intent intent, int flags, int startId) {
        Log.i("TimeService", "------onStartCommand------");
        return super.onStartCommand(intent, flags, startId);
    }

    @Override
    public IBinder onBind(Intent intent) {
        Log.i("TimeService", "------onBind------");
        return null;
    }

    @Override
    public boolean onUnbind(Intent intent) {
        Log.i("TimeService", "------onUnbind------");
        return super.onUnbind(intent);
    }
```

```
    /**
     * 相关变量初始化
     */
    private void init() {
        timer = new Timer();
        sdf = new SimpleDateFormat("yyyy-MM-dd HH:mm:ss.SSS");
        timeIntent = new Intent();
        bundle = new Bundle();
    }

    /**
     * 发送广播,通知UI层时间已改变
     */
    private void sendTimeChangedBroadcast() {
        bundle.putString("time", getTime());
        timeIntent.putExtras(bundle);
        timeIntent.setAction(TIME_CHANGED_ACTION);
        //发送广播,通知UI层时间改变了
        sendBroadcast(timeIntent);
    }

    /**
     * 获取最新系统时间
     */
    private String getTime() {
        return sdf.format(new Date());
    }

    @Override
    public ComponentName startService(Intent service) {
        Log.i("TimeService", "------startService------");
        return super.startService(service);
    }

    @Override
    public void onDestroy() {
        super.onDestroy();
        Log.i("TimeService", "------onDestroy------");
        flag = false;
    }
}
```

代码文件：codes\07\7.1\CountDemo\cn\edu\hstc\countdemo\service\TimeService.java

上面程序实现了一个可运行于后台的Service组件,该Service负责通过一个定时器Timer每一秒钟发送一个广播,并且该广播在传播过程中将带着系统当前最新的时间。当该Service被销毁的时候,该定时器将不再发送广播。

在实现TimeService类的源码之后,还需要在Android全局配置文件中对该Service进行配置,如下:

```
<?xml version = "1.0" encoding = "utf - 8"?>
< manifest xmlns:android = "http://schemas.android.com/apk/res/android"
    package = "cn.edu.hstc.countdemo.activity"
    android:versionCode = "1"
    android:versionName = "1.0" >

    < uses - sdk
        android:minSdkVersion = "8"
        android:targetSdkVersion = "8" />

    < application
        android:allowBackup = "true"
        android:icon = "@drawable/ic_launcher"
        android:label = "@string/app_name"
        android:theme = "@style/AppTheme" >
        < activity
            android:name = "cn.edu.hstc.countdemo.activity.MainActivity"
            android:theme = "@android:style/Theme.Dialog" >
            < intent - filter >
                < action android:name = "android.intent.action.MAIN" />

                < category android:name = "android.intent.category.LAUNCHER" />
            </intent - filter >
        </activity >
        < activity android:name = "cn.edu.hstc.countdemo.activity.SecondActivity" />
        < service android:name = "cn.edu.hstc.countdemo.service.TimeService">
            < intent - filter android: name = "cn.edu.hstc.service.action.TIME_CHANGED_
ACTION"></intent - filter >
        </service >
    </application >

</manifest >
```

代码文件：codes\07\7.1\CountDemo\res\AndroidManifest.xml

有了该Service不断记录当前最新时间，为了实现不断告诉用户该时间和一开始单击“开始计时”按钮时获取的时间的时间差，必须采用一个Activity与用户进行交互。新建一个MainActivity类，实现代码如下：

```
package cn.edu.hstc.countdemo.activity;

import java.text.DateFormat;
import java.text.ParseException;
import java.text.SimpleDateFormat;
import java.util.Date;

import android.app.Activity;
import android.content.BroadcastReceiver;
import android.content.Context;
import android.content.Intent;
import android.content.IntentFilter;
```

```
import android.graphics.Color;
import android.os.Bundle;
import android.view.View;
import android.view.Window;
import android.view.WindowManager;
import android.widget.Button;
import android.widget.EditText;
import cn.edu.hstc.countdemo.service.TimeService;

public class MainActivity extends Activity {
    private static String TIME_CHANGED_ACTION = "cn.edu.hstc.service.action.TIME_CHANGED_
ACTION";
    private EditText timeEdit = null;
    private Button button = null;
    private Intent serviceIntent = null;
    private DateFormat df = new SimpleDateFormat("yyyy-MM-dd HH:mm:ss.SSS");
    private Date startDate = null, currentDate = null;

    @Override
    public void onCreate(Bundle savedInstanceState) {
        super.onCreate(savedInstanceState);
          this.getWindow().setFlags(WindowManager.LayoutParams.FLAG_FULLSCREEN,
WindowManager.LayoutParams.FLAG_FULLSCREEN);
        this.requestWindowFeature(Window.FEATURE_NO_TITLE);
        setContentView(R.layout.activity_main);
        //初始化布局上的控件 Widget
        initWidget();
        //注册广播,监听后台 Service 发送过来的广播
        registerBroadcastReceiver();
    }

    private class TimeReceiver extends BroadcastReceiver {
        @Override
        public void onReceive(Context context, Intent intent) {
            String action = intent.getAction();
            if (action!= null && TIME_CHANGED_ACTION.equals(action)) {
                String currentTime = intent.getExtras().getString("time");
                try {
                    currentDate = df.parse(currentTime);
                } catch (ParseException e) {
                    e.printStackTrace();
                }
                long diff = currentDate.getTime() - startDate.getTime();
                                                          //这样得到的差值是微秒级别
                long days = diff / (1000 * 60 * 60 * 24);
                long hours = diff / (60 * 60 * 1000) - days * 24;
                long mins = (diff / (60 * 1000)) - days * 24 * 60 - hours * 60;
                long s = diff / 1000 - days * 24 * 60 * 60 - hours * 60 * 60 - mins * 60;
                String diffTimeStr = days + "天" + hours + "小时" + mins + "分" + s + "秒";
                timeEdit.setText(diffTimeStr);
            }
```

```
        }
    }

    /**
     * 初始化 UI
     */
    private void initWidget() {
        timeEdit = (EditText) findViewById(R.id.edit_time);
        timeEdit.setTextColor(Color.RED);
        timeEdit.setTextSize(13);
        button = (Button) findViewById(R.id.button);
        button.setOnClickListener(new View.OnClickListener() {
            @Override
            public void onClick(View v) {
                if (button.getText().toString().equals("开始计时")) {
                    //启动服务,时间改变后发送广播,通知 UI 层修改时间
                    try {
                        String nowDateStr = df.format(new Date());
                        startDate = df.parse(nowDateStr);
                    } catch (ParseException e) {
                        e.printStackTrace();
                    }
                    startTimeService();
                    button.setText("停止计时");
                } else if (button.getText().toString().equals("停止计时")) {
                    stopService(serviceIntent);
                    serviceIntent = null;
                    startDate = null;
                    currentDate = null;
                    button.setText("开始计时");
                }
            }
        });
    }

    /**
     * 注册广播
     */
    private void registerBroadcastReceiver() {
        TimeReceiver receiver = new TimeReceiver();
        IntentFilter filter = new IntentFilter(TIME_CHANGED_ACTION);
        registerReceiver(receiver, filter);
    }

    /**
     * 启动服务
     */
    private void startTimeService() {
        serviceIntent = new Intent(this, TimeService.class);
        this.startService(serviceIntent);
    }
```

```
        @Override
        protected void onDestroy() {
            super.onDestroy();
            //停止服务
            if (serviceIntent!= null) {
                stopService(serviceIntent);
                serviceIntent = null;
            }
        }
    }
```

代码文件：codes\07\7.1\CountDemo\cn\edu\hstc\countdemo\activity\MainActivity.java

上面程序实现当用户单击 Activitiy 中的"开始计时"按钮时，记录下此时系统当前时间并启动服务 TimeService，TimeService 开始将更新的时间通过广播发送出去，MainActivity 中的注册的广播接收者开始不断监听 TimeService 发来的广播并将接收到的广播 Intent 中所带的时间和记录的开始时间进行比较，算出时间差并将该时间差显示在页面中的文本输入框中。当用户单击"停止计时"按钮时，将停止 TimeServie，触发 TimeService 的销毁事件控制 TimeService 中的定时器不再发送广播。

为了使该 Activity 以对话框 Dialog 的形式启动，需要在 Android 的全局配置文件中配置该 Activity 的主题样式。

```
<?xml version = "1.0" encoding = "utf - 8"?>
<manifest xmlns:android = "http://schemas.android.com/apk/res/android"
    package = "cn.edu.hstc.countdemo.activity"
    android:versionCode = "1"
    android:versionName = "1.0" >

    <uses - sdk
        android:minSdkVersion = "8"
        android:targetSdkVersion = "8" />

    <! -- 通过 android:theme = "@android:style/Theme.Dialog"指定 Activity 的主题资源为对话
框主题 -->
    <application
        android:allowBackup = "true"
        android:icon = "@drawable/ic_launcher"
        android:label = "@string/app_name"
        android:theme = "@style/AppTheme" >
        <activity
            android:name = "cn.edu.hstc.countdemo.activity.MainActivity"
            android:theme = "@android:style/Theme.Dialog" >
            <intent - filter>
                <action android:name = "android.intent.action.MAIN" />

                <category android:name = "android.intent.category.LAUNCHER" />
            </intent - filter>
        </activity>
        <activity android:name = "cn.edu.hstc.countdemo.activity.SecondActivity" />
```

```
        <service android:name = "cn.edu.hstc.countdemo.service.TimeService">
            <intent - filter android:name = "cn.edu.hstc.service.action.TIME_CHANGED_
ACTION"></intent - filter>
        </service>
    </application>

</manifest>
```

代码文件：codes\07\7.1\CountDemo\res\AndroidManifest.xml

将应用部署在 Android 模拟器上，运行效果如图 7.1 所示。

用户单击“开始计时”按钮，计时器开始计时，效果图如图 7.2 所示。

图 7.1 启动计时器

图 7.2 开始计时

至此，计时器的全部实现代码已介绍完毕，接下来将对该计时器进行剖析，介绍计时器中所使用到的 Service 和 BroadcastReceiver 的一些特性。

7.2 剖析计时器

Service 作为 Android 组件，与 window 中的服务是类似的，Service 一般没有用户操作界面，运行于系统中不易被用户发现，可以使用它开发监控程序等。一旦 Service 被启动起来，它就与 Activity 一样有了自己的生命周期。

广播接收者 BroadcastReceiver 用于接收广播 Intent，通常一个广播可以被订阅了该广播 Intent 的多个广播接收者接收。BroadcastReceiver 就如一个全局的事件监听器，监听着系统发出的 Broadcast。

接下来，通过剖析计时器的源代码中关于 TimeService 和 BroadcastReceiver 的开发来介绍这两者的相关特性。

7.2.1　计时服务 TimeService 的创建、配置

查看计时器中 TimeService 的实现源码，可以发现该类继承自 Service 基类，并重写了 onCreate()方法和 onDestroy()方法。这说明开发 Service 组件的第一个必要的步骤就是定义一个继承 Service 类的子类。

TimeService 重写了 onCreate()方法和 onDestroy()方法，这两个方法其实都是属于 Service 的生命周期方法，关于 Service 的生命周期，将会在后面单独介绍，此处不再赘述。

接下来，查看 AndroidManifest. xml 文件源码，可以发现其中有这样一行代码：<service android:name="cn. edu. hstc. countdemo. service. TimeService"></service>，该行代码配置了 TimeService。这正是开发 Service 的第二个必要的步骤——在 AndroidManifest. xml 中配置 Service。

事实上，配置 Service 时也可以为<service/>元素配置<intent-filter/>子元素，用于说明该 Service 可被哪些 Intent 启动。如下配置片段：

```
<service android:name = "cn.edu.hstc.countdemo.service.TimeService">
    <intent - filter android:name = "cn.edu.hstc.service.action.TIME_CHANGED_ACTION"></
intent - filter>
</service>
```

7.2.2　计时服务 TimeService 的启动和停止

一个服务开发好了，比如由访问者去启动该服务。在计时器这个案例中，是通过 MainActivity 去启动服务的，单击 MainActivity 中的“开始计时”按钮，将会启动 TimeService 服务。查看 MainActivity 类，有如下代码片段：

```
/**
 * 启动服务
 */
private void startTimeService() {
    serviceIntent = new Intent(this, TimeService.class);
    this.startService(serviceIntent);
}
```

该方法通过上下文对象 Context. startService()方法启动 TimeService 服务。需要强调的是，通过该方法启动的 Service 与访问者没有关联，即使访问者退出了，Service 仍然运行。

实际上，还可以通过 Context. bindService()方法来启动一个 Service，这时访问者与 Service 绑定在一起，访问者一旦退出，Service 也就终止。接下来将继续介绍通过 bindService()方法来启动 Service 的情况。

当用户单击“停止计时”按钮时，程序将会终止 TimeService 服务的运行，实现停止计

时。在我们的代码中,停止 TimeService 十分简单,通过代码行 stopService(serviceIntent)便可停止 TimeService 计时服务的运行。所以,当想要停止一个通过 Context. startService()方法启动的服务时,只需调用 Context. stopService()方法。通过 Context. bindService()方法启动服务,则需要通过 Context. unbindService()方法来解绑服务,这在下面将会有所介绍。

7.2.3 计时器里的广播接收者(BroadcastReceiver)的创建、配置、启动

BroadcastReceiver 用于接收包括用户开发的程序和系统内建的程序所发出的 Broadcast Intent。在此,我们暂不关心计时器里的广播接收者 BroadcastReceiver 接收的广播来自哪里,此时关心的是如何接收广播。

为了能够接收程序所发出的广播 Intent,我们需要自定义自己的 BroadcastReceiver。本例中,MainActivity 类中有以下代码片段:

```
private class TimeReceiver extends BroadcastReceiver {
    @Override
    public void onReceive(Context context, Intent intent) {
        String action = intent.getAction();
        if (action!= null && TIME_CHANGED_ACTION.equals(action)) {
            String currentTime = intent.getExtras().getString("time");
            try {
                currentDate = df.parse(currentTime);
            } catch (ParseException e) {
                e.printStackTrace();
            }
            long diff = currentDate.getTime() - startDate.getTime();
                                                        //这样得到的差值是微秒级别
            long days = diff / (1000 * 60 * 60 * 24);
            long hours = diff / (60 * 60 * 1000) - days * 24;
            long mins = (diff / (60 * 1000)) - days * 24 * 60 - hours * 60;
            long s = diff / 1000 - days * 24 * 60 * 60 - hours * 60 * 60 - mins * 60;
            String diffTimeStr = days + "天" + hours + "小时" + mins + "分" + s + "秒";
            timeEdit.setText(diffTimeStr);
        }
    }
}
```

从以上程序中可以看出,自定义广播接收者类继承自 BroadcastReceiver 基类并实现 onReceive(Context context, Intent intent)方法。

为了让开发出来的 BroadcastReceiver 能够监听特定的广播 Intent,需要为其注册广播。如 MainActivity 类中的代码片段:

```
/**
 * 注册广播
 */
private void registerBroadcastReceiver() {
    TimeReceiver receiver = new TimeReceiver();
    IntentFilter filter = new IntentFilter(TIME_CHANGED_ACTION);
```

```
        registerReceiver(receiver, filter);
    }
```

从以上代码片段可以看出，最终调用的是 registerReceiver(BroadcastReceiver receiver, IntentFilter filter)方法为广播接收者注册广播。

实际上，我们还有另一种注册广播的方式供开发者选择，在本例子中的广播接收者类是作为 MainActivity 类的内部类的，如果自定义 BroadcastReceiver 时采用的是外部类方式，也可按照以下配置片段实现广播接收者的注册：

```
<receiver android:name = ".TimeReceiver">
    <intent-filter>
        <action android:name = "cn.edu.hstc.service.action.TIME_CHANGED_ACTION" />
    </intent-filter>
</receiver>
```

BroadcastReceiver 本质上是一个系统级的监听器，拥有自己的进程，只要存在与之匹配的广播 Intent 被广播出来，BroadcastReceiver 便会被激发。每次系统广播事件发生后，系统就会创建 BroadcastReceiver 的实例，并自动触发它的 onReceive(Context context, Intent intent)方法，该方法执行完后，BroadcastReceiver 的实例就会被销毁。

因此，BroadcastReceiver 无须开发者控制启动，而是在特定的时间点，由 Android 系统自身激发。

如果 onReceive(Context context, Intent intent)方法在 10s 内未被执行完成，Android 会认为该程序无响应。所以不要在该方法中执行一些耗时的操作，否则会弹出 Application No Response 的对话框。如果确实需要由 BroadcastReceiver 来完成一项耗时的操作，则可以在 onReceive(Context context, Intent intent)方法中启动一个 Service 来完成。但不要考虑用新线程去完成耗时操作，因为 BroadcastReceiver 本身的生命周期很短，可能出现子线程还没有结束，BroadcastReceiver 就已经退出的情况。如果 BroadcastReceiver 所在的进程结束，虽然该进程还有用户启动的新线程，但由于该进程不包含任何活动组件，因此系统可能在内存紧张时优先结束该进程，这样就可能导致 BroadcastReceiver 启动的子线程不能执行完成。

7.2.4　发送广播以及广播类型

当用户单击“开始计时”按钮时，将会在后台启动计时服务，计时服务开启了一个定时器不断的发送广播 Intent，该 Intent 带着最新的当前时间，由上节所介绍的广播接收者 BroadcastReceiver 接收。也就是说，该广播是由服务发出的，于是我们在 TimeService 类中可以看到以下代码片段：

```
/**
 * 发送广播,通知 UI 层时间已改变
 */
private void sendTimeChangedBroadcast() {
    bundle.putString("time", getTime());
    timeIntent.putExtras(bundle);
    timeIntent.setAction(TIME_CHANGED_ACTION);
```

```
        //发送广播,通知 UI 层时间改变了
        sendBroadcast(timeIntent);
    }
```

以上程序中的 TIME_CHANGED_ACTION 为一个字符串常量,在 MainActivity 的广播接收者也必须是 IntentFilter 为该常量的才能监听到该广播。如下代码片段:

```
<receiver android:name = ".TimeReceiver">
    <intent-filter>
        <action android:name = "cn.edu.hstc.service.action.TIME_CHANGED_ACTION" />
    </intent-filter>
</receiver>
```

由于该 BroadcastReceiver 所注册的 IntentFilter 的 action 属性值与上面的广播 Intent 的 action 值一样,故该 BroadcastReceiver 可被该广播所激活。

由以上代码可以看出,发送广播非常简单,只需要调用 Context. sendBroadcast(Intent intent)方法。这条广播将会启动 intent 参数所对应的 BroadcastReceiver。

事实上,通过 Context. sendBroadcast(Intent intent)方法所发送的广播为普通广播,而广播还有另外一种类型叫有序广播。发送有序广播调用的是另外的一个方法,这也就涉及接下来所要介绍的广播的类型。

Broadcast 分为如下两种类型:

- Normal Broadcast(普通广播)——异步广播,可以在同一个时刻被多个广播接收者所接收,消息传递效率较高。不足之处是接收者不能将处理结果传递给下一个接收者,并且无法终止广播 Intent 的传播。
- Ordered Broadcast(有序广播)——接收者按照预先声明的优先级依次接收有序广播,优先级高的优先接收到有序广播,并将该广播传递到下一个优先级的接收者。优先级声明配置在<intent-filter/>元素的 android:priority 属性中,数值越大,优先级别越高,值的范围为－1000～1000,优先级别也可以调用 IntentFilter 对象的 setPriority()方法进行设置。有序广播的接收者可以终止该广播的传播,使后面的接收者将不再接收到该 Broadcast。另外,有序广播的接收者可以将数据传递到下一个接收者。

在以上的程序片段中,发送普通广播所调用的方法为 Context. sendBroadcast(Intent intent),而发送有序广播所调用的方法则为 Context. sendOrderedBroadcast(Intent intent)。

有序广播接收者通过 BroadcastReceiver 的 abortBroadcast()方法终止广播 Intent 的继续传播。优先接收到广播的接收者通过 setResultExtras(Bundle bundle)方法将处理结果存入广播 Broadcast 中,下一个接收者可通过代码 Bundle bundle ＝ getResultExtras(true)获取上一个接收者所存入的数据。

7.3 建立与访问者相互通信的本地 Service

当程序通过 startService()启动服务和通过 stopService()关闭服务时,Service 与访问者之间基本不存在太多的关联,因此 Service 和访问者之间无法进行通信或数据交换。因此，Android提供了另一种方式即使用bindService()和unbindService()的方法来启动和关闭服务，以使Service和访问者之间可以进行方法调用或数据交换。

Context的bindService()方法的完整方法签名为bindService(Intent service, ServiceConnection conn, int flag)，该方法的三个参数含义如下：

- service——该参数通过Intent指定要启动的Service。
- conn——该参数是一个ServiceConnection对象，该对象用于监听访问者与Service之间的连接情况。当访问者与Service之间连接成功时，将回调ServiceConnection对象的onServiceConnected(ComponentName name, IBinder service)方法；当访问者与Service之间断开连接时，将回调该ServiceConnection对象的onServiceDisconnected(ComponentName name)方法。
- flag——指定绑定时是否自动创建Service(如果Service还未创建)。该参数可指定为0(不自动创建)或BIND_AUTO_CREATE(自动创建)。

注意，ServiceConnection对象的onServiceConnected方法中有一个IBinder对象，该对象即可实现与被绑定Service之间的通信。

当开发Service类时，该Service类必须提供一个IBinder onBind(Intent intent)方法，在绑定本地Service的情况下，onBind(Intent intent)方法所返回的IBinder对象将会传给ServiceConnection对象中onServiceConnected(ComponentName name, IBinder service)方法的service参数，这样访问者就可通过该IBinder对象与Service进行通信。

实际开发中通常会采用继承Binder(IBinder的实现类)的方式实现自己的IBinder对象。

下面的程序示范了如何在Activity中绑定本地Service，并获取Service的运行状态。该程序的Service类需要真正实现onBind()方法，并让该方法返回一个有效的IBinder对象，该Service类的代码如下：

```
import android.app.Service;

import android.content.Intent;
import android.os.Binder;
import android.os.IBinder;

public class BindService extends Service {
    private int count;
    private boolean quit;
    //定义 onBinder 方法所返回的对象
    private MyBinder binder = new MyBinder();

    //通过继承 Binder 来实现 IBinder 类
    public class MyBinder extends Binder {
        public int getCount() {
            //获取 Service 的运行状态: count
            return count;
        }
    }
```

```
//必须实现的方法
@Override
public IBinder onBind(Intent intent) {
    System.out.println("Service is Binded");
    //返回 IBinder 对象
    return binder;
}

//Service 被创建时回调该方法。
@Override
public void onCreate() {
    super.onCreate();
    System.out.println("Service is Created");
    //启动一条线程、动态地修改 count 状态值
    new Thread() {
        @Override
        public void run() {
            while (!quit) {
                try {
                    Thread.sleep(1000);
                } catch (InterruptedException e) {
                }
                count++;
            }
        }
    }.start();
}

//Service 被断开连接时回调该方法
@Override
public boolean onUnbind(Intent intent) {
    System.out.println("Service is Unbinded");
    return true;
}

//Service 被关闭之前回调
@Override
public void onDestroy() {
    super.onDestroy();
    this.quit = true;
    System.out.println("Service is Destroyed");
}

@Override
public void onRebind(Intent intent) {
    super.onRebind(intent);
    this.quit = true;
```

```
            System.out.println("Service is ReBinded");
        }
    }
```

上面的Service类实现了onBind()方法,该方法返回一个可访问该Service状态数据(count值)的IBinder对象,该对象将被传给该Service的访问者。

接下来定义一个Activity来绑定该Service,并在该Activity中通过MyBinder对象访问Service的内部状态。该Activity的界面上包含三个按钮:第一个按钮用于绑定Service,第二个按钮用于解除绑定,第三个按钮用于获取Service的运行状态。该Activity的代码如下:

```
import android.app.Activity;
import android.app.Service;
import android.content.ComponentName;
import android.content.Intent;
import android.content.ServiceConnection;
import android.os.Bundle;
import android.os.IBinder;
import android.view.View;
import android.view.View.OnClickListener;
import android.widget.Button;
import android.widget.Toast;

public class MainActivity extends Activity {
    Button bind, unbind, getServiceStatus;
    //保持所启动的Service的IBinder对象
    BindService.MyBinder binder;
    //定义一个ServiceConnection对象
    private ServiceConnection conn = new ServiceConnection() {
        //当该Activity与Service连接成功时回调该方法
        @Override
        public void onServiceConnected(ComponentName name, IBinder service) {
            System.out.println("--Service Connected--");
            //获取Service的onBind方法所返回的MyBinder对象
            binder = (BindService.MyBinder) service;
        }

        //当该Activity与Service断开连接时回调该方法
        @Override
        public void onServiceDisconnected(ComponentName name) {
            System.out.println("--Service Disconnected--");
        }
    };

    @Override
    public void onCreate(Bundle savedInstanceState) {
        super.onCreate(savedInstanceState);
        setContentView(R.layout.main);
```

```
        //获取程序界面中的 start、stop、getServiceStatus 按钮
        bind = (Button) findViewById(R.id.bind);
        unbind = (Button) findViewById(R.id.unbind);
        getServiceStatus = (Button) findViewById(R.id.getServiceStatus);
        //创建启动 Service 的 Intent
        final Intent intent = new Intent();
        //为 Intent 设置 Action 属性
        intent.setAction("org.crazyit.service.BIND_SERVICE");
        bind.setOnClickListener(new OnClickListener() {
            @Override
            public void onClick(View source) {
                //绑定指定 Serivce
                bindService(intent, conn, Service.BIND_AUTO_CREATE);
            }
        });
        unbind.setOnClickListener(new OnClickListener() {
            @Override
            public void onClick(View source) {
                //解除绑定 Serivce
                unbindService(conn);
            }
        });
        getServiceStatus.setOnClickListener(new OnClickListener() {
            @Override
            public void onClick(View source) {
                //获取、并显示 Service 的 count 值
                Toast.makeText(MainActivity.this, "Serivce 的 count 值为: " + binder.
getCount(), 4000).show();
            }
        });
    }
}
```

上面的程序用于在该 Activity 与 Service 连接成功时获取 Service 的 onBind()方法所返回的 MyBinder 对象并通过 MyBinder 对象来访问 Service 的运行状态。

运行该程序，单击程序界面中的“绑定 Service”按钮，即可看到 DDMS 的 LogCat 有如图 7.3 所示的输出。

I	09-28 15:32:55.138	921	921	org.crazyit.service	System.out	Service is Created
I	09-28 15:32:55.178	921	921	org.crazyit.service	System.out	Service is Binded
I	09-28 15:32:55.188	921	921	org.crazyit.service	System.out	--Service Connected--

图 7.3　绑定 Service

单击界面上的“获取 Service 状态”按钮可看到如图 7.4 所示效果。

单击界面上的“解除绑定”按钮，即可在 DDMS 的 LogCat 中看到如图 7.5 所示的输出。

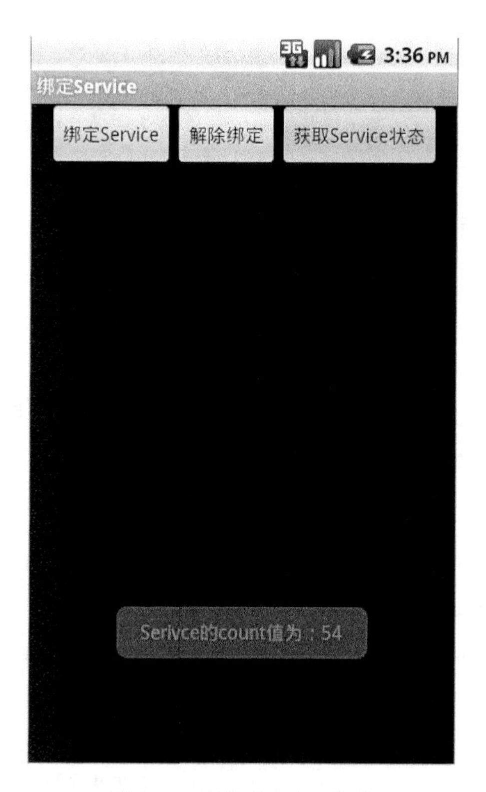

图 7.4　获取 Service 状态

```
I   09-28 15:39:15.138   950   950   org.crazyit.service   System.out   Service is Unbinded
I   09-28 15:39:15.138   950   950   org.crazyit.service   System.out   Service is Destroyed
```

图 7.5　解除绑定 Service

7.4　Service 的生命周期

在运行 7.3 节所附带的例子时，通过在 DDMS 面板上的输出也大致了解了启动服务方式为 bindService 以及停止服务方式为 unbindService 时的生命周期形态，那么如果程序是通过 startService 和 stopService 方式来启动和停止服务，其生命周期又是如何呢？

为了让读者有直观的感受，我们对 7.1 节所实现的计时器服务类 TimeService 稍做修改，在 Service 类所有的生命周期方法中都打印出相应的日志，例如，在 TimeService 类中的 onCreate() 方法中加入 Log. i(" TimeService", "------onCreate------") 代码行。在 TimeService 生命周期期间将会自动回调这些方法并打印出相应日志。

接着，启动计时器应用，单击“开始计时”按钮，这时在 DDMS 面板中可以看到如图 7.6 所示的输出日志。

```
I   10-01 06:12:25.097   277   277   cn.edu.hstc.count...   TimeService   ------onCreate------
I   10-01 06:12:25.117   277   277   cn.edu.hstc.count...   TimeService   ------onStartCommand------
I   10-01 06:12:25.117   277   277   cn.edu.hstc.count...   TimeService   ------onStart------
```

图 7.6　startService 时回调方法

然后，单击程序界面中的“停止计时”按钮，这时在 DDMS 面板中可以看到如图 7.7 所示的打印日志。

I	10-01 06:28:19.946	277	277	cn.edu.hstc.count...	TimeService	------onCreate------
I	10-01 06:28:19.946	277	277	cn.edu.hstc.count...	TimeService	------onStartCommand------
I	10-01 06:28:19.956	277	277	cn.edu.hstc.count...	TimeService	------onStart------
I	10-01 06:28:22.236	277	277	cn.edu.hstc.count...	TimeService	------onDestroy------

stopService时回调方法

图 7.7　stopService 时回调方法

从上面的输入日志可以了解到，程序使用 startService 启动服务以及 stopService 停止服务时，Service 的生命周期为 onCreate() --> onStartCommand() --> onStart() --> Service running　--> (如果调用 Context. stopService())onDestroy() --> Service shut down。

在这里需要注意的是，onStart 方法是在 Android 2.0 之前的平台使用的。在 Android 2.0 及其之后，则需重写 onStartCommand 方法，同时，旧的 onStart 方法不会再被直接调用，而是外部调用 onStartCommand 方法，onStartCommand 方法会再调用 onStart 方法。在 Android 2.0 之后，推荐覆盖 onStartCommand 方法，而为了向前兼容，在 onStartCommand 依然会调用 onStart 方法。

当程序使用 bindService 绑定服务以及 unbindService 解绑服务时，Service 的生命周期为 onCreate() --> onBind() --> Service running --> (如果调用 Context. unbindService()) onUnbind() --> onDestroy() --> Service stop。

在此，通过如图 7.8 所示的流程图来描述上面两种 Service 生命周期形态。

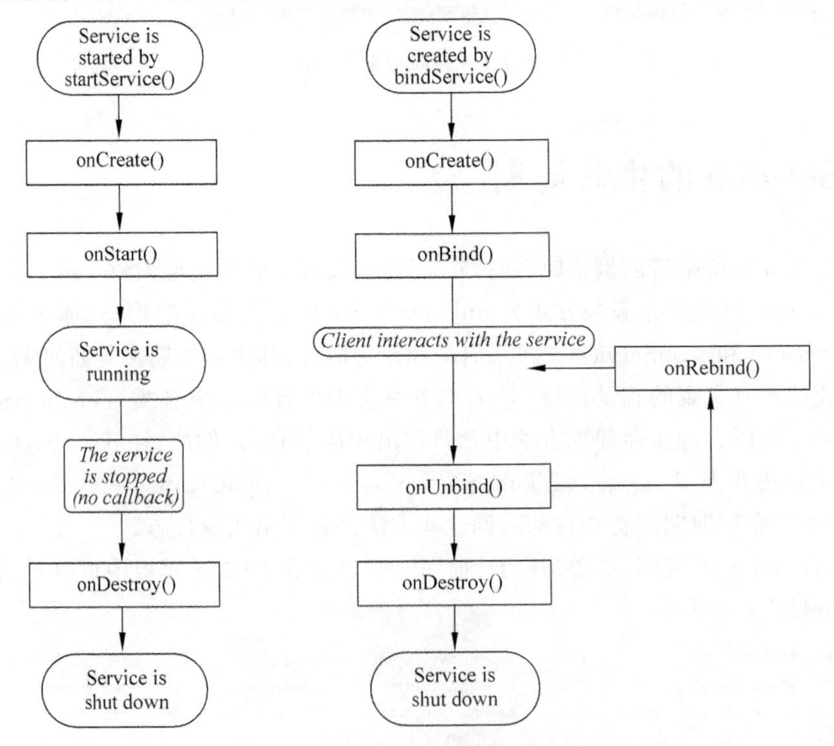

图 7.8　两种 Service 生命周期过程

在如图7.8所示的流程图中，左边为startService启动Service时所经过的生命周期，右边为bindService绑定Service时所经过的生命周期。

7.5　接收系统广播消息

现在，假定一个场景，当用户的手机电量发生改变或者电量过低时，会有一个消息提示用户。想要做到这一点，就需要利用Android系统对外发送的电池电量广播。接下来，通过一个手机电池电量监听器的应用来了解Android如何处理系统广播消息。

开发该手机电量监听程序的核心代码是开发一个广播接收者，该广播接收者监听系统发送的手机电量广播，当电量发生变化时，提示用户。核心代码如下：

```
package cn.edu.hstc.batteryreceiverdemo.broadcast;

import android.content.BroadcastReceiver;
import android.content.Context;
import android.content.Intent;
import android.os.Bundle;
import android.widget.Toast;

public class BatteryReceiver extends BroadcastReceiver {
    @Override
    public void onReceive(Context context, Intent intent) {
        if (Intent.ACTION_BATTERY_OKAY.equals(intent.getAction())) {
            Toast.makeText(context, "电量已恢复,可以使用!", Toast.LENGTH_LONG).show();
        }
        if (Intent.ACTION_BATTERY_LOW.equals(intent.getAction())) {
            Toast.makeText(context, "电量过低,请尽快充电!", Toast.LENGTH_LONG).show();
        }
        if (Intent.ACTION_BATTERY_CHANGED.equals(intent.getAction())) {
            Bundle bundle = intent.getExtras();
            //获取当前电量
            int current = bundle.getInt("level");
            //获取总电量
            int total = bundle.getInt("scale");
            StringBuffer sb = new StringBuffer();
            sb.append("当前电量为:" + current * 100 / total + "%" + "  ");
            //如果当前电量小于总电量的15%
            if (current * 1.0 / total < 0.15) {
                sb.append("电量过低,请尽快充电!");
            } else {
                sb.append("电量足够,请放心使用!");
            }
            Toast.makeText(context, sb.toString(), Toast.LENGTH_LONG).show();
        }
    }
}
```

代码文件：codes\04\4.1\BatteryReceiverDemo\cn\edu\hstc\batteryreceiverdemo\broadcast\BatteryReceiver.java

上面的程序实现通过接收系统广播消息获取手机电池总电量、当前电量，当电量发生变化时，如果当前电量与总电量百分比小于15%，则提示用户“电量过低，请尽快充电！”，否则提示“电量足够，请放心使用！”。当电量从不足状态恢复时，提示用户“电量已恢复，可以使用！”。

接着，需要在全局配置文件配置开发好的BoradcastReceiver，代码片段如下：

```
<?xml version = "1.0" encoding = "utf - 8"?>
< manifest xmlns:android = "http://schemas.android.com/apk/res/android"
    package = "com.example.batteryreceiverdemo"
    android:versionCode = "1"
    android:versionName = "1.0" >

    < uses - sdk
        android:minSdkVersion = "8"
        android:targetSdkVersion = "8" />

    <! -- 配置广播接收者 -->
    < application
        android:allowBackup = "true"
        android:icon = "@drawable/ic_launcher"
        android:label = "@string/app_name"
        android:theme = "@style/AppTheme" >
            < receiver android: name = " cn. edu. hstc. batteryreceiverdemo. broadcast.
BatteryReceiver" >
            < intent - filter >
                < action android:name = "android.intent.action.BATTERY_CHANGED" />
                < action android:name = "android.intent.action.BATTERY_OKAY" />
                < action android:name = "android.intent.action.BATTERY_LOW" />
            </intent - filter >
        </receiver >
    </application >

</manifest >
```

上面配置文件中为BatteryReceiver配置了intent-filter，该intent-filter的三个action属性值分别为BATTERY_CHANGED(电量发生改变时，系统对外发送Intent的Action为ACTION_BATTERY_CHANGED常量的广播)、BATTERY_OKAY(电量从不足状态恢复时，系统对外发送Intent的Action为ACTION_BATTERY_ OKAY常量的广播)、BATTERY_LOW(电量过低时，系统对外发送Intent的Action为ACTION_BATTERY_LOW常量的广播)。

从上面小程序的开发介绍中可以了解到，BroadcastReceiver除了用于接收用户自定义的广播之外，还用于接收系统的广播，这也是其重要的用途之一。Android的大量系统事件都会对外发送标准广播。下面是Android常见的广播Action常量：

- ACTION_TIME_CHANGED——系统时间被改变。
- ACTION_DATE_CHANGED——系统日期被改变。
- ACTION_TIMEZONE_CHANGED——系统时区被改变。

- ACTION_BOOT_COMPLETED——系统启动完成。
- ACTION_BATTERY_CHANGED——电池电量改变。
- ACTION_BATTERY_ LOW——电池电量过低。
- ACTION_POWER_ CONNECTED——系统连接电源。
- ACTION_POWER_ DISCONNECTED——系统与电源断开。

7.6 本章小结

本章通过实现计时器介绍了开发 Service 以及 BoradcastReceiver 的流程，读者应将学习重点放在创建和配置 Service 组件以及如何启动和停止 Service 上；学习 BoradcastReceiver 时，需要掌握创建、配置 BoradcastReceiver 组件以及如何在程序中发送 Broadcast。

第8章 Android的数据存储以及文件读写

当想要设置应用程序的某些参数、保存应用退出时的运行状态、将大量的歌曲放在手机中时，这些数据都需要存储在外部存储器上，如 SD 卡(SD Card)，这样在系统关机之后数据并不会丢失。以上这些都将涉及 Android 系统一些专门操作输入输出的 API。

当然，如果仅仅只是想将少量的数据保存起来，那么使用普通文件就可以了；但是如果应用程序有大量数据需要存储和访问，就需要借助 Android 系统内置的一个轻量级数据库 SQLite 了。SQLite 内置于 Android 中，没有后台进程，一个数据库就对应一个文件，做到真正的轻量级。本章将会详细介绍如何利用 Android 提供的大量的操作 SQLite 数据库的 API 来使用 SQLite 数据库。

8.1 使用 SharedPreferences

应用程序有时需要保存的数据数量非常少并且数据格式非常简单，如是否打开音效、QQ 软件打开夜间模式、收到微信消息时是否打开振动效果、记住登录账号和密码等操作时存储的配置信息。这些都是一些普通的字符串或者键值对(key-value)，对于这种数据，Android 使用 SharedPreferences 进行保存。

8.1.1 通过密码记住功能学习使用 SharedPreferences

本节将实现一个带有密码记住功能的登录界面。该登录界面比较简单，从上至下放置了账号输入框、密码输入框、记住密码的多选按钮、登录按钮。XML 代码如下：

```
<?xml version = "1.0" encoding = "utf - 8"?>
< LinearLayout xmlns:android = "http://schemas.android.com/apk/res/android"
    android:layout_width = "match_parent"
    android:layout_height = "match_parent"
    android:background = "#f1f0f0"
    android:orientation = "vertical" >

    < TableLayout
        android:layout_width = "fill_parent"
        android:layout_height = "wrap_content"
        android:paddingTop = "30dp"
        android:stretchColumns = "2" >
```

```
<TableRow>

    <TextView
        android:layout_width="10dp"
        android:layout_height="1dp" />

    <TextView
        android:id="@+id/txt_account"
        android:layout_width="wrap_content"
        android:layout_height="wrap_content"
        android:text="账号：" />

    <EditText
        android:id="@+id/edit_account"
        android:layout_width="wrap_content"
        android:layout_height="wrap_content"
        android:background="@drawable/shurukuang"
        android:paddingLeft="5dp" />

    <TextView
        android:layout_width="10dp"
        android:layout_height="1dp" />
</TableRow>

<TextView
    android:layout_width="fill_parent"
    android:layout_height="10dp" />

<TableRow>

    <TextView
        android:layout_width="10dp"
        android:layout_height="1dp" />

    <TextView
        android:id="@+id/txt_pwd"
        android:layout_width="wrap_content"
        android:layout_height="wrap_content"
        android:text="密码：" />

    <EditText
        android:id="@+id/edit_pwd"
        android:layout_width="wrap_content"
        android:layout_height="wrap_content"
        android:background="@drawable/shurukuang"
        android:password="true"
        android:paddingLeft="5dp" />

    <TextView
        android:layout_width="10dp"
        android:layout_height="1dp" />
```

```
        </TableRow>
    </TableLayout>

    <LinearLayout
        android:layout_width="fill_parent"
        android:layout_height="wrap_content"
        android:gravity="center_horizontal"
        android:orientation="horizontal" >

        <TextView
            android:layout_width="wrap_content"
            android:layout_height="wrap_content"
            android:text="*记住账号密码,方便下次登录"
            android:textColor="@android:color/black" />

        <CheckBox
            android:id="@+id/cbx_remember"
            android:layout_width="wrap_content"
            android:layout_height="wrap_content"
            style="@style/CustomCheckboxTheme"
            android:layout_marginLeft="10dp"
            android:checked="false" />
    </LinearLayout>

    <Button
        android:id="@+id/btn_login"
        android:layout_width="fill_parent"
        android:layout_height="wrap_content"
        android:background="@drawable/style_button"
        android:layout_gravity="center_horizontal"
        android:layout_marginLeft="10dp"
        android:layout_marginRight="10dp"
        android:text="登录" />

</LinearLayout>
```

当用户单击“登录”按钮时,程序将会自动识别用户是否选中了记住密码功能,如有,则将账号和密码写入 SharedPreferences 中并跳转到欢迎页面。在登录界面所对应的主程序 LoginActivity 类中实现了该核心功能。

```
package net.edu.hstc.rememberpwd.activity;

import java.util.Timer;
import java.util.TimerTask;

import android.app.Activity;
import android.app.AlertDialog;
import android.content.Context;
import android.content.DialogInterface;
import android.content.DialogInterface.OnCancelListener;
import android.content.Intent;
```

```
import android.content.SharedPreferences;
import android.content.SharedPreferences.Editor;
import android.os.Bundle;
import android.view.View;
import android.widget.Button;
import android.widget.CheckBox;
import android.widget.CompoundButton;
import android.widget.CompoundButton.OnCheckedChangeListener;
import android.widget.EditText;
import android.widget.LinearLayout;
import android.widget.TextView;
import android.widget.Toast;

public class LoginActivity extends Activity {
    private EditText accountEdit, pwdEdit;
    private CheckBox rememberCbx;
    private Button loginBtn;
    //声明 SharedPreferences
    private SharedPreferences sp;
    //声明 SharedPreferences.Editor
    private Editor editor;
    private AlertDialog prompt;
    private Timer timer;
    private boolean flag = true;

    @Override
    protected void onCreate(Bundle savedInstanceState) {
        super.onCreate(savedInstanceState);
        setContentView(R.layout.activity_login);
        //初始化组件
        initView();
        //获得 SharedPreferences 实例对象并指定该 SharedPreferences 数据能被其他应用程序
读,但不能写
        sp = this.getSharedPreferences("userInfo", Context.MODE_WORLD_READABLE);
        editor = sp.edit();
        timer = new Timer();
        //当获取 SharedPreferences 中 remember 值为 true 时,将选项按钮置为选中状态
        //将 account 值跟 pwd 值分别设在账号输入框跟密码输入框中
        if (sp.getBoolean("remember", false)) {
            if (!sp.getString("account", "").equals("") && !sp.getString("pwd", "").equals
("")) {
                rememberCbx.setChecked(true);
                accountEdit.setText(sp.getString("account", ""));
                pwdEdit.setText(sp.getString("pwd", ""));
            }
        }
    }

    /**
     * 通过 id 加载,初始化登录界面中各个组件
     */
```

```
private void initView() {
    accountEdit = (EditText) findViewById(R.id.edit_account);
    pwdEdit = (EditText) findViewById(R.id.edit_pwd);
    rememberCbx = (CheckBox) findViewById(R.id.cbx_remember);
    loginBtn = (Button) findViewById(R.id.btn_login);

    //为多选按钮设置监听器
    rememberCbx.setOnCheckedChangeListener(new OnCheckedChangeListener() {
        @Override
        public void onCheckedChanged(CompoundButton buttonView, boolean isChecked) {
            if (rememberCbx.isChecked()) {
                editor.putBoolean("remember", true).commit();
            } else if (!rememberCbx.isChecked()) {
                editor.putBoolean("remember", false);
                editor.remove("account");
                editor.remove("pwd");
                editor.commit();
            }
        }
    });

    loginBtn.setOnClickListener(new View.OnClickListener() {
        @Override
        public void onClick(View v) {
            String accountStr = accountEdit.getText().toString();
            String pwdStr = pwdEdit.getText().toString();
            if (accountStr.equals("hstc") && pwdStr.equals("123456")) {
                if (rememberCbx.isChecked()) {  //当用户记住密码时
                    editor.putString("account", accountStr);
                    editor.putString("pwd", pwdStr);
                    editor.commit();
                }
                prompt = prompt(LoginActivity.this, "登录成功,正在跳转...");
                prompt.setOnCancelListener(new OnCancelListener() {
                    @Override
                    public void onCancel(DialogInterface dialog) {
                        if (prompt.isShowing()) {
                            prompt.dismiss();
                        }
                    }
                });
                timer.schedule(new TimerTask() {
                    @Override
                    public void run() {
                        if (flag) {
                            if (prompt.isShowing()) {
                                prompt.dismiss();
                            }
                            flag = false;
                            timer.cancel();
                            timer.purge();
```

```
                        timer = null;
                        Intent intent = new Intent(LoginActivity.this,
                        MainActivity.class);
                        LoginActivity.this.startActivity(intent);
                        LoginActivity.this.finish();
                    }
                }
            }, 3000, 5000);                    //弹窗提示3秒
        } else if (accountStr.equals("") || pwdStr.equals("")) {
                Toast.makeText(LoginActivity.this, "账号或密码不能为空!", Toast.
LENGTH_LONG).show();
        } else {
                Toast.makeText(LoginActivity.this, "账号或密码错误,请重新登录!",
Toast.LENGTH_LONG).show();
        }
    }
  });
}

/**
  * 提示对话框
  */
private AlertDialog prompt(Activity context, String content) {
    AlertDialog.Builder builder = new AlertDialog.Builder(context);
    LinearLayout layout = (LinearLayout) context.getLayoutInflater().inflate(R.layout.
prompt, null);
    ((TextView) layout.findViewById(R.id.message)).setText(content);
    builder.setView(layout);
    AlertDialog dialog = builder.show();
    dialog.setCanceledOnTouchOutside(false);
    return dialog;
}
}
```

上面程序实现当用户选择记住密码登录时，将会把账号与对应的密码一起写入SharedPreferences中，在程序退出后再次被打开时，在登录界面中将会自动把存放在SharedPreferences中的账号与密码获取出来并显示在相应的文本输入框中。程序中所对应的欢迎界面MainActivity类比较简单，此处不再赘述。

将程序部署在Android模拟器中，出现了如图8.1所示的登录界面，在"账号"输入框中输入hstc并在"密码"输入框中输入123456，选中记住密码功能，然后单击"登录"按钮，图8.2是当用户记住密码登录后重启应用时的登录界面效果图，图8.3为用户正在登录中界面，3秒钟后跳到如图8.4所示的欢迎页面，按下硬键盘上的返回键，退出程序，重新启动程序，将会看到此时的登录页面中的账号输入框以及密码输入框已经自动有了值。

本应用为了记住用户密码，使用了SharedPreferences。SharedPreferences实现Android平台的轻量级存储，主要用来保存一些常用的配置参数，如保存了用户的登录信息。SharedPreferences采用XML文件存放数据，文件存放在/data/data/< packagename >/shared_prefs目录下。

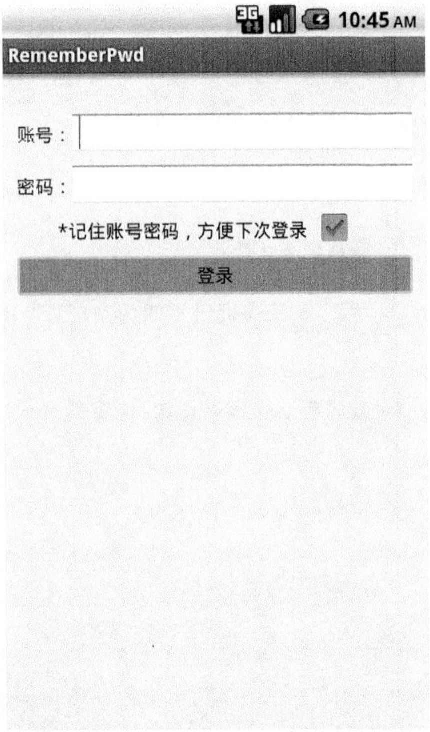

10:45 AM
RememberPwd
账号：
密码：
*记住账号密码，方便下次登录
登录

图 8.1　应用初次打开时登录界面

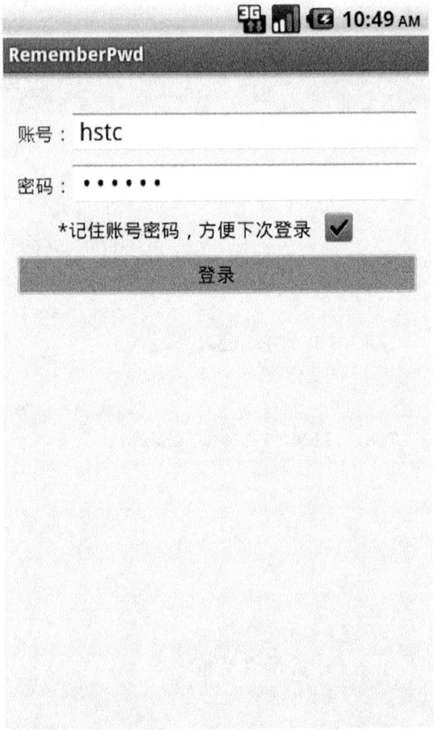

10:49 AM
RememberPwd
账号： hstc
密码： ••••••
*记住账号密码，方便下次登录
登录

图 8.2　输入账号密码并选中记住密码

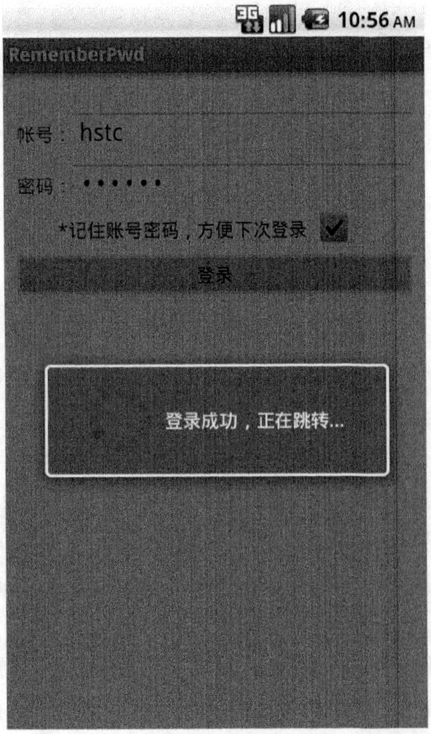

10:56 AM
RememberPwd
帐号： hstc
密码： ••••••
*记住账号密码，方便下次登录
登录
登录成功，正在跳转...

图 8.3　登录中

11:03 AM
RememberPwd
欢迎使用本软件

图 8.4　欢迎页面

从本应用的 LoginActivity 类中的源代码可以看出，程序并没有直接创建 SharedPreferences 实例，而是通过 Context 提供的 getSharedPreferences(String name, int mode)方法来获取 SharedPreferences 实例，如代码行 getSharedPreferences("userInfo", Context.MODE_WORLD_READABLE)。实际上，该方法的第二个参数通常会指定以下三个值：

➢ Context.MODE_PRIVATE——指定该 SharedPreferences 数据只能被本应用程序所读、写。
➢ Context.MODE_WORLD_READABLE——指定该 SharedPreferences 数据能被其他应用程序读，但不能写。
➢ Context.MODE_WORLD_WRITEABLE——指定该 SharedPreferences 数据能被其他应用程序所读、写。

SharedPreferences 是以键值对(key-value)的形式保存数据的，因此是通过 SharedPreferences 本身的 getXxx(String key, xxx defValue)方法来获取相关的数值的。如程序中代码行 accountEdit.setText(sp.getString("account", ""))。实际上，SharedPreferences 提供了如下常用方法来访问应用程序的 Preferences 数据。

➢ boolean contains(String key)：判断 SharedPreferences 是否包含特定 key 的数据。
➢ abstract Map < String, ? > getAll()：获取 SharedPreferences 数据里全部的键值对。
➢ xxx getXxx(String key, xxx defValue)：获取 SharedPreferences 数据里指定 key 对应的 value。如果该 key 不存在，则返回默认值 defValue。其中 xxx 可以是 boolean、float、int、long、String 等各种基本数据类型。

由于 SharedPreferences 接口本身并没有写入数据的操作能力，故需要通过 SharedPreferences 的内部接口 Editor 提供的方法来向 SharedPreferences 中写入数据。如代码行 editor = sp.edit()，SharedPreferences 调用本身的 edit()方法即可获取它所对应的 Editor 对象，进而写入数据。SharedPreferences.Editor 提供了如下写操作数据方法：

➢ SharedPreferences.Editor clear()——清空 SharedPreferences 中的所有数据。
➢ SharedPreferences.Editor putXxx(String key, xxx value)——向 SharedPreferences 中存入指定 key 对应的数据。其中 xxx 可以是 boolean、float、int、long、String 等各种基本数据类型。
➢ SharedPreferences.Editor remove(String key)——删除 SharedPreferences 中指定 key 所对应的数据项。
➢ boolean commit()——操作前面的所有写操作方法的最后都需要调用该方法提交。

如 LoginAction 类中为多选按钮设置监听器的代码片段中，当用户反选该多选按钮时，向 SharedPreferences.Editor 的对象 editor 中存放一个值为 false 的 boolean 类型数据，并将调用 remove(String key)方法删除掉 SharedPreferences 数据中的 account、pwd 的值，最后调用 commit()方法提交所有操作。代码片段如下：

```
//为多选按钮设置监听器
rememberCbx.setOnCheckedChangeListener(new OnCheckedChangeListener() {
    @Override
    public void onCheckedChanged(CompoundButton buttonView, boolean isChecked) {
        if (rememberCbx.isChecked()) {
            editor.putBoolean("remember", true).commit();
```

```
            } else if (!rememberCbx.isChecked()) {
                editor.putBoolean("remember", false);
                editor.remove("account");
                editor.remove("pwd");
                editor.commit();
            }
        }
    });
```

8.1.2 SharedPreferences 的存储位置和格式

运行 8.1.1 节中的记住密码功能的应用程序，选中记住密码选项按钮并单击“登录”按钮，此时打开 DDMS 面板的 File Explorer，然后展开文件浏览树，可以看到如图 8.5 所示的窗口。

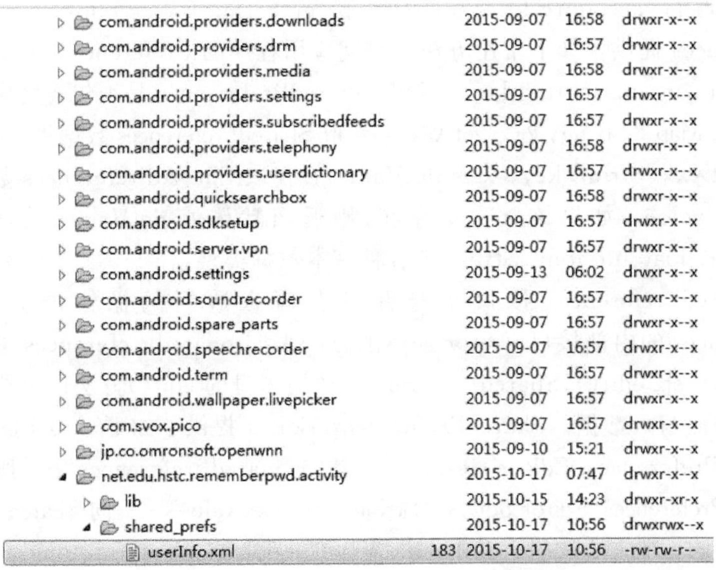

Name	Size	Date	Time	Permissions
▷ com.android.providers.downloads		2015-09-07	16:58	drwxr-x--x
▷ com.android.providers.drm		2015-09-07	16:57	drwxr-x--x
▷ com.android.providers.media		2015-09-07	16:58	drwxr-x--x
▷ com.android.providers.settings		2015-09-07	16:57	drwxr-x--x
▷ com.android.providers.subscribedfeeds		2015-09-07	16:57	drwxr-x--x
▷ com.android.providers.telephony		2015-09-07	16:58	drwxr-x--x
▷ com.android.providers.userdictionary		2015-09-07	16:57	drwxr-x--x
▷ com.android.quicksearchbox		2015-09-07	16:58	drwxr-x--x
▷ com.android.sdksetup		2015-09-07	16:57	drwxr-x--x
▷ com.android.server.vpn		2015-09-07	16:57	drwxr-x--x
▷ com.android.settings		2015-09-13	06:02	drwxr-x--x
▷ com.android.soundrecorder		2015-09-07	16:57	drwxr-x--x
▷ com.android.spare_parts		2015-09-07	16:57	drwxr-x--x
▷ com.android.speechrecorder		2015-09-07	16:57	drwxr-x--x
▷ com.android.term		2015-09-07	16:57	drwxr-x--x
▷ com.android.wallpaper.livepicker		2015-09-07	16:57	drwxr-x--x
▷ com.svox.pico		2015-09-07	16:57	drwxr-x--x
▷ jp.co.omronsoft.openwnn		2015-09-10	15:21	drwxr-x--x
◢ net.edu.hstc.rememberpwd.activity		2015-10-17	07:47	drwxr-x--x
▷ lib		2015-10-15	14:23	drwxr-xr-x
◢ shared_prefs		2015-10-17	10:56	drwxrwx--x
userInfo.xml	183	2015-10-17	10:56	-rw-rw-r--

图 8.5 SharedPreferences 数据文件存储路径

在密码记住功能的程序代码中，在 LoginActivity 类中有代码行 sp = this. getSharedPreferences("userInfo", Context. MODE_WORLD_READABLE)，该行代码实际上已经实现在/data/data/< package name >/shared _ prefs 目录下建立了一个名为 userInfo. xml 的文件，该文件用于保存该应用的 SharedPreferences 数据。通过其文件名后缀，可以想到，SharedPreferences 数据是以 XML 格式写入文件的。

通过 File Explorer 面板中的导出文件按钮将 userInfo. xml 文件导出到电脑桌面，通过 Notepad++文档阅读器打开该文件，可以看到以下内容：

```
<?xml version='1.0' encoding='utf-8' standalone='yes' ?>
<map>
<string name="account">hstc</string>
<boolean name="remember" value="true" />
<string name="pwd">123456</string>
</map>
```

可以看出，该文件中 account 所对应的值正是 8.1.1 节应用中所对应的登录账号，remember 所对应的值正是密码的值，因此，上面文件保存了用户账号与密码信息以及是否记住密码的标识。

从文件内容可以看出，SharedPreferences 数据文件的 XML 根元素是< map/>，该元素包含的每个子元素正代表了每个保存到文件中的键值对。根据保存的基本数据类型，这些子元素可以是< Boolean/>、< float/>、< int/>、< long/>、< String/>等等。

根据代码，当用户反选了记住密码功能的选项按钮，将会删除掉 SharedPreferences 数据中 key 为 account 和 pwd 的键值对，此时再次将 File Explorer 面板中的 userInfo. xml 文件导出到电脑桌面并打开，可以看到如下内容：

```
<?xml version='1.0' encoding='utf-8' standalone='yes' ?>
<map>
<boolean name="remember" value="false" />
</map>
```

通过以上的介绍，不难发现，SharedPreferences 数据总是保存在/data/data/< package name >/shared_prefs 目录下并且以 XML 格式保存。

8.2 文件(File)存储

前面介绍的 SharedPreferences 存储方式非常简单方便，但其只适合存储比较简单的数据，如果需要存储更多的数据，可行选择的方式有好几种，这里要给读者介绍的是文件存储的方式。

8.2.1 文件的保存与读取

与传统的 Java IO 访问磁盘文件类似，Android 同样提供了自己的文件操作的 API 供用户来访问手机存储器上的文件。

下面通过一个简易型备忘录来介绍文件的保存和读取。应用启动界面放置一个文本输入框以及一个按钮，在文本输入框中输入要记录的文字后单击按钮，将会把文本输入框中的文字保存到指定的文件中并跳转到查看备忘录页面，在该页面读取指定的文件内容并展示在页面的输入框中，该页面同时放置了两个按钮：一个用于清空备忘录文件，一个用于退出整个应用程序。软件所需要用到的两个界面都比较简单，此处不再给出源代码，读者可以通过本书附带源码查看这两个界面布局代码。接下介绍的是保存内容到指定文件以及读取指定文件的内容并展示到页面的核心程序代码。

保存内容到指定文件的核心程序如下：

```
package cn.edu.hstc.fileiodemo.activity;

import java.io.BufferedWriter;
import java.io.FileOutputStream;
import java.io.OutputStreamWriter;

import android.app.Activity;
```

```
import android.content.Intent;
import android.os.Bundle;
import android.view.View;
import android.view.Window;
import android.view.WindowManager;
import android.widget.Button;
import android.widget.EditText;

public class InputActivity extends Activity {
    //定义要保存的文件名
    final String FILE_NAME = "memorandum.txt";
    private EditText edit;
    private Button button, reset;

    @Override
    protected void onCreate(Bundle savedInstanceState) {
        super.onCreate(savedInstanceState);
        //下面两句实现界面全屏化
        this.getWindow().setFlags(WindowManager.LayoutParams.FLAG_FULLSCREEN,
        WindowManager.LayoutParams.FLAG_FULLSCREEN);
        this.requestWindowFeature(Window.FEATURE_NO_TITLE);
        setContentView(R.layout.activity_input);
        initView();
    }

    private void initView() {
        edit = (EditText) findViewById(R.id.edit);
        button = (Button) findViewById(R.id.button);
        reset = (Button) findViewById(R.id.reset);
        //为记录按钮绑定事件监听器
        button.setOnClickListener(new View.OnClickListener() {
            @Override
            public void onClick(View v) {
                //将文本输入框中的内容保存到文件 memorandum.txt 中
                saveDataToFile(FILE_NAME, edit.getText().toString());
                //实现跳转到查看备忘录页面
                Intent intent = new Intent(InputActivity.this, OutputActivity.class);
                startActivity(intent);
            }
        });

        //为重置按钮添加事件监听器
        reset.setOnClickListener(new View.OnClickListener() {
            @Override
            public void onClick(View v) {
                edit.setText("");
            }
        });
    }

    /**
```

```
     * 向 File 中保存数据
     * @param fileName
     * @param data
     */
    private void saveDataToFile(String fileName, String data) {
        FileOutputStream fileOutputStream = null;
        OutputStreamWriter outputStreamWriter = null;
        BufferedWriter bufferedWriter = null;
        try {
            fileOutputStream = openFileOutput(fileName, MODE_APPEND);
                                                            //以追加方式打开文件
            outputStreamWriter = new OutputStreamWriter(fileOutputStream);
            bufferedWriter = new BufferedWriter(outputStreamWriter);
            bufferedWriter.write(data);
        } catch (Exception e) {
            e.printStackTrace();
        } finally {
            try {
                if (bufferedWriter != null) {
                    bufferedWriter.close();
                }
            } catch (Exception e) {
                e.printStackTrace();
            }
        }
    }
}
```

代码文件：codes\08\8.2\FileIODemo\cn\edu\hstc\activity\InputActivity.java

从指定的文件中读取内容并显示在页面中的核心程序代码如下：

```
package cn.edu.hstc.fileiodemo.activity;

import java.io.BufferedReader;
import java.io.FileInputStream;
import java.io.InputStreamReader;

import android.app.Activity;
import android.content.Intent;
import android.os.Bundle;
import android.view.View;
import android.view.Window;
import android.view.WindowManager;
import android.widget.Button;
import android.widget.EditText;
import android.widget.LinearLayout;

public class OutputActivity extends Activity {
    //定义要读取的文件名
    final String FILE_NAME = "memorandum.txt";
    private LinearLayout backlayout;
```

```
private EditText show;
private Button clear, exit;

@Override
protected void onCreate(Bundle savedInstanceState) {
    super.onCreate(savedInstanceState);
    //以下两句实现页面全屏化
    this.getWindow().setFlags(WindowManager.LayoutParams.FLAG_FULLSCREEN,
    WindowManager.LayoutParams.FLAG_FULLSCREEN);
    this.requestWindowFeature(Window.FEATURE_NO_TITLE);
    setContentView(R.layout.activity_output);
    initView();
}

private void initView() {
    backlayout = (LinearLayout) findViewById(R.id.layout_back);
    show = (EditText) findViewById(R.id.show);
    clear = (Button) findViewById(R.id.btn_clear);
    exit = (Button) findViewById(R.id.btn_exit);
    //将读取的文件内容显示在文本输入框中
    show.setText(getDataFromFile(FILE_NAME));
    backlayout.setOnClickListener(new View.OnClickListener() {
        @Override
        public void onClick(View v) {
            OutputActivity.this.finish();
        }
    });
    clear.setOnClickListener(new View.OnClickListener() {
        @Override
        public void onClick(View v) {
            //清空备忘录,可以将文件内容置为空,也可以直接将文件删掉,这里采用的是直
              接删除文件
            deleteFile(FILE_NAME);
            show.setText("");
        }
    });
    exit.setOnClickListener(new View.OnClickListener() {
        @Override
        public void onClick(View v) {
            Intent intent = new Intent(Intent.ACTION_MAIN);
            intent.addCategory(Intent.CATEGORY_HOME);
            intent.setFlags(Intent.FLAG_ACTIVITY_CLEAR_TOP);
            startActivity(intent);
            android.os.Process.killProcess(android.os.Process.myPid());
        }
    });
}

/**
 * 从File中读取数据
 * @param fileName
```

```
     * @return
     */
    private String getDataFromFile(String fileName) {
        FileInputStream fileInputStream = null;
        InputStreamReader inputStreamReader = null;
        BufferedReader bufferedReader = null;
        StringBuilder stringBuilder = null;
        String line = null;
        try {
            stringBuilder = new StringBuilder();
            fileInputStream = openFileInput(fileName);
            inputStreamReader = new InputStreamReader(fileInputStream);
            bufferedReader = new BufferedReader(inputStreamReader);
            while ((line = bufferedReader.readLine()) != null) {
                stringBuilder.append(line);
            }
        } catch (Exception e) {
            e.printStackTrace();
        } finally {
            try {
                if (bufferedReader != null) {
                    bufferedReader.close();
                }
            } catch (Exception e) {
                e.printStackTrace();
            }
        }
        if (stringBuilder != null) {
            return stringBuilder.toString();
        } else {
            return "";
        }
    }
}
```

代码文件：codes\08\8.2\FileIODemo\cn\edu\hstc\activity\OutputActivity.java

上面两段程序实现将用户输入的内容保存在一个叫 memorandum.txt 的备忘录文件中并跳转到查看备忘录内容页面，在查看页面中读取 memorandum.txt 文件的内容并展示在页面输入框中，当用户单击“清空备忘录”按钮时，删除 memorandum.txt 文件。将程序部署在 Android 模拟器上，并按照上面的操作步骤，效果如图 8.6 和图 8.7 所示。

此时展开 File Explorer 面板的文件浏览树，可以看到如图 8.8 所示窗口。

从图 8.8 可以看出，应用程序的数据文件保存路径是/data/data/< package name >/files。

从本应用示例可以看出，Android 提供了如下两个方法来访问应用程序的文件 IO 流：

➢ FileInputStream openFileInput(String name)——打开应用程序的数据文件夹下的 name 文件对应输入流。

➢ FileOutputStream openFileOutput(String name, int mode)——打开应用程序的数据文件夹下的 name 文件对应输出流。

实际上，Android 还提供了如下方法来访问应用程序的数据文件夹或文件夹下的文件：

图 8.6 输入内容并单击“记录”按钮保存文件

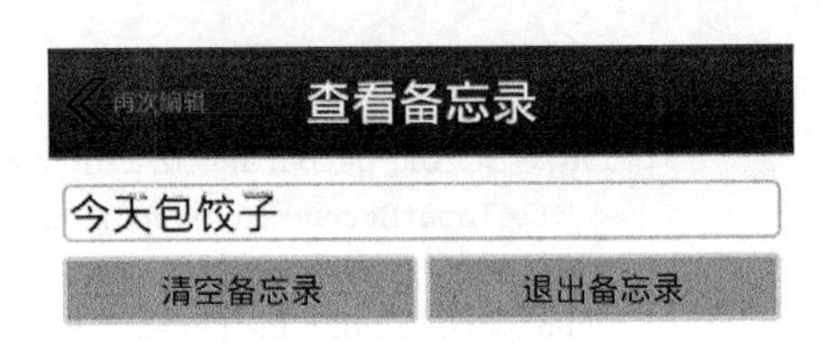

图 8.7 读取文件内容并展示

Name	Size	Date	Time	Permissions
data		2015-09-07	16:57	drwxrwx--x
app		2015-10-18	12:29	drwxrwx--x
app-private		2015-09-07	16:56	drwxrwx--x
backup		2015-09-07	16:57	drwx------
dalvik-cache		2015-10-18	12:29	drwxrwx--x
data		2015-10-18	07:12	drwxrwx--x
android.tts		2015-09-07	16:57	drwxr-x--x
cn.edu.hstc.fileiodemo.activity		2015-10-18	07:12	drwxr-x--x
files		2015-10-18	13:05	drwxrwx--x
memorandum.txt	15	2015-10-18	13:05	-rw-rw----

图 8.8 数据文件保存路径

- getDir(String name, int mode)——在应用程序的数据文件夹下获取或创建 name 对应的子目录。
- File getFilesDir()——获取应用程序的数据文件夹的绝对路径。
- String[] fileList()——返回应用程序的数据文件夹下的全部文件。
- deleteFile(String name)——删除应用程序的数据文件夹下的指定文件。

8.2.2 文件的操作模式

在上节介绍的简易型备忘录的程序代码中有这样一句代码：fileOutputStream = openFileOutput(fileName, MODE_APPEND)，这句代码的功能也正是上节所介绍的打开应用程序的数据文件夹下的 fileName 文件对应的输出流。该句代码所对应的方法的第二个参数 MODE_APPEND 指的是应用程序可以以追加方式打开该文件。实际上该方法第二个参数指定的是打开文件的模式,该模式可以指定如下值:

- MODE_PRIVATE——设置文件只能被当前程序读写。
- MODE_APPEND——设置应用程序可以向该文件追加内容。
- MODE_WORLD_READABLE——设置该文件可以被其他应用程序读取。
- MODE_WORLD_WRITEABLE——设置该文件可以被其他应用程序读、写。

8.2.3　通过图片下载器实现操作 SD 卡

由于设备内置存储空间是有限的,所以,仅通过 openFileInput 或 openFileOutput 打开应用程序的数据文件夹里的文件输入流、输出流来操作文件数据的存储是远远不够的,有时为了更好地存、取大文件数据,应用程序需要访问到设备的外部存储器,如 SD 卡,这将大大扩充 Android 设备的存储能力。

本节将通过一个简单的图片下载器应用来介绍程序如何将从网络下载的图片保存在 SD 卡上以及从 SD 卡上将存储的图片显示在页面,以此掌握访问 SD 卡的方法。

该图片下载器应用界面十分简单,从上至下分别放置了一个文本输入框用于显示图片下载地址,一个 Button 按钮用来单击以便将网络图片下载并保存在手机 SD 卡中,另一个 Button 按钮供用户单击从 SD 卡中获取下载的图片并显示在界面底部的图像视图 ImageView 中。

由于该界面布局比较简单,故此处不再给出源码,读者可以通过阅读本书所附带源码获得。接下来重点学习的是本实例中需要用到的操作 SD 卡的工具类 FileUtil 类的源码,通过学习该工具类掌握如何访问 Android 设备的 SD 卡。该工具类源码如下:

```
package cn.edu.hstc.sdcarddemo.util;
import java.io.File;
import java.io.IOException;
import android.os.Environment;

public class FileUtil {
    /**
     * 判断是否 Android 设备插入了 SD 卡并且应用程序具有读写 SD 卡的权限
     */
    public static Boolean isHaveSDCard() {
        return Environment.getExternalStorageState().equals(Environment.MEDIA_MOUNTED);
    }

    /**
     * 获取设备 SD 卡绝对路径
     */
    public static String getSDCardPath() {
        return Environment.getExternalStorageDirectory().getAbsolutePath();
    }

    /**
     * 在 SD 卡上创建目录
```

```
     */
    public static File createDIR(String dirPath) {
        File dir = new File(dirPath);
        if (!dir.exists()) {                          //若不存在文件夹,则创建
            dir.mkdir();
        }
        return dir;
    }

    /**
     * 在 SD 卡上创建文件
     */
    public static File createFile(String filePath) throws IOException {
        File file = new File(filePath);
        if (!file.exists()) {                         //若文件不存在,则创建
            try {
                file.createNewFile();
            } catch (Exception e) {
                e.printStackTrace();
            }
        }
        return file;
    }

    /**
     * 判断文件是否存在
     */
    public static boolean isFileExists(String filePath) {
        File file = new File(filePath);
        return file.exists();                 //返回 boolean 类型值,true: 存在; false: 不存在
    }
}
```

代码文件: codes\08\8.2\SDCardDemo\cn\edu\hstc\sdcarddemo\util\FileUtil.java

该工具类中对 SD 卡的各种读写操作在源码中已经做了详细的注释,相信读者通过阅读注释已经对 Android 操作 SD 卡有了初步的认知,在将本实例的源码介绍完整后会对其作进一步的总结,接下来,需要编写本应用的 Activity 以实现与用户的交互。该 Activity 类的源码如下:

```
package cn.edu.hstc.sdcarddemo.activity;

import java.io.BufferedInputStream;
import java.io.File;
import java.io.FileOutputStream;
import java.io.InputStream;
import java.io.OutputStream;
import java.net.URL;
import java.net.URLConnection;

import android.app.Activity;
```

```
import android.app.AlertDialog;
import android.content.DialogInterface;
import android.content.DialogInterface.OnCancelListener;
import android.graphics.Bitmap;
import android.graphics.BitmapFactory;
import android.os.AsyncTask;
import android.os.Bundle;
import android.view.View;
import android.view.View.OnClickListener;
import android.widget.Button;
import android.widget.ImageView;
import android.widget.LinearLayout;
import android.widget.TextView;
import android.widget.Toast;
import cn.edu.hstc.sdcarddemo.util.DateUtil;
import cn.edu.hstc.sdcarddemo.util.FileUtil;

public class MainActivity extends Activity {
    //声明布局中的各个组件
    private Button startBtn, showBtn;
    private ImageView image;
    //图片下载中提示对话框
    private AlertDialog prompt;
    //是否有插入 SD 卡标识,true: 下载完毕; false: 未插入 SD 卡
    private boolean flag;
    private File file;

    @Override
    protected void onCreate(Bundle savedInstanceState) {
        super.onCreate(savedInstanceState);
        setContentView(R.layout.activity_main);
        //通过组件 ID 加载界面布局中的各个组件
        image = (ImageView) findViewById(R.id.image);
        startBtn = (Button) findViewById(R.id.startBtn);
        showBtn = (Button) findViewById(R.id.showBtn);
        startBtn.setOnClickListener(new OnClickListener() {
            public void onClick(View v) {
                startDownload();                    //开始下载图片
            }
        });
        showBtn.setOnClickListener(new View.OnClickListener() {
            @Override
            public void onClick(View v) {
                if (file != null && file.exists()) {
                    Bitmap bm = BitmapFactory.decodeFile(file.getAbsolutePath());
                    image.setImageBitmap(bm);       //将位图加载到图像视图
                } else {
                    Toast.makeText(MainActivity.this, "请先下载图片", 3000).show();
                }
            }
        });
```

```
    }

    private void startDownload() {
        String pictureUrl = "http://pic1.nipic.com/2008-12-25/2008122510134038_2.jpg";
        //创建轻量级异步任务,启动下载图片
        new DownloadFileAsync().execute(pictureUrl);
    }

    //内部类,轻量级异步任务类,继承 AsyncTask
    class DownloadFileAsync extends AsyncTask<String, String, String> {
        @Override
        protected void onPreExecute() {
            super.onPreExecute();
            prompt = prompt(MainActivity.this, "图片下载中...");
            prompt.setOnCancelListener(new OnCancelListener() {
                @Override
                public void onCancel(DialogInterface dialog) {
                    cancel(true);
                }
            });
        }

        @Override
        protected String doInBackground(String... params) {
            int count;
            try {
                URL url = new URL(params[0]);
                URLConnection conexion = url.openConnection();
                conexion.connect();
                int lenghtOfFile = conexion.getContentLength();
                InputStream input = new BufferedInputStream(url.openStream());
                if (FileUtil.isHaveSDCard()) {      //判断设备是否插入 SD 卡
                    //创建文件夹
                    File dir = FileUtil.createDIR(FileUtil.getSDCardPath() + File.
                    separator + "photos");
                    //在创建的文件夹下创建文件,以当前时间命名图片文件
                    file = FileUtil.createFile(dir.getAbsolutePath() + File.separator +
DateUtil.getCurrentDate() + ".jpg");
                    //创建文件输出流,将流写入文件中
                    OutputStream output = new FileOutputStream(file);
                    byte data[] = new byte[1024];
                    long total = 0;
                    while ((count = input.read(data)) != -1) {
                        total += count;
                        publishProgress("" + (int) ((total * 100) / lenghtOfFile));
                        output.write(data, 0, count);
                    }
                    output.flush();
                    output.close();
                    input.close();
                    flag = true;
```

```
                } else {
                    flag = false;
                }
            } catch (Exception e) {
                System.out.println(e.getMessage().toString());
            }
            return null;
        }

        @Override
        protected void onPostExecute(String unused) {
            prompt.dismiss();                          //关闭提示对话框
            if (flag == true) {
                Toast.makeText(MainActivity.this, "下载完毕!", 3000).show();
            } else if (flag == false) {
                Toast.makeText(MainActivity.this, "设备未插入 SD 卡", 3000).show();
            }
        }
    }

    /**
     * 返回提示对话框
     */
    private AlertDialog prompt(Activity context, String content) {
        AlertDialog.Builder builder = new AlertDialog.Builder(context);
        //prompt.xml 为对话框布局,布局中放置了 ProgressBar 进度条以及显示提示信息的 TextView
        LinearLayout layout = (LinearLayout) context.getLayoutInflater().inflate(R.layout.
prompt, null);
        ((TextView) layout.findViewById(R.id.message)).setText(content);
        builder.setView(layout);
        AlertDialog dialog = builder.show();
        dialog.setCanceledOnTouchOutside(false);
        return dialog;
    }
}
```

代码文件：codes\08\8.2\SDCardDemo\cn\edu\hstc\sdcarddemo\activity\MainActivity.java

上面的 Activity 类实现了与用户的交互，由于下载的文件保存时会以当前日期命名，故其中用到了一个叫 DateUtil 的工具类，该工具类用于获取当前日期并格式化，该工具类源码如下：

```
package cn.edu.hstc.sdcarddemo.util;

import java.text.SimpleDateFormat;
import java.util.Date;

public class DateUtil {
    public static String getCurrentDate() {
        Date nowDate = new Date();
        //日期格式化
        SimpleDateFormat format = new SimpleDateFormat("yyyyMMddHHmmss");
```

```
            return format.format(nowDate);
        }
    }
```

代码文件: codes\08\8.2\SDCardDemo\cn\edu\hstc\sdcarddemo\util\DateUtil.java

接下来，需要在应用程序的清单文件中配置添加读写 SD 卡的权限以及访问网络的权限，配置片段如下所示：

```
<!-- 访问网络权限 -->
<uses-permission android:name="android.permission.INTERNET" />
<!-- 在 SD 卡中创建与删除文件权限 -->
<uses-permission android:name="android.permission.MOUNT_UNMOUNT_FILESYSTEMS" />
<!-- 往 SD 卡写入数据权限 -->
<uses-permission android:name="android.permission.WRITE_EXTERNAL_STORAGE" />
```

至此，简易型图片下载器的主要源码已经全部实现了。将应用部署在模拟器上，单击"开始下载"按钮，运行效果如图 8.9 所示。

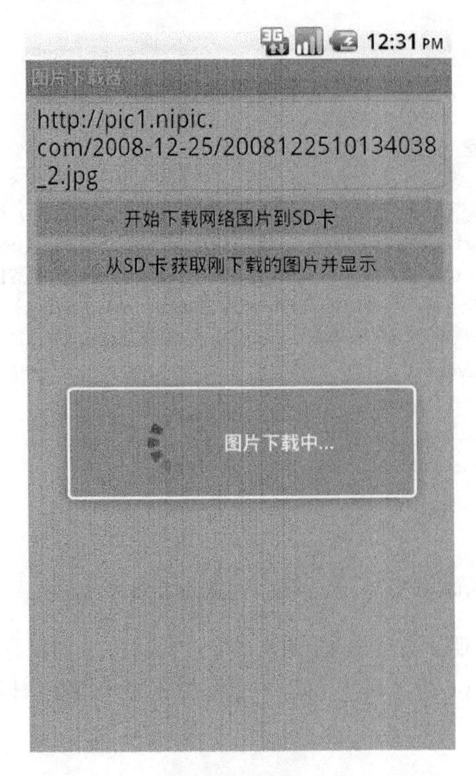

图 8.9　从网络上下载图片至 SD 卡中

当图片下载完毕后，打开 File Explorer 面板，可以看到如图 8.10 所示界面。

由图 8.10 可以看出，SD 卡的路径为/mnt/sdcard，由于在程序中实现了在 SD 卡目录下创建了一个叫 photos 的文件夹并将下载的图片写入该文件夹下，故可以看到图片的保存路径为/mnt/sdcard/photos/。

单击界面中的"显示图片"按钮，此时会将/mnt/sdcard/photos/目录下刚才保存的图片文件读取出来并显示在界面的图像视图中，可以看到如图 8.11 所示界面。

◢ mnt		2015-10-30	12:29	drwxrwxr-x
▷ asec		2015-10-30	12:29	drwxr-xr-x
◢ sdcard		1970-01-01	00:00	d---rwxr-x
▷ DCIM		2015-10-19	14:55	d---rwxr-x
▷ LOST.DIR		2015-09-07	16:58	d---rwxr-x
◢ photos		2015-10-30	12:35	d---rwxr-x
20151030123152.jpg	223067	2015-10-30	12:31	----rwxr-x

图 8.10　图片保存路径

图 8.11　从 SD 卡中读取图片并显示

总结以上开发过程，可以看出，操作 Android 设备 SD 卡中的文件可以按照如下步骤进行：

① 调用 Environment 的 getExternalStorageState()方法判断手机上是否插入了 SD 卡，并且应用程序具有读、写 SD 卡的权限。例如本 Demo 中所用到的 FileUtil 类中的代码片段：

```
/**
 * 判断 Android 设备是否插入了 SD 卡并且应用程序具有读写 SD 卡的权限
 */
public static Boolean isHaveSDCard() {
    return Environment.getExternalStorageState().equals(Environment.MEDIA_MOUNTED);
}
```

如果手机已插入 SD 卡，且应用程序具有读写 SD 卡的能力，则该方法返回 true。

② 调用 Environment.getExternalStorageDirectory()方法来获取外部存储器，如

FileUtil 工具类中的代码片段：

```
/**
 * 获取设备 SD 卡绝对路径
 */
public static String getSDCardPath() {
    return Environment.getExternalStorageDirectory().getAbsolutePath();
}
```

以上方法返回 SD 卡的目录。

③ 使用 FileInputStream、FileOutputStream、FileReader、FileWriter 读、写 SD 卡里的文件。在本 Demo 中正是使用了 FileOutputStream 将网络资源写入 SD 卡中的文件的。

8.3 通过简易旅游记录仪详细介绍 SQLite 数据库

利用外存储设备如 SD 卡存储数据毕竟需要依赖外来硬件，这有时会给没有 SD 卡的用户带来不便，并不是所有数据都需要存储在外部存储设备的。Android 系统集成了一个真正轻量级的数据库：SQLite，SQLite 作为一个嵌入式的数据库引擎，非常适合适量数据的存取，多用于手机、平板等移动端设备上，一个 SQLite 数据库就是一个文件。

8.3.1 实现简易旅游记录仪

简易旅游记录仪能帮助用户记录旅途中美好的瞬间，用户通过该应用能将风景或者其他所看到的事物拍下来，并对其进行简单描述，然后保存起来，当用户遇到能歌善舞的游伴时，可以利用该应用将其美妙的声音记录起来并进行简单描述，然后保存起来，过后，用户可以通过翻阅历史查看过去的美好点滴。

该简易旅游记录仪将用户记录的每一个美好的事物作为一个实体 History 保存在一张数据库表中，这也是实现该记录仪时所需要掌握的知识点：Android 操作 SQLite 数据库。接下来，让我们一起来实现该旅游记录仪的源代码。首先，新建名为 SQLiteDemo 的 Android Application Project，然后实现应用启动界面。应用主界面下方为三个菜单项，分别是查看照片类型的历史、查看音频类型的历史、设置(清空数据库)。单击不同的菜单项展示不同的界面，应用采用 Fragment 技术实现该功能。界面布局代码如下：

```
<?xml version = "1.0" encoding = "utf - 8"?>
<LinearLayout xmlns:android = "http://schemas.android.com/apk/res/android"
    android:layout_width = "fill_parent"
    android:layout_height = "fill_parent"
    android:background = "#F5F5F5"
    android:orientation = "vertical" >

    <FrameLayout
        android:id = "@ + id/realtabcontent"
        android:layout_width = "fill_parent"
        android:layout_height = "0dip"
        android:layout_weight = "1" />
```

```
    <android.support.v4.app.FragmentTabHost
        android:id = "@android:id/tabhost"
        android:layout_width = "fill_parent"
        android:layout_height = "wrap_content"
        android:background = "@drawable/buttom_back" >

        <FrameLayout
            android:id = "@android:id/tabcontent"
            android:layout_width = "0dp"
            android:layout_height = "0dp"
            android:layout_weight = "0" />
    </android.support.v4.app.FragmentTabHost>

</LinearLayout>
```

代码文件：codes\08\8.3\SQLiteDemo\res\layout\activity_main.xml

接着，实现各个 Fragment 的布局，以展示照片类型历史列表的 Fragment 为例，代码如下：

```
<?xml version = "1.0" encoding = "utf - 8"?>
<LinearLayout xmlns:android = "http://schemas.android.com/apk/res/android"
    android:layout_width = "match_parent"
    android:layout_height = "match_parent"
    android:orientation = "vertical" >

    <RelativeLayout
        android:id = "@ + id/layout_takephoto"
        android:layout_width = "fill_parent"
        android:layout_height = "wrap_content"
        android:background = "@color/gainsboro"
        android:padding = "10dp" >

        <TextView
            android:layout_width = "wrap_content"
            android:layout_height = "wrap_content"
            android:layout_centerInParent = "true"
            android:text = "Take Photo"
            android:textSize = "22sp" />
    </RelativeLayout>

    <ListView
        android:id = "@ + id/listView_camera"
        android:layout_width = "fill_parent"
        android:layout_height = "wrap_content"
        android:divider = "@color/black"
        android:dividerHeight = "1px" />

</LinearLayout>
```

代码文件：codes\08\8.3\SQLiteDemo\res\layout\ fragment_camera.xml

上面的布局从上而下依次放置了一个 RelativeLayout 布局，该布局相当于一个按钮，单击将会调用系统相机，一个 ListView 组件用于展示历史数据。展示音频类历史列表的 Fragment 的布局代码与上面的代码基本一样，在此不再重复给出，单击底部第三个菜单项时跳转到设置的 Fragment，该 Fragment 的布局更加简单，只是在 Fragment 中放置了一个用于清空数据库的按钮，在此也不再重复给出，可以通过查看本书所附带源码进行获得该部分源码。

接着，为了实现底部菜单项样式，我们新建了一个叫 tab_item_view. xml 的 Layout 文件，在该布局文件中实现菜单项按钮样式，该文件源代码如下所示：

```
<?xml version = "1.0" encoding = "utf - 8"?>
< LinearLayout xmlns:android = "http://schemas.android.com/apk/res/android"
    android:layout_width = "wrap_content"
    android:layout_height = "wrap_content"
    android:gravity = "center"
    android:orientation = "vertical" >

    < ImageView
        android:id = "@ + id/imageview"
        android:layout_width = "wrap_content"
        android:layout_height = "wrap_content"
        android:focusable = "false"
        android:padding = "3dp"
        android:src = "@drawable/camera_tab_btn" />

</LinearLayout >
```

代码文件：codes\08\8.3\SQLiteDemo\res\layout\ tab_item_view.xml

可以看到，上面的代码对 TextView 组件作了注释操作，即在底部菜单按钮中屏蔽了图片下方文字的展示，若需要编写一个底部菜单按钮图片下方带文字的功能，可将该部分注释掉的代码恢复。

接下来，由于展示历史列表时用到了 ListView 组件，如 fragment_camera. xml 中所示，故我们需要布局 ListView 组件的样式，源代码如下：

```
<?xml version = "1.0" encoding = "utf - 8"?>
< RelativeLayout xmlns:android = "http://schemas.android.com/apk/res/android"
    android:layout_width = "match_parent"
    android:layout_height = "match_parent"
    android:paddingBottom = "2dp"
    android:paddingLeft = "20dp"
    android:paddingRight = "20dp"
    android:paddingTop = "2dp" >

    < TextView
        android:id = "@ + id/txt_title"
        android:layout_width = "wrap_content"
        android:layout_height = "wrap_content"
        android:layout_alignParentLeft = "true"
        android:layout_centerVertical = "true" />
```

```
    <TextView
        android:id = "@ + id/txt_ge"
        android:layout_width = "wrap_content"
        android:layout_height = "wrap_content"
        android:layout_centerVertical = "true"
        android:layout_toRightOf = "@id/txt_title"
        android:text = "|" />

    <TextView
        android:id = "@ + id/txt_time"
        android:layout_width = "wrap_content"
        android:layout_height = "wrap_content"
        android:layout_centerVertical = "true"
        android:layout_toRightOf = "@id/txt_ge" />

    <ImageView
        android:id = "@ + id/image_icon"
        android:layout_width = "wrap_content"
        android:layout_height = "wrap_content"
        android:layout_alignParentRight = "true"
        android:layout_centerVertical = "true"
        android:background = "@drawable/right_icon" />

</RelativeLayout>
```

代码文件：codes\08\8.3\SQLiteDemo\res\layout\listview_item.xml

上面的布局设定了 ListView 每一个 Item 项的样式。展示图片类型的历史数据时，单击列表上方的 Take Photo 按钮将会调用系统相机，拍摄照片后将会回到列表页面，此时会弹出一个对话框供用户完善资料信息，比如写下对该照片的简单描述等，该自定义对话框亦对应一个独立的布局文件，该部分源代码如下：

```
<?xml version = "1.0" encoding = "utf - 8"?>
<LinearLayout xmlns:android = "http://schemas.android.com/apk/res/android"
    android:layout_width = "match_parent"
    android:layout_height = "match_parent"
    android:orientation = "vertical"
    android:paddingBottom = "5dp"
    android:paddingLeft = "10dp"
    android:paddingRight = "10dp"
    android:paddingTop = "5dp" >

    <LinearLayout
        android:layout_width = "fill_parent"
        android:layout_height = "wrap_content"
        android:orientation = "horizontal" >

        <TextView
            android:layout_width = "wrap_content"
            android:layout_height = "wrap_content"
```

```
            android:text = "标题"
            android:textColor = "@color/black" />

        <EditText
            android:id = "@ + id/edt_title"
            android:layout_width = "fill_parent"
            android:layout_height = "wrap_content"
            android:layout_marginLeft = "5dp"
            android:background = "@drawable/contact_edit_edittext_normal"
            android:paddingLeft = "5dp" />
    </LinearLayout>

    <LinearLayout
        android:layout_width = "fill_parent"
        android:layout_height = "wrap_content"
        android:orientation = "horizontal"
        android:paddingTop = "5dp" >

        <TextView
            android:layout_width = "wrap_content"
            android:layout_height = "wrap_content"
            android:text = "描述"
            android:textColor = "@color/black" />

        <EditText
            android:id = "@ + id/edt_content"
            android:layout_width = "fill_parent"
            android:layout_height = "wrap_content"
            android:layout_marginLeft = "5dp"
            android:background = "@drawable/contact_edit_edittext_normal"
            android:lines = "4"
            android:paddingLeft = "5dp" />
    </LinearLayout>

    <LinearLayout
        android:layout_width = "wrap_content"
        android:layout_height = "wrap_content"
        android:orientation = "horizontal"
        android:paddingTop = "5dp"
        android:layout_gravity = "center_horizontal" >

        <TextView
            android:layout_width = "wrap_content"
            android:layout_height = "wrap_content"
            android:text = "描述"
            android:textColor = "@color/black"
            android:visibility = "invisible" />

        <Button
            android:id = "@ + id/btn_over"
            android:layout_width = "wrap_content"
```

```
            android:layout_height = "wrap_content"
            android:background = "@drawable/style_button"
            android:layout_marginLeft = "5dp"
            android:paddingLeft = "30dp"
            android:paddingRight = "30dp"
            android:text = "完成" />
    </LinearLayout>

</LinearLayout>
```

该布局从上而下放置了两个输入框用于填写标题和描述，一个Button按钮用于单击完成数据插入。

至此，所有页面布局文件的源码实现已经全部完成，其他辅助样式如自定义按钮的样式等文件可以通过本书附带源码中获得(项目路径为/res/drawable/)。

接下来，需要实现的是本应用的主程序代码。首先，根据面向对象程序设计思想，每一个存入数据库表中的数据行，都对应了一个具体的实体类对象。在本应用中，该实体类为History类，该类实现代码如下：

```
package cn.edu.hstc.sqlitedemo.entity;

import java.io.Serializable;

public class History implements Serializable {
    private static final long serialVersionUID = 1L;
    public int _id;
    public String title;
    public String content;
    public String filePath;
    public String time;
    public String type;

    public History() {
    }

    public History(String title, String content, String filePath, String time, String type) {
        this.title = title;
        this.content = content;
        this.filePath = filePath;
        this.time = time;
        this.type = type;
    }

    public History(int _id, String title, String content, String filePath, String time, String
type) {
        this._id = _id;
        this.title = title;
        this.content = content;
        this.filePath = filePath;
        this.time = time;
```

```
        this.type = type;
    }
}
```

代码文件: codes\08\8.3\SQLiteDemo\cn\edu\hstc\sqlitedemo\entity\History.java

细心的读者会发现,该类实现了 Serializable 接口,之所以这么做,是因为这样可以将整个 History 对象放入 Bundle 对象中传递到其他页面中。

接下来要实现的是本应用的重点所在,也是掌握 SQLite 数据库开发的重点所在,主要包括数据库工具类 DBHelper 类以及操作数据库的工具类 DBManager 类,所有有关数据库对基础操作都放在这两个工具类中实现了。这两个类中所包含的源码便是 SQLite 数据库开发的知识点所在。在此,只给出源码,当然源码中会有附带相关注释,更详细的知识点将在下一节介绍。

首先,先来看看实际操作创建数据库文件的类 DBHelper 的源代码:

```
package cn.edu.hstc.sqlitedemo.util;

import android.content.Context;
import android.database.sqlite.SQLiteDatabase;
import android.database.sqlite.SQLiteOpenHelper;

public class DBHelper extends SQLiteOpenHelper {
    //数据库文件名
    private static final String DATABASE_NAME = "sqlitedemo.db";
    private static final int DATABASE_VERSION = 1;

    public DBHelper(Context context) {
        //CursorFactory设置为null,使用默认值
        super(context, DATABASE_NAME, null, DATABASE_VERSION);
    }

    //数据库第一次被创建时onCreate会被调用
    @Override
    public void onCreate(SQLiteDatabase db) {
        db.execSQL("create table if not exists history" + "(_id integer primary key autoincrement, title varchar, content varchar, filePath varchar, time varchar, type varchar)");
    }

    //如果DATABASE_VERSION值被改为2,系统发现现有数据库版本不同,即会调用onUpgrade
    @Override
    public void onUpgrade(SQLiteDatabase db, int oldVersion, int newVersion) {
        db.execSQL("alter table history add column other string");
    }
}
```

代码文件: codes\08\8.3\SQLiteDemo\cn\edu\hstc\sqlitedemo\util\DBHelper.java

上面的工具类做了创建数据库文件的实际操作,在下一节将会详细讲解该类,此处只给出源码。接下来,需要实现另一个工具类 DBManager 类,用于实现数据库的数据操作功能,源代码如下:

```
package cn.edu.hstc.sqlitedemo.util;

import java.util.ArrayList;
import java.util.List;

import android.content.Context;
import android.database.Cursor;
import android.database.sqlite.SQLiteDatabase;
import cn.edu.hstc.sqlitedemo.entity.History;

public class DBManager {
    private DBHelper helper;
    private SQLiteDatabase db;

    public DBManager(Context context) {
        helper = new DBHelper(context);
        //因为 getWritableDatabase 内部调用了 mContext.openOrCreateDatabase(mName, 0,
mFactory);
        //所以要确保 context 已初始化,可以把实例化 DBManager 的步骤放在 Activity 的
onCreate 里
        db = helper.getWritableDatabase();
    }

    public void add(History history) {
        db.beginTransaction();                          //开始事务
        try {
            db.execSQL("insert into history values(null, ?, ?, ?, ?, ?)", new Object[]
            {history.title, history.content, history.filePath, history.time, history.type});
            db.setTransactionSuccessful();              //设置事务成功完成
        } catch (Exception e) {
        } finally {
            db.endTransaction();                        //结束事务
        }
    }

    public void add(List<History> histories) {
        db.beginTransaction();                          //开始事务
        try {
            for (History history : histories) {
                db.execSQL("insert into history values(null, ?, ?, ?, ?, ?)", new Object[]
{history.title, history.content, history.filePath, history.time, history.type});
            }
            db.setTransactionSuccessful();              //设置事务成功完成
        } catch (Exception e) {
        } finally {
            db.endTransaction();                        //结束事务
        }
    }

    /* public void addContact(List<Contact> contacts) {
        db.beginTransaction();                          //开始事务
```

```
        try {
            for (Contact contact : contacts) {
                db.execSQL("insert into contact values(null, ?, ?, ?, ?, ?, ?, ?, ?)", new
Object[ ] {contact.name, contact.phone, contact.email, contact.company, contact.title,
contact.address, contact.website, contact.theId});
            }
            db.setTransactionSuccessful();          //设置事务成功完成
        } catch (Exception e) {
        } finally {
            db.endTransaction();                    //结束事务
        }
    } */

    public void delete(int theId) {
        db.beginTransaction();
        try {
            db.execSQL("delete from history where _id = " + theId);
            db.setTransactionSuccessful();
        } catch (Exception e) {
        } finally {
            db.endTransaction();
        }
    }

    public void getLastId(String table_name) {
    }

    public void deleteTableHistory() {
        db.delete("history", null, null);
    }

    public List<History> query(String[ ] type) {
        ArrayList<History> histories = new ArrayList<History>();
        Cursor cursor = queryTheCursor(type);
        while (cursor.moveToNext()) {
            History history = new History();
            history._id = cursor.getInt(cursor.getColumnIndex("_id"));
            history.title = cursor.getString(cursor.getColumnIndex("title"));
            history.content = cursor.getString(cursor.getColumnIndex("content"));
            history.filePath = cursor.getString(cursor.getColumnIndex("filePath"));
            history.time = cursor.getString(cursor.getColumnIndex("time"));
            history.type = cursor.getString(cursor.getColumnIndex("type"));
            histories.add(history);
        }
        cursor.close();
        return histories;
    }

    public Cursor queryTheNew(String[ ] time) {
        Cursor c = db.rawQuery("select * from history where time = ?", time);
        return c;
```

```
    }

    public Cursor queryTheCursor(String[] type) {
        Cursor c = db.rawQuery("select * from history where type = ? order by _id desc",
type);
        return c;
    }

    public Cursor queryTheCursor1(String[] whereStr) {
        Cursor c = db.rawQuery("select * from history where title = ? or time = ?", whereStr);
        return c;
    }

    public void deleteTheTable(String tableName) {
        if (hasTable(tableName)) {
            db.execSQL("delete from " + tableName);
        }
    }

    public void closeDB() {
        db.close();
    }

    public boolean hasTable(String table_name) {
        boolean result = false;
        Cursor cur = null;
        try {
            String sql_table = "select count( * ) as c from Sqlite_master where type = 'table'
and name = '" + table_name.trim() + "'";
            cur = db.rawQuery(sql_table, null);
            if (cur.moveToNext()) {
                int count = cur.getInt(0);
                if (count > 0) {
                    result = true;
                }
            }
            cur.close();
        } catch (Exception e) {
            return result;
        }
        return result;
    }
}
```

代码文件：codes\08\8.3\SQLiteDemo\cn\edu\hstc\sqlitedemo\util\DBManager.java

上述源码做了操作数据库表的数据的实际操作，具体就是数据库的增、删、改、查。实际上，该工具类还可继续完善，随着APP业务需求的增多以及繁杂度的加大，可以向该类添加更多的数据库操作的方法。接下来，介绍本应用中涉及的另一个工具类——自定义ListView适配器MyListViewAdapter类，源代码如下所示：

```
package cn.edu.hstc.sqlitedemo.util;

import java.util.ArrayList;
import java.util.List;

import android.content.Context;
import android.view.LayoutInflater;
import android.view.View;
import android.view.ViewGroup;
import android.widget.BaseAdapter;
import android.widget.TextView;
import cn.edu.hstc.sqlitedemo.activity.R;
import cn.edu.hstc.sqlitedemo.entity.History;

public class MyListViewAdapter extends BaseAdapter {
    //填充 ListView 的数据源
    private List<History> listHistories = new ArrayList<History>();
    private LayoutInflater inflater;

    //在构造器中注入数据源 listHistories
    public MyListViewAdapter(Context context, List<History> listHistories) {
        this.listHistories = listHistories;
        inflater = (LayoutInflater) context.getSystemService(Context.LAYOUT_INFLATER_
        SERVICE);
    }

    @Override
    public int getCount() {
        return listHistories.size();
    }

    @Override
    public Object getItem(int position) {
        return listHistories.get(position);
    }

    @Override
    public long getItemId(int position) {
        return position;
    }

    //自定义列表项
    @Override
    public View getView(int position, View convertView, ViewGroup parent) {
        ViewHolder holder = null;
        if (convertView == null) {
            //加载列表项布局文件
            convertView = inflater.inflate(R.layout.listview_item, null);
            holder = new ViewHolder();
            holder.title = (TextView) convertView.findViewById(R.id.txt_title);
            holder.time = (TextView) convertView.findViewById(R.id.txt_time);
```

```
                convertView.setTag(holder);
            } else {
                holder = (ViewHolder)convertView.getTag();
            }
            holder.title.setText(listHistories.get(position).title);
            holder.time.setText(listHistories.get(position).time);
            return convertView;
        }

        //更新数据,刷新 ListView
        public void addItem(final History history) {
            listHistories.add(history);
            notifyDataSetChanged();
        }

        public static class ViewHolder {
            public TextView title;
            public TextView time;
        }
    }
```

代码文件：codes\08\8.3\SQLiteDemo\cn\edu\hstc\sqlitedemo\util\MyListViewAdapter.java

上述工具类作为 ListView 组件的自定义适配器，在调用时，只需创建该类对象并传入 List 数组的数据源，便可将数据填充到 ListView 中，非常方便。这在本书第 4 章中介绍 ListView 列表组件时就有所介绍，对这部分知识点不熟悉的读者可以翻阅之前的章节进行巩固。

当用户使用该应用拍摄照片或者录制音频文件后，会自动将所拍摄的照片或录制的音频文件进行保存到本地的操作，此时会使用精确到秒的当前系统时间作为该文件的文件名，故在本应用中还涉及到另外两个工具类——DateUtil 类以及 FileUtil 类。DateUtil 类主要是返回任意格式的系统当前时间，FileUtil 类主要是操作文件以及 SD 卡，例如，将任意文件复制到另一个任意目录下。读者可以通过阅读本书所附带源码来获取这两个类的源码。

本应用采用 Fragment 技术实现底部三个菜单的界面切换，因此需要实现三个菜单各自的 Fragment 组件。首先，实现第一个菜单按钮所对应的 Fragment 及其功能。源码如下：

```
package cn.edu.hstc.sqlitedemo.fragment;

import java.io.File;
import java.io.FileNotFoundException;
import java.io.FileOutputStream;
import java.io.IOException;
import java.util.ArrayList;
import java.util.List;

import android.app.Activity;
import android.app.AlertDialog;
import android.content.Context;
import android.content.Intent;
```

```
import android.database.Cursor;
import android.graphics.Bitmap;
import android.os.Bundle;
import android.os.Environment;
import android.provider.MediaStore;
import android.support.v4.app.Fragment;
import android.view.LayoutInflater;
import android.view.View;
import android.view.ViewGroup;
import android.widget.AdapterView;
import android.widget.AdapterView.OnItemClickListener;
import android.widget.Button;
import android.widget.EditText;
import android.widget.ListView;
import android.widget.RelativeLayout;
import android.widget.Toast;
import cn.edu.hstc.sqlitedemo.activity.MainActivity;
import cn.edu.hstc.sqlitedemo.activity.R;
import cn.edu.hstc.sqlitedemo.activity.ShowCameraActivity;
import cn.edu.hstc.sqlitedemo.entity.History;
import cn.edu.hstc.sqlitedemo.util.DateUtil;
import cn.edu.hstc.sqlitedemo.util.FileUtil;
import cn.edu.hstc.sqlitedemo.util.MyListViewAdapter;

public class CameraHistroyFragment extends Fragment {
    private MyListViewAdapter cameraAdapter;   //自定义 ListView 适配器
    private ListView cameraListView;           //ListView 组件,用于显示照片类型的历史列表
    private RelativeLayout takephotoLayout;    //在顶部单击调用系统照相机的按钮
    private List<History> historyList;         //List 集合,存放数据库表对应的实体

    @Override
    public View onCreateView(LayoutInflater inflater, ViewGroup container, Bundle
    savedInstanceState) {
        return inflater.inflate(R.layout.fragment_camera, null);
    }

    @Override
    public void onActivityCreated(Bundle savedInstanceState) {
        super.onActivityCreated(savedInstanceState);
        //初始化 ListView
        initListView();
    }

    @Override
    public void onActivityResult(int requestCode, int resultCode, Intent data) {
        super.onActivityResult(requestCode, resultCode, data);
        if (requestCode == 1 && resultCode == Activity.RESULT_OK) {
            boolean flag = true;
            String sdStatus = Environment.getExternalStorageState();
            if (!sdStatus.equals(Environment.MEDIA_MOUNTED)) { //检测 SD 卡是否可用
                Toast.makeText(getActivity(), "存储设备未插入", 3000).show();
```

```
                return;
            }
            if (data != null) {
                Bundle bundle = data.getExtras();
                Bitmap bitmap = (Bitmap) bundle.get("data");
                                    //获取相机返回的数据,并转换为 Bitmap 图片格式
                FileOutputStream b = null;
                File dir1 = FileUtil.createDIR(FileUtil.getSDCardPath() + File.separator +
"sqlitedemo");
                File dir2 = FileUtil.createDIR(dir1.getAbsolutePath() + File.separator +
"images");
                final String fileName = dir2.getAbsolutePath() + File.separator +
                DateUtil.getCurrentDate() + ".jpg";
                try {
                    b = new FileOutputStream(fileName);
                    bitmap.compress(Bitmap.CompressFormat.JPEG, 100, b); //把数据写入文件
                    Toast.makeText(getActivity(), "照片已保存", 3000).show();
                } catch (FileNotFoundException e) {
                    e.printStackTrace();
                    flag = false;
                    Toast.makeText(getActivity(), "照片保存出错", 3000).show();
                } finally {
                    try {
                        b.flush();
                        b.close();
                    } catch (IOException e) {
                        e.printStackTrace();
                    }
                }
                if (flag == true) {
                    AlertDialog.Builder builder = new AlertDialog.Builder(getActivity());
                    LayoutInflater inflater = (LayoutInflater) getActivity().
                    getSystemService(Context.LAYOUT_INFLATER_SERVICE);
                    View layout = inflater.inflate(R.layout.dialog_input, null);
                    builder.setView(layout);
                    builder.setTitle("完善此刻想说的话");
                    final AlertDialog dialog = builder.show();
                    Button button = (Button) layout.findViewById(R.id.btn_over);
                    final EditText titleEdt = (EditText) layout.findViewById(R.id.edt_title);
                    final EditText contentEdt = (EditText) layout.findViewById(R.id.edt_
content);
                    button.setOnClickListener(new View.OnClickListener() {
                        @Override
                        public void onClick(View v) {
                            if (dialog != null) {
                                String title = titleEdt.getText().toString();
                                String content = contentEdt.getText().toString();
                                String time = DateUtil.getCurrentDate1();
                                String type = "camera";
                                MainActivity.dbManager.add(new History(title, content,
                                fileName, time, type));
```

```
                            dialog.dismiss();
                            String timeRaw[] = {time};
                            History history = queryNew(timeRaw);
                            historyList.add(0, history);
                            cameraAdapter.notifyDataSetChanged();
                        }
                    }
                });
            }
        }
    }
}

/**
 * 查找数据库,将数据填充在 ListView 中
 */
private void initListView() {
    takephotoLayout = (RelativeLayout) getActivity().findViewById(R.id.layout_
    takephoto);
    takephotoLayout.setOnClickListener(new View.OnClickListener() {
        @Override
        public void onClick(View v) {
            Intent intent = new Intent(MediaStore.ACTION_IMAGE_CAPTURE);
            startActivityForResult(intent, 1);
        }
    });

    String[] typeRaw = { "camera" };
    historyList = handleDB(typeRaw);  //查询表,并将查找出来的数据拼装到 List 集合中
    //将查询出来的数据作为数据源填充到 ListView 适配器中
    cameraAdapter = new MyListViewAdapter(CameraHistroyFragment.this.getActivity(),
historyList);
    cameraListView = (ListView) CameraHistroyFragment.this.getActivity().findViewById
(R.id.listView_camera);
    cameraListView.setAdapter(cameraAdapter);
    cameraListView.setOnItemClickListener(new OnItemClickListener() {
        @Override
        public void onItemClick(AdapterView<?> arg0, View arg1, int arg2, long arg3) {
            History history = historyList.get(arg2);
            Intent intent = new Intent(CameraHistroyFragment.this.getActivity(),
            ShowCameraActivity.class);
            Bundle bundle = new Bundle();
            bundle.putSerializable("history", history);
            intent.putExtras(bundle);
            startActivity(intent);
        }
    });
}

/**
 * 操作数据库,查询表
```

```
     */
    private List<History> handleDB(String[] typeRaw) {
        List<History> historyList = new ArrayList<History>();
        Cursor cursor = null;
        //使用 dbManager 数据库操作工具查询满足条件的数据填充游标
        cursor = MainActivity.dbManager.queryTheCursor(typeRaw);
        //托付给 activity 根据自己的生命周期去管理 Cursor 的生命周期
        CameraHistroyFragment.this.getActivity().startManagingCursor(cursor);
        while (cursor.moveToNext()) {          //开始将查询出来的数据添加在 List 数组中
            int _id = cursor.getInt(0);
            String title = cursor.getString(1);
            String content = cursor.getString(2);
            String filePath = cursor.getString(3);
            String time = cursor.getString(4);
            String type = cursor.getString(5);
            History history = new History(_id, title, content, filePath, time, type);
            historyList.add(history);
        }
        return historyList;
    }

    /**
     * 查询最新一条数据
     */
    private History queryNew(String[] timeRaw) {
        Cursor cursor = null;
        //根据时间字段查询单条数据
        cursor = MainActivity.dbManager.queryTheNew(timeRaw);
        //托付给 activity 根据自己的生命周期去管理 Cursor 的生命周期
        CameraHistroyFragment.this.getActivity().startManagingCursor(cursor);
        while (cursor.moveToNext()) {
            int _id = cursor.getInt(0);
            String title = cursor.getString(1);
            String content = cursor.getString(2);
            String filePath = cursor.getString(3);
            String time = cursor.getString(4);
            String type = cursor.getString(5);
            History history = new History(_id, title, content, filePath, time, type);
            return history;
        }
        return null;
    }
}
```

代码文件：codes\08\8.3\SQLiteDemo\cn\edu\hstc\sqlitedemo\fragment\CameraHistroyFragment.java

上面文件中的程序实现了单击顶部按钮调用系统照相机，照相之后返回该 Fragment 并将所拍摄的照片保存在手机 SD 卡的指定文件夹中，并记住所保存的绝对路径，然后弹出一个对话框供用户填写标题以及对该照片的简单描述，单击对话框中的完成按钮将会生成一条新的 History 数据，该 History 数据的 title 属性和 content 属性为用户在对话框中所填写的标题和简述，fileName 属性为照片的保存路径，time 属性为系统当前时间，type 属性为 camera 字符串，最后调用数据库操作工具类实现将该条数据作为数据库表的一条记录保存到数据库中并刷新 ListView 列表。

当然，上面程序必然会涉及数据库查询操作，将查询到的数据作为 ListView 数据源填充到 ListView 中。单击 ListView 列表中的某一列表项，将会把该列表项所对应的 History 实体类放在 Bundle 中并传递到历史记录详情页面。该详情页面的实现并不是本章的学习重点，在此不提供代码，读者可以通过阅读本书所附带项目源码获取该页面实现源码。

单击底部菜单中的第二个菜单项，界面将会显示其对应的第二个 Fragment——AudioHistoryFragment。该 Fragment 的实现与 CameraHistoryFragment 的实现类似，只不过调用的系统工具为设备的录音功能，保存的文件也不再是照片而是一段音频，当然保存到数据库所对应的 History 数据中的 type 字段也将保存为 audio，页面中展示历史数据也是用 ListView 组件，单击其某个列表项也会跳转到对应的详情页面，在该详情页面展示的不再是照片而是一段音频，单击即可实现播放与停止。因此，AudioHistoryFragment 类以及对应的详情页面的实现源码都可通过本书附带源码获取。

底部菜单中提供了第三个菜单供用户单击显示第三个 Fragment——SettingFragment。该 Fragment 放置了一个按钮供用户单击以清空数据库，实现源码如下：

```
package cn.edu.hstc.sqlitedemo.fragment;

import android.app.AlertDialog;
import android.content.DialogInterface;
import android.content.DialogInterface.OnClickListener;
import android.os.Bundle;
import android.support.v4.app.Fragment;
import android.view.LayoutInflater;
import android.view.View;
import android.view.ViewGroup;
import android.widget.RelativeLayout;
import android.widget.Toast;
import cn.edu.hstc.sqlitedemo.activity.MainActivity;
import cn.edu.hstc.sqlitedemo.activity.R;

public class SettingFragment extends Fragment {
    private RelativeLayout settingLayout;

    @Override
    public View onCreateView(LayoutInflater inflater, ViewGroup container, Bundle
    savedInstanceState) {
        return inflater.inflate(R.layout.fragment_setting, null);
    }

    @Override
    public void onActivityCreated(Bundle savedInstanceState) {
        super.onActivityCreated(savedInstanceState);
        initView();
```

```
    }

    private void initView() {
        settingLayout = (RelativeLayout) getActivity().findViewById(R.id.layout_cleardb);
        settingLayout.setOnClickListener(new View.OnClickListener() {
            @Override
            public void onClick(View v) {
                AlertDialog.Builder builder = new AlertDialog.Builder(getActivity());
                builder.setMessage("确认清空所有数据吗?");
                builder.setTitle("提示");
                builder.setPositiveButton("确认", new OnClickListener() {
                    @Override
                    public void onClick(DialogInterface dialog, int which) {
                        //调用数据库操作工具类删除表数据
                        MainActivity.dbManager.deleteTheTable("history");
                        dialog.dismiss();
                        Toast.makeText(getActivity(), "清空数据完毕", 3000).show();
                    }
                });
                builder.setNegativeButton("取消", new OnClickListener() {
                    @Override
                    public void onClick(DialogInterface dialog, int which) {
                        dialog.dismiss();
                    }
                });
                builder.create().show();
            }
        });
    }
}
```

代码文件：codes\08\8.3\SQLiteDemo\cn\edu\hstc\sqlitedemo\fragment\SettingFragment.java

上面程序最核心的一行代码就是 MainActivity.dbManager.deleteTheTable("history")，通过该行代码调用数据库操作工具类来删除 history 表中的全部数据，实现了清空所有数据的功能。当然底层的数据库操作依旧是在上面所介绍的 DBManager 工具类中实现的。该工具类的实现与完善才是本章的重点，Android 操作 SQLite 数据库的实际工作便是在该工具类中实现的。

接下来，需要将上面的这三个 Fragment 整合到 Activity 中才能实现与用户的交互。该 Activity 也是应用的入口 Activity，源码实现如下：

```
package cn.edu.hstc.sqlitedemo.activity;

import android.content.Intent;
import android.os.Bundle;
import android.support.v4.app.FragmentActivity;
import android.support.v4.app.FragmentTabHost;
import android.view.LayoutInflater;
```

```
import android.view.View;
import android.view.Window;
import android.widget.ImageView;
import android.widget.TabHost.TabSpec;
import cn.edu.hstc.sqlitedemo.fragment.AudioHistoryFragment;
import cn.edu.hstc.sqlitedemo.fragment.CameraHistroyFragment;
import cn.edu.hstc.sqlitedemo.fragment.SettingFragment;
import cn.edu.hstc.sqlitedemo.util.DBManager;

public class MainActivity extends FragmentActivity {
    private FragmentTabHost tabHost;
    private LayoutInflater layoutInflater;
    //定义数组来存放 Fragment 界面
    private Class<?> fragmentArr[] = { CameraHistroyFragment.class, AudioHistoryFragment.
class, SettingFragment.class };
    //定义数组来存放底部按钮图片
    private int tabImageArr[] = { R.drawable.camera_tab_btn, R.drawable.audio_tab_btn, R.
drawable.setting_tab_btn };
    //定义数组来存放底部选项卡文字
    private String tabTextArr[] = { "Camera", "Audio", "Setting"};
    public static DBManager dbManager;

    @Override
    protected void onCreate(Bundle savedInstanceState) {
        super.onCreate(savedInstanceState);
        this.requestWindowFeature(Window.FEATURE_NO_TITLE);  //去掉标题栏
        setContentView(R.layout.activity_main);
        //创建 DBManager 对象
        dbManager = new DBManager(MainActivity.this);
        initView();
    }

    @Override
    protected void onActivityResult(int arg0, int arg1, Intent arg2) {
        super.onActivityResult(arg0, arg1, arg2);
    }

    private void initView() {
        //实例化布局对象
        layoutInflater = LayoutInflater.from(this);
        tabHost = (FragmentTabHost) this.findViewById(android.R.id.tabhost);
        tabHost.setup(this, getSupportFragmentManager(), R.id.realtabcontent);
        int count = fragmentArr.length;
        for (int i = 0; i < count; i++) {
            //为每一个 Tab 按钮设置图标、文字
            TabSpec tabSpec = tabHost.newTabSpec(tabTextArr[i]).setIndicator
            (getTabItemView(i));
            tabHost.addTab(tabSpec, fragmentArr[i], null);
```

```
            tabHost.getTabWidget().getChildAt(i).setBackgroundResource(R.drawable.
            selector_tab_background);
        }
    }

    /**
     * 给 Tab 按钮设置图标和文字
     */
    private View getTabItemView(int index) {
        View view = layoutInflater.inflate(R.layout.tab_item_view, null);
        ImageView imageView = (ImageView) view.findViewById(R.id.imageview);
        imageView.setImageResource(tabImageArr[index]);
        return view;
    }
}
```

代码文件：codes\08\8.3\SQLiteDemo\cn\edu\hstc\sqlitedemo\activity\MainActivity.java

至此，简易旅游记录仪的所有代码实现便已介绍完毕。接下来主要探讨的是该应用源码中的 DBHelper 类以及 DBManager 类，以便读者更牢固地掌握 Android SQLite 知识。将程序部署在 Android 设备上并使用，图 8.12～图 8.15 是一些操作界面。

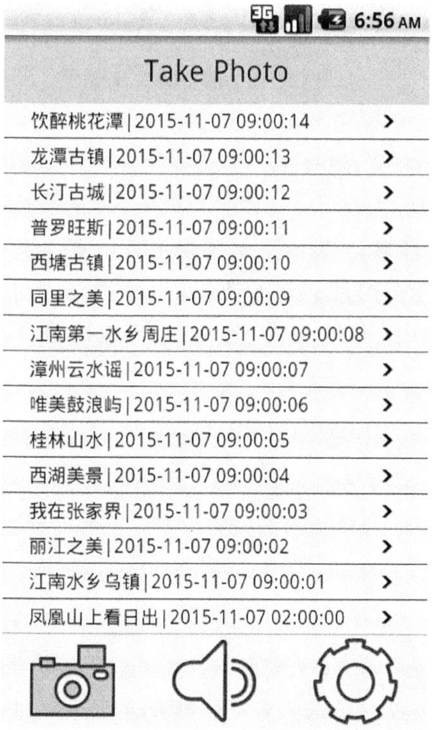

6:56 AM
Take Photo
饮醉桃花潭|2015-11-07 09:00:14
龙潭古镇|2015-11-07 09:00:13
长汀古城|2015-11-07 09:00:12
普罗旺斯|2015-11-07 09:00:11
西塘古镇|2015-11-07 09:00:10
同里之美|2015-11-07 09:00:09
江南第一水乡周庄|2015-11-07 09:00:08
漳州云水谣|2015-11-07 09:00:07
唯美鼓浪屿|2015-11-07 09:00:06
桂林山水|2015-11-07 09:00:05
西湖美景|2015-11-07 09:00:04
我在张家界|2015-11-07 09:00:03
丽江之美|2015-11-07 09:00:02
江南水乡乌镇|2015-11-07 09:00:01
凤凰山上看日出|2015-11-07 02:00:00

图 8.12　历史列表

6:57 AM
Record your message
Recording
00:07

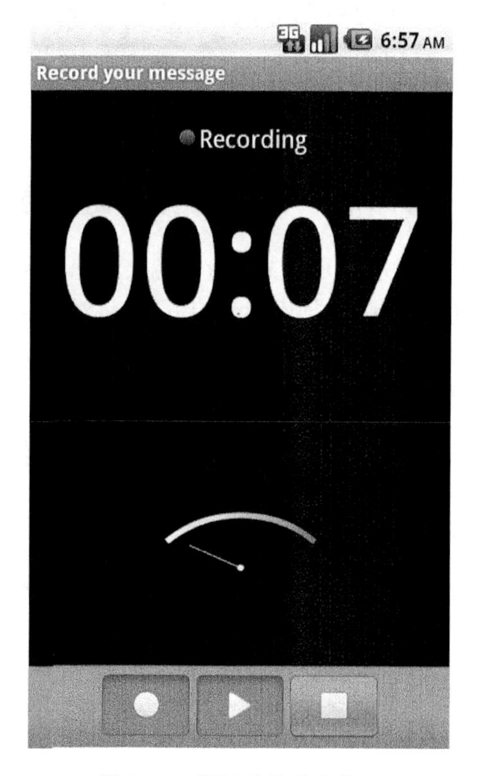

图 8.13　调用系统录音设备

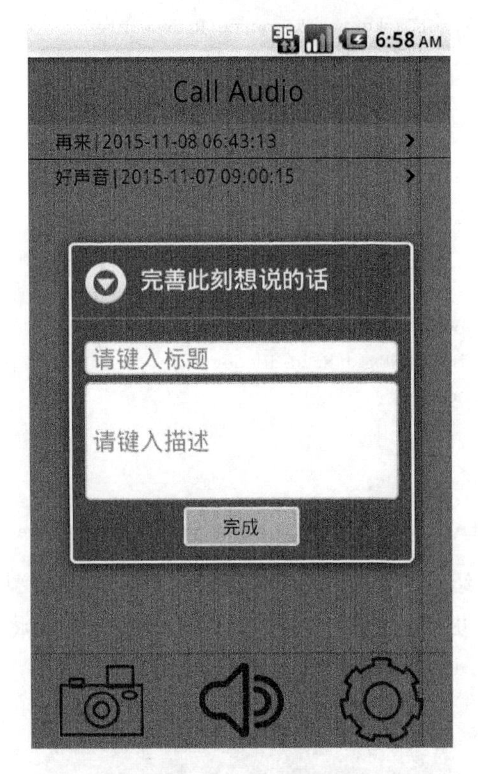

图 8.14　完善数据

图 8.15　清空数据库

8.3.2　剖析简易旅游记录仪

通过了 8.3.1 节对简易旅游记录仪的实现，相信读者已基本了解了 Android 是如何操作 SQLite 数据库的。接下来，将对该应用的两个关键类的源码进行剖析，以帮助读者加深对 SQLite 数据库操作的理解。首先，再次将创建数据库的类 DBHelper 的源码在下面贴出：

```
package cn.edu.hstc.sqlitedemo.util;

import android.content.Context;
import android.database.sqlite.SQLiteDatabase;
import android.database.sqlite.SQLiteOpenHelper;

public class DBHelper extends SQLiteOpenHelper {
    //数据库文件名
    private static final String DATABASE_NAME = "sqlitedemo.db";
    private static final int DATABASE_VERSION = 1;

    public DBHelper(Context context) {
        //CursorFactory 设置为 null,使用默认值
        super(context, DATABASE_NAME, null, DATABASE_VERSION);
    }
```

```
    //数据库第一次被创建时 onCreate 会被调用
    @Override
    public void onCreate(SQLiteDatabase db) {
        db.execSQL("create table if not exists history" + "(_id integer primary key autoincrement,
title varchar, content varchar, filePath varchar, time varchar, type varchar)");
    }

    //如果 DATABASE_VERSION 值被改为 2,系统发现现有数据库版本不同,即会调用 onUpgrade
    @Override
    public void onUpgrade(SQLiteDatabase db, int oldVersion, int newVersion) {
        db.execSQL("alter table history add column other string");
    }
}
```

通过阅读上面程序源码可以发现,帮助我们创建了 SQLite 数据库的类继承了 SQLiteOpenHelper 类。该类是 SQLiteDatabase 的一个帮助类,用来管理数据库的创建和版本的更新,一般是建立一个类继承它并实现它的 onCreate 和 onUpgrade 方法。SQLiteOpenHelper 类具有如表 8.1 所示的一些方法。

表 8.1　SQLiteOpenHelper 类的方法

方法名	方法描述
SQLiteOpenHelper (Context context, String name, SQLiteDatabase.CursorFactory factory, int version)	构造方法,一般把要创建的数据库的名称作为参数进行传递
onCreate(SQLiteDatabase db)	创建数据库时调用
onUpgrade (SQLiteDatabase db, int oldVersion, int newVersion)	版本更新时调用
getReadableDatabase()	创建或打开一个只读数据库
getWritableDatabase()	创建或打开一个读写数据库

在此,姑且先不去研究程序是怎么创建 sqlitedemo.db 这个数据库文件的。从自定义的 DBHelper 类中,可以看到,我们是在 onCreate 方法中调用了 SQLiteDatabase 对象的 execSQL 方法来创建了一张名为 history 的数据库表。SQLiteDatabase 类提供了很多实际操作数据库的方法,较为常用的方法如表 8.2 所示。

表 8.2　SQLiteDatabase 类提供的常用方法

(返回值)方法名	方法描述
(int) delete(String table, String whereClause, String[] whereArgs)	删除数据行的便捷方法
(long) insert (String table, String nullColumnHack, ContentValues values)	添加数据行的便捷方法
(int) update (String table, ContentValues values, String whereClause, String[] whereArgs)	更新数据行的便捷方法
(void) execSQL(String sql)	执行一个 SQL 语句,可以是一个 select 或其他的 SQL 语句
(void) close()	关闭数据库

续表

(返回值)方法名	方法描述
(Cursor) query(String table, String[] columns, String selection, String[] selectionArgs, String groupBy, String having, String orderBy, String limit)	查询指定的数据表返回一个带游标的数据集
(Cursor) rawQuery(String sql, String[] selectionArgs)	运行一个预置的 SQL 语句,返回带游标的数据集(与上面的语句最大的区别就是防止 SQL 注入)

在 DBHelper 类中正是使用了 SQLiteDatabase 类中的 execSQL(String sql)方法来执行一段创建数据库表的 SQL 语句。

实际上,SQLiteDatabase 类代表了一个数据库,一旦应用程序获得了代表指定数据库的 SQLiteDatabase 对象,接下来就可以通过其对象来管理和操作数据库了。可是,程序是如何创建 SQLiteDatabase 数据库的呢?这也要归功于 SQLiteDatabase。SQLiteDatabase 提供了如表 8.3 所示的静态方法来打开或创建数据库文件。

表 8.3　SQLiteDatabase 类提供的打开或创建数据库文件的方法

(返回值)方法名	方法描述
static SQLiteDatabase openDatabase(String path, SQLiteDatabase.CursorFactory factory, int flags)	打开 path 文件所代表的 SQLite 数据库
static SQLiteDatabase openOrCreateDatabase(File file, SQLiteDatabase.CursorFactory factory)	打开或创建 file 文件所代表的 SQLite 数据库
static SQLiteDatabase openOrCreateDatabase(String path, SQLiteDatabase.CursorFactory factory)	打开或创建 path 路径下文件所代表的 SQLite 数据库

从上面的 DBHelper 类的源码中,我们并没有看到以上打开或创建数据库文件的代码,那么程序在哪里调用了 SQLiteDatabase 对象来创建数据库呢?答案就在 DBManager 类源码中。在此,将 DBManager 类的源码再次贴出:

```
package cn.edu.hstc.sqlitedemo.util;

import java.util.ArrayList;
import java.util.List;

import android.content.Context;
import android.database.Cursor;
import android.database.sqlite.SQLiteDatabase;
import cn.edu.hstc.sqlitedemo.entity.History;

public class DBManager {
    private DBHelper helper;
    private SQLiteDatabase db;

    public DBManager(Context context) {
        helper = new DBHelper(context);
        //因为 getWritableDatabase 内部调用了 mContext.openOrCreateDatabase(mName, 0, mFactory);
        //所以要确保 context 已初始化,可以把实例化 DBManager 的步骤放在 Activity 的 onCreate 中
```

```
        db = helper.getWritableDatabase();
    }

    public void add(History history) {
        db.beginTransaction();              //开始事务
        try {
            db.execSQL("insert into history values(null, ?, ?, ?, ?, ?)", new Object[]
{history.title, history.content, history.filePath, history.time, history.type});
            db.setTransactionSuccessful();  //设置事务成功完成
        } catch (Exception e) {
        } finally {
            db.endTransaction();            //结束事务
        }
    }

    public void add(List<History> histories) {
        db.beginTransaction();              //开始事务
        try {
            for (History history : histories) {
                db.execSQL("insert into history values(null, ?, ?, ?, ?, ?)", new Object[]
{history.title, history.content, history.filePath, history.time, history.type});
            }
            db.setTransactionSuccessful();  //设置事务成功完成
        } catch (Exception e) {
        } finally {
            db.endTransaction();            //结束事务
        }
    }

    /* public void addContact(List<Contact> contacts) {
        db.beginTransaction();              //开始事务
        try {
            for (Contact contact : contacts) {
                db.execSQL("insert into contact values(null, ?, ?, ?, ?, ?, ?, ?, ?)", new
Object[]{contact.name, contact.phone, contact.email, contact.company, contact.title,
contact.address, contact.website, contact.theId});
            }
            db.setTransactionSuccessful();  //设置事务成功完成
        } catch (Exception e) {
        } finally {
            db.endTransaction();            //结束事务
        }
    } */

    public void delete(int theId) {
        db.beginTransaction();
        try {
            db.execSQL("delete from history where _id = " + theId);
            db.setTransactionSuccessful();
        } catch (Exception e) {
        } finally {
```

```
            db.endTransaction();
        }
    }

    public void getLastId(String table_name) {
    }

    public void deleteTableHistory() {
        db.delete("history", null, null);
    }

    public List<History> query(String[] type) {
        ArrayList<History> histories = new ArrayList<History>();
        Cursor cursor = queryTheCursor(type);
        while (cursor.moveToNext()) {
            History history = new History();
            history._id = cursor.getInt(cursor.getColumnIndex("_id"));
            history.title = cursor.getString(cursor.getColumnIndex("title"));
            history.content = cursor.getString(cursor.getColumnIndex("content"));
            history.filePath = cursor.getString(cursor.getColumnIndex("filePath"));
            history.time = cursor.getString(cursor.getColumnIndex("time"));
            history.type = cursor.getString(cursor.getColumnIndex("type"));
            histories.add(history);
        }
        cursor.close();
        return histories;
    }

    public Cursor queryTheNew(String[] time) {
        Cursor c = db.rawQuery("select * from history where time=?", time);
        return c;
    }

    public Cursor queryTheCursor(String[] type) {
        Cursor c = db.rawQuery("select * from history where type=? order by _id desc", type);
        return c;
    }

    public Cursor queryTheCursor1(String[] whereStr) {
        Cursor c = db.rawQuery("select * from history where title=? or time=?", whereStr);
        return c;
    }

    public void deleteTheTable(String tableName) {
        if (hasTable(tableName)) {
            db.execSQL("delete from " + tableName);
        }
    }

    public void closeDB() {
        db.close();
```

```
    }

    public boolean hasTable(String table_name) {
        boolean result = false;
        Cursor cur = null;
        try {
            String sql_table = "select count( * ) as c from Sqlite_master where type = 'table'
and name = '" + table_name.trim() + "'";
            cur = db.rawQuery(sql_table, null);
            if (cur.moveToNext()) {
                int count = cur.getInt(0);
                if (count > 0) {
                    result = true;
                }
            }
            cur.close();
        } catch (Exception e) {
            return result;
        }
        return result;
    }
}
```

阅读DBManager类源码发现，其构造方法中创建继承了SQLiteOpenHelper类的DBHelper类对象，并调用其getWritableDatabase()方法获得了一个SQLiteDatabase对象，程序正是通过调用该方法创建了SQLite数据库的。下面是SQLiteOpenHelper中的getWritableDatabase()的源码：

```
public synchronized SQLiteDatabase getWritableDatabase() {
        if (mDatabase != null && mDatabase.isOpen() && !mDatabase.isReadOnly()) {
            return mDatabase;  //The database is already open for business //如果已经打开
                               //了，则直接返回
        }
        if (mIsInitializing) {
            throw new IllegalStateException("getWritableDatabase called recursively");

        }
        //If we have a read-only database open, someone could be using it
        //(though they shouldn't), which would cause a lock to be held on
        //the file, and our attempts to open the database read-write would
        //fail waiting for the file lock.  To prevent that, we acquire the
        //lock on the read-only database, which shuts out other users.
        boolean success = false;
        SQLiteDatabase db = null;
        if (mDatabase != null) mDatabase.lock();
        try {
            mIsInitializing = true;
            if (mName == null) {
                db = SQLiteDatabase.create(null);  //以上没多大意义，可以不管
            } else {
```

```
                db = mContext.openOrCreateDatabase(mName, 0, mFactory);//关键是这行代码
            }
            int version = db.getVersion();
            if (version != mNewVersion) {
                db.beginTransaction();
                try {
                    if (version == 0) {
                            onCreate(db);  //在子类中重写了该方法(其实什么也没做,要想在
                              //第一次创建时做一些操作就自己在子类的方法中实现,如创建表)
                    } else {
                        onUpgrade(db, version, mNewVersion);
                                                        //在子类中重写该方法(版本发生变化)
                    }
                    db.setVersion(mNewVersion);
                    db.setTransactionSuccessful();
                } finally {
                    db.endTransaction();
                }
            }
            onOpen(db);                                 //在子类中重写该方法
            success = true;
            return db;
        } finally {
            mIsInitializing = false;
            if (success) {
                if (mDatabase != null) {
                    try { mDatabase.close(); } catch (Exception e) { }
                    mDatabase.unlock();
                }
                mDatabase = db;
            } else {
                if (mDatabase != null) mDatabase.unlock();
                if (db != null) db.close();
            }
        }
    }
```

阅读上面源码,可以发现,在getWritableDatabase()方法中有代码行 db = mContext.openOrCreateDatabase(mName, 0, mFactory),正是该句代码实现了打开或创建了对应数据库文件名的数据库。所以,我们通常是不用直接去调用SQLiteDatabase的openOrCreateDatabase方法的,而是通过SQLiteDatabase的一个帮助类SQLiteOpenHelper来实现数据库的打开或创建。

当主程序中创建DBManager对象时,实际上便是调用DBManager类的构造方法,通过该构造方法实现了数据库的打开或创建。通常把实例化DBManager的步骤放在Activity的onCreate方法中。

在DBManager类中,通过新增方法,实现了对数据库的底层的增、删、改、查操作。而在业务层面中,比如Activity或Fragment中,只需要调用实例化后的DBManager对象的这些方法,便可以实现实际上的数据库操作。可以看出,在本应用的DBManager数据库管理类

中，对数据库的增、删、改等非查询的操作的方法都是调用 SQLiteDatabase 的 execSQL 方法来实现的。实际上，正如上面介绍 SQLiteDatabase 类常用方法的表格中所提到的，SQLiteDatabase 还提供了如 insert、update、delete、query 等方法来实现操作数据库，完全可以不必通过执行 SQL 语句来完成，但 Android 考虑到部分开发者对 SQL 语法的不熟悉，故提供了这些方法帮助开发者以更简单的方式来操作数据库表的数据。

可以发现，DBManager 类中查询数据库表数据的基础方法都是返回一个 Cursor 对象，Android 中的 Cursor 类似于 JDBC 的 ResultSet，Cursor 同样提供了如表 8.4 所示的方法来移动查询结果的记录指针。

表 8.4　移动查询结果的记录指针的方法

方　法　名	方 法 描 述
move(int offset)	将记录指针向上或向下移动指定的行数。offset 为正数就是向下移动，为负数就是向上移动
boolean moveToFirst()	将记录指针移动到第一行，如果成功则返回 true
boolean moveToLast()	将记录指针移动到最后一行，如果成功则返回 true
boolean moveToNext()	将记录指针移动到下一行，如果成功则返回 true
boolean moveToPosition(int position)	将记录指针移动到指定的行，如果成功则返回 true
boolean moveToPrevious()	将记录指针移动到上一行，如果成功则返回 true

一旦将记录指针移动到指定行之后，接下来就可以调用 Cursor 的 getXxx()方法来获取该行的指定列的数据。

8.4　本章小结

本章主要介绍了 Android 的数据存储以及文件的读写。利用 SharedPreferences 可以非常方便地读、写应用程序的参数、选项。Android 还提供了 openFileOutput 和 openFileInput 两个便捷的方法来实现文件的 IO 操作。除此之外，Android 操作 SQLite 数据库也是本章的重点，读者需要学习如何利用 Android 提供的大量方便的工具类来访问 SQLite 数据库。

第9章 使用ContentProvider

Android 官方指出的数据存储方式总共有五种，分别是 Shared Preferences、网络存储、文件存储、外储存储、SQLite。但是一般这些存储都只是在单独的一个应用程序之中达到一个数据的共享，因为在不同应用之间共享数据时使用这种方式显得太杂乱了，不利于操作，但用户依然可能选择这么做。因此，为了适应一个应用程序操作另一个应用程序的一些数据，例如，需要操作系统里的媒体库、通讯录等，这时就可能通过 ContentProvider 来满足人们的需求了。

ContentProvider 提供了在应用程序之间共享数据的一种机制，是不同应用程序之间数据交换的标准 API，当一个应用程序需要把自己的数据暴露给其他程序使用时，该应用程序就可通过提供 ContentProvider 来实现，其他应用程序就可通过 ContentResolver 来操作 ContentProvider 暴露的数据。

9.1 实现通过 ContentProvider 共享数据的应用

本节通过实现一个用于对图书进行管理的 APP，但其并没有直接进行数据库操作，而是提供一个 ContentProvider，共享数据，然后再实现另一个 APP，通过 ContentResolver 来操作第一个应用所暴露的数据。

新建一个图书实体类 BookBean，该实体类对应了数据库表 book，源码如下：

```
package cn.edu.hstc.contentproviderdemo.entity;

public class BookBean {
    public int _id;
    public String number;
    public String name;
    public String press;

    public BookBean() {
    }

    public BookBean(String number, String name, String press) {
        this.number = number;
        this.name = name;
        this.press = press;
    }
```

```
    public BookBean(int _id, String number, String name, String press) {
        this._id = _id;
        this.number = number;
        this.name = name;
        this.press = press;
    }
}
```

代码文件：codes\09\9.1\ContentProviderDemo\cn\edu\hstc\util\BookBean.java

由于需要操作 SQLite 数据库，还需要两个数据库管理与操作的工具类，这与第 8 章中实现简易旅游记录仪 APP 时所用到的数据库工具类类似，并且 SQLite 的知识并不是本章的重点，在此不给出源码，读者可以通过本书所附带的源码获得。实现本应用需要用到另一个常量工具类，源码如下：

```
package cn.edu.hstc.contentproviderdemo.util;

import android.net.Uri;
import android.provider.BaseColumns;

public class Constant {
    //定义 ContentProvider 的 Authority
    public static final String AUTHORITY = "cn.edu.hstc.providers.bookprovider";

    //定义一个静态内部类
    public static final class Book implements BaseColumns {
        //定义 ContentProvider 所允许操作的 4 个数据列
        public final static String _ID = "_id";
        public final static String NUMBER = "number";
        public final static String NAME = "name";
        public final static String PRESS = "press";
        //定义 ContentProvider 提供服务的两个 Uri
        public final static Uri BOOKS_CONTENT_URI = Uri.parse("content://" + AUTHORITY +
"/books");
        public final static Uri BOOK_CONTENT_URI = Uri.parse("content://" + AUTHORITY + "/
book");
    }
}
```

代码文件：codes\09\9.1\ContentProviderDemo\cn\edu\hstc\util\Constant.java

接着，需要实现本应用的重点，即实现能共享数据的 ContentProvider，该 ContentProvider 的源码如下：

```
package cn.edu.hstc.contentproviderdemo.util;

import android.content.ContentProvider;
import android.content.ContentUris;
import android.content.ContentValues;
import android.content.UriMatcher;
import android.database.Cursor;
```

```
import android.net.Uri;

public class BookProvider extends ContentProvider {
    private static UriMatcher matcher = new UriMatcher(UriMatcher.NO_MATCH);
    private static final int BOOKS = 1;
    private static final int BOOK = 2;
    private DBManager dbManager;
    static {
        //为 UriMatcher 注册两个 Uri
        matcher.addURI(Constant.AUTHORITY, "books", BOOKS);
        matcher.addURI(Constant.AUTHORITY, "book/#", BOOK);
    }

    /**
     * 第一次调用该 ContentProvider 时,系统先创建 BookProvider 对象,并回调该方法 */
    @Override
    public boolean onCreate() {
        dbManager = new DBManager(this.getContext());
        return false;
    }

    @Override
    public String getType(Uri uri) {
        switch (matcher.match(uri)) {
        //如果操作的数据是多项记录
        case BOOKS:
            return "vnd.android.cursor.dir/cn.edu.hstc.providers.book";
        //如果操作的数据是单项记录
        case BOOK:
            return "vnd.android.cursor.item/cn.edu.hstc.providers.book";
        default:
            throw new IllegalArgumentException("未知 Uri:" + uri);
        }
    }

    @Override
    public Uri insert(Uri uri, ContentValues values) {
        //插入数据,返回行 ID
        long rowId = dbManager.insert(Constant.Book._ID, values);
        //如果插入成功返回 uri
        if (rowId > 0) {
            //在已有的 Uri 的后面追加 ID 数据
            Uri wordUri = ContentUris.withAppendedId(uri, rowId);
            //通知数据已经改变
            getContext().getContentResolver().notifyChange(wordUri, null);
            return wordUri;
        }
        return null;
    }

    @Override
```

```
    public int delete(Uri uri, String selection, String[] selectionArgs) {
        //记录所删除的记录数
        int num = 0;
        //对于 uri 进行匹配。
        switch (matcher.match(uri)) {
        case BOOKS:
            num = dbManager.delete(selection, selectionArgs);
            break;
        case BOOK:
            //解析出所需要删除的记录 ID
            long id = ContentUris.parseId(uri);
            String where = Constant.Book._ID + " = " + id;
            //如果原来的 where 子句存在,拼接 where 子句
            if (selection != null && !selection.equals("")) {
                where = where + " and " + selection;
            }
            num = dbManager.delete(where, selectionArgs);
            break;
        default:
            throw new IllegalArgumentException("未知 Uri:" + uri);
        }
        //通知数据已经改变
        getContext().getContentResolver().notifyChange(uri, null);
        return num;
    }

    @Override
     public int update ( Uri uri, ContentValues values, String selection, String [ ]
selectionArgs) {
        //记录所修改的记录数
        int num = 0;
        switch (matcher.match(uri)) {
        case BOOKS:
            num = dbManager.update(values, selection, selectionArgs);
            break;
        case BOOK:
            //解析出想修改的记录 ID
            long id = ContentUris.parseId(uri);
            String where = Constant.Book._ID + " = " + "?";
            //如果原来的 where 子句存在,拼接 where 子句
            if (selection != null && !selection.equals("")) {
                where = where + " and " + selection;
            }
            num = dbManager.update(values, where, selectionArgs);
            break;
        default:
            throw new IllegalArgumentException("未知 Uri:" + uri);
        }
        //通知数据已经改变
        getContext().getContentResolver().notifyChange(uri, null);
        return num;
```

```
    }

    @Override
    public Cursor query (Uri uri, String[ ] projection, String selection, String[ ]
selectionArgs, String sortOrder) {
        switch (matcher.match(uri)) {
        case BOOKS:
            //执行查询
            return dbManager.query(projection, selection, selectionArgs, sortOrder);
        case BOOK:
            //解析出想查询的记录 ID
            long id = ContentUris.parseId(uri);
            String where = Constant.Book._ID + " = " + id;
            //如果原来的 where 子句存在,拼接 where 子句
            if (selection != null && !"".equals(selection)) {
                where = where + " and " + selection;
            }
            return dbManager.query(projection, where, selectionArgs, sortOrder);
        default:
            throw new IllegalArgumentException("未知 Uri:" + uri);
        }
    }
}
```

代码文件:codes\09\9.1\ContentProviderDemo\cn\edu\hstc\util\BookProvider.java

该应用并没有实现自身操作 SQLite 数据库,故没有提供让用户操作图书数据的交互界面,启动该应用完成的是新建数据库 book.db 并建立数据库表 book,同时注册了 BookProvider。只要启动一次该应用,就会将数据通过 BookProvider 暴露给其他 APP,其他 APP 便可操作其数据了。

这里并没有通过自身去操作 SQLite 数据库,所以并不用提供与用户操作交互的界面,而是在启动界面中提示用户数据库表已存在以及 ContentProvider 已存在就可以了。此部分的实现所涉及的布局以及 Activity 源码非常简单,故不贴出源码,读者可以通过本书附带源码获得。

接下来,需要实现另一个 APP,即利用 ContentResolver 操作刚才已经实现的 APP 通过 ContentProvider 暴露的数据。

由于现在要实现的这个应用是需要操作数据的,所以需要提供与用户交互的界面,该界面提供展示图书编号、图书名称、出版社的文本输入框,以及对应增、删、改、查四个操作的 Button 按钮。该界面布局比较简单,读者可通过学习本书附带源码获得。

实现该应用还需要提供一个与暴露数据的应用一样的常量类 Constant,同样定义了 ContentProvider 所允许操作的四个数据列,以及 ContentProvider 提供服务的两个 Uri。

接下来所要介绍的与用户交互的各个按钮的核心功能代码将是需要重点掌握的。该交互 Activity 的源码如下:

```
package cn.edu.hstc.contentresolverdemo.activity;

import android.app.Activity;
```

```
import android.content.ContentResolver;
import android.content.ContentUris;
import android.content.ContentValues;
import android.database.Cursor;
import android.os.Bundle;
import android.view.View;
import android.view.View.OnClickListener;
import android.widget.Button;
import android.widget.EditText;
import android.widget.Toast;
import cn.edu.hstc.contentresolverdemo.util.Constant;

public class MainActivity extends Activity implements OnClickListener {
    ContentResolver contentResolver;
    private Button insertBtn, deleteBtn, updateBtn, queryBtn;
    private int _id;

    @Override
    protected void onCreate(Bundle savedInstanceState) {
        super.onCreate(savedInstanceState);
        setContentView(R.layout.activity_main);
        //获取系统的 ContentResolver 对象
        contentResolver = getContentResolver();
        initView();
    }

    private void initView() {
        insertBtn = (Button) findViewById(R.id.btn_add);
        insertBtn.setOnClickListener(this);
        deleteBtn = (Button) findViewById(R.id.btn_delete);
        deleteBtn.setOnClickListener(this);
        updateBtn = (Button) findViewById(R.id.btn_update);
        updateBtn.setOnClickListener(this);
        queryBtn = (Button) findViewById(R.id.btn_query);
        queryBtn.setOnClickListener(this);
    }

    @Override
    public void onClick(View v) {
        switch (v.getId()) {
        case R.id.btn_add:
            insertBook();
            break;
        case R.id.btn_delete:
            deleteBook();
            break;
        case R.id.btn_update:
            updateBook();
            break;
        case R.id.btn_query:
            queryBook();
```

```
            break;
        default:
            break;
        }
    }

    private void insertBook() {
        String number = ((EditText) findViewById(R.id.book_number)).getText().toString();
        String name = ((EditText) findViewById(R.id.book_name)).getText().toString();
        String press = ((EditText) findViewById(R.id.book_press)).getText().toString();
        //插入图书
        ContentValues values = new ContentValues();
        values.put(Constant.Book.NUMBER, number);
        values.put(Constant.Book.NAME, name);
        values.put(Constant.Book.PRESS, press);
        contentResolver.insert(Constant.Book.BOOK_CONTENT_URI, values);
        //显示提示信息
        Toast.makeText(MainActivity.this, "添加图书成功!", 20000).show();
    }

    private void deleteBook() {
        contentResolver.delete(Constant.Book.BOOKS_CONTENT_URI, "_id = ?", new String[]
{String.valueOf(_id)});
        ((EditText) findViewById(R.id.book_number)).setText("");
        ((EditText) findViewById(R.id.book_name)).setText("");
        ((EditText) findViewById(R.id.book_press)).setText("");
        Toast.makeText(MainActivity.this, "删除图书成功!", 20000).show();
    }

    private void updateBook() {
        String number = ((EditText) findViewById(R.id.book_number)).getText().toString();
        String name = ((EditText) findViewById(R.id.book_name)).getText().toString();
        String press = ((EditText) findViewById(R.id.book_press)).getText().toString();
        //修改图书
        ContentValues values = new ContentValues();
        values.put(Constant.Book.NUMBER, number);
        values.put(Constant.Book.NAME, name);
        values.put(Constant.Book.PRESS, press);
        //contentResolver.update(Constant.Book.BOOKS_CONTENT_URI, values, "_id = ?", new
String[]{String.valueOf(_id)});
        contentResolver.update(ContentUris.withAppendedId(Constant.Book.BOOK_CONTENT_URI,
_id), values, null, new String[]{String.valueOf(_id)});
        Toast.makeText(MainActivity.this, "更新图书成功!", 20000).show();
    }

    private void queryBook() {
        //获取用户输入
        String number = ((EditText) findViewById(R.id.book_number)).getText().toString();
        //执行查询
        Cursor cursor = contentResolver.query(Constant.Book.BOOKS_CONTENT_URI, null,
"number like ?", new String[]{"%" + number + "%"}, null);
        while (cursor.moveToNext()) {
            _id = cursor.getInt(0);
```

```
                ((EditText) findViewById(R.id.book_name)).setText(cursor.getString(2));
                ((EditText) findViewById(R.id.book_press)).setText(cursor.getString(3));
            }
        }
    }
```

代码文件：codes\09\9.1\ContentResolverDemo\cn\edu\hstc\activity\MainActivity.java

将 ContentProviderDemo 项目部署在 Android 模拟器上并启动应用，可看到如图 9.1 所示界面。

接着，将 ContentResolverDemo 项目部署在 Android 模拟器上并启动应用，在页面中输入图书编号、图书名称以及出版社名称，单击"增"按钮，将会为 ContentProviderDemo 项目中的 book.db 数据库中的 book 表添加一行数据，如图 9.2 所示。

3:00 PM

ContentProviderDemo

数据库表已存在！ContentProvider已注册！

图 9.1 启动应用，暴露数据给其他应用

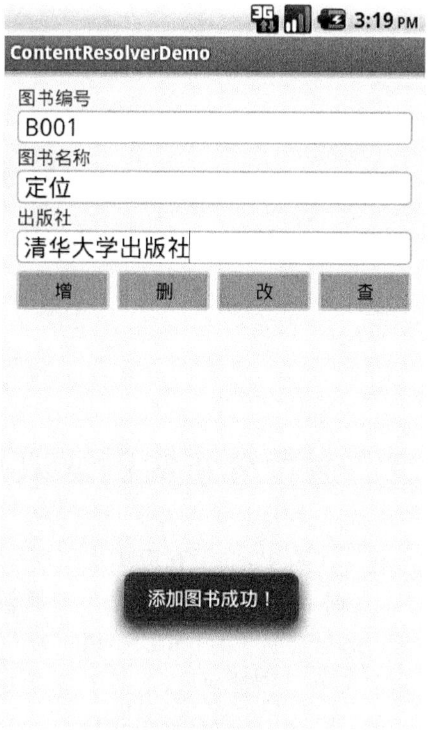

3:19 PM

ContentResolverDemo

图书编号

B001

图书名称

定位

出版社

清华大学出版社

增 删 改 查

添加图书成功！

图 9.2 新增图书信息

将图书名称修改为《中国经济论》，单击"改"按钮，出现如图 9.3 所示效果。

将图书名称以及出版社的输入框留空，只剩下图书编号写入"B001"，然后单击"查"按钮，将会看到与修改图书信息时一样的界面，这是因为根据编号查出了刚才的图书信息，然后，当单击"删"按钮时，将会提示用户删除图书成功，再次以 B001 的编号去查找图书，已经查不到任何信息了。

至此，自定义 ContentProvider 暴露自身数据以及通过 ContentResolver 操作其他应用的数据的开发流程以及项目演示已全部介绍完毕。接下来将会对以上的开发程序进行剖析，帮助读者更进一步地理解 ContentProvider 以及 ContentResolver 两个组件。

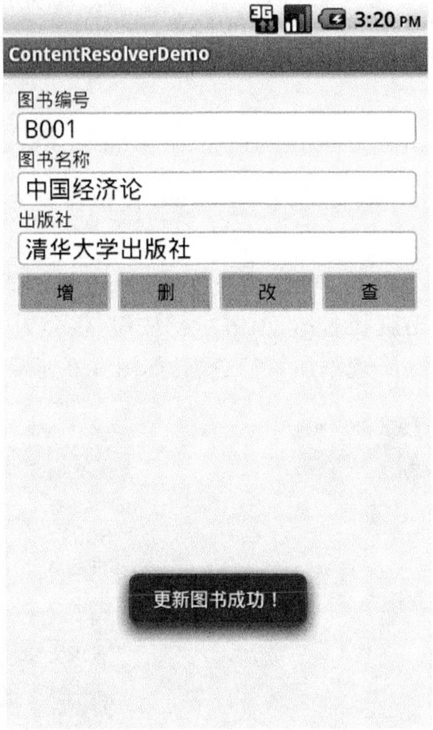

图 9.3 修改图书信息

9.2 通过分析实例认识 ContentProvider

在 9.1 节中，已经在应用中实现了自己的 ContentProvider 组件，成功将数据暴露出来，并且，实现了另一个应用程序。即通过 ContentResolver 对暴露接口的应用进行数据管理操作。通过分析 ContentProviderDemo 的程序，可以发现，实现自定义 ContentProvider 其实非常简单，只需按照以下两个步骤去操作即可：

① 定义自己的 ContentProvider 类，该类继承自 Android 提供的 ContentProvider 基类。

② 向 Android 系统注册开发好的 ContentProvider 组件，即在全局配置文件 AndroidManifest.xml 中声明该 ContentProvider，同时为其绑定一个 authorities 属性，相当于对外开放了一个访问接口地址。

在 AndroidManifest.xml 注册 ContentProvider 的方式为在< application/>元素中添加一个< provider/>子元素。代码片段如下所示：

```
<!-- 注册一个 ContentProvider -->
<provider
    android:name="cn.edu.hstc.contentproviderdemo.util.BookProvider"
    android:authorities="cn.edu.hstc.providers.bookprovider" />
```

从 BookProvider 的开发实现过程中，可以看出，BookProvider 是通过实现其基类中的以下几个方法来实现将程序内部数据暴露出来的。这些方法包括了程序对数据的 CRUD 操作。通过阅读 BookProvider 类的源码,可以总结得到表 9.1。

表 9.1 ContentProvider 需要实现的方法

方 法 名	描 述
public void onCreate()	当其他程序第一次访问 ContentProvider 时,该方法在 ContentProvider 创建后会被立即调用
public Uri insert(Uri uri, ContentValues values)	根据 Uri 插入 values 对应的数据
public int delete (Uri uri, String selection, String[] selectionArgs)	根据 Uri 删除 select 条件所匹配的全部记录
public int update(Uri uri, ContentValues values, String selection, String[] selectionArgs)	根据 Uri 修改 select 条件所匹配的全部记录
public Cursor query(Uri uri, String[] projection, String selection, String[] selectionArgs, String sortOrder) {	根据 Uri 查询出 select 条件所匹配的全部记录,其中 projection 就是一个列名列表,表示只选择指定的数据列
public String getType(Uri uri)	该方法用于返回当前 Uri 所代表的数据的 MIME 类型。如果该 Uri 对应数据可能包含多条记录,那么 MIME 类型字符串应该以 vnd. android. cursor. dir/开头;如果该 Uri 对应的数据只包含一条记录,那么返回的 MIME 类型字符串应该以 vnd. android. cursor. item/开头

通过表 9.1 中的介绍中不难发现,对于 ContentProvider 而言,Uri 是一个非常重要的概念。实际上,ContentProvider 也是通过该 Uri 向其他应用暴露了自己的数据操作接口。这个模式就像通过网址去访问网站一样。下面将对 Uri 这个概念进行介绍。

打开 ContentProviderDemo 应用源码中的 Constant 类,发现其中定义了两个可以为外界提供服务的 Uri,这两个 Uri 是通过 Android 提供的 Uri 工具类中的 parse()静态方法将字符串转换成一个真正的 Uri 地址后。例如,Constant 类中的以下代码片段:

```
//定义 ContentProvider 提供服务的两个 Uri
public final static Uri BOOKS_CONTENT_URI = Uri.parse("content://" + AUTHORITY + "/books");
public final static Uri BOOK_CONTENT_URI = Uri.parse("content://" + AUTHORITY + "/book");
```

阅读上下文代码,不难发现,这两个 Uri 地址所对应的字符串形式如下:

① content://cn. edu. hstc. providers. bookprovider/book

② content://cn. edu. hstc. providers. bookprovider/books

对于以上的两个 Uri 地址,可以分成以下三个部分:

- content://——这个部分是 Android 系统规定的,是固定的,相当于一种协议;
- cn. edu. hstc. providers. bookprovider——这个部分就是 AndroidManifest. xml 配置文件中注册 ContentProvider 时所绑定的 authorities 属性的值。系统就是通过这个部分来找到操作哪个 ContentProvider;
- books——资源部分(数据部分),这个部分是需要动态改变的,因为访问者需要访问到不同资源。

例如,如果想访问 book 数据中_id 为 1 的记录,则可以通过 content://cn. edu. hstc.

providers. bookprovider/book/1 这样的 Uri 来访问。

如果想访问 book 数据中_id 为 1 的记录的 number 字段，则可以通过 content://cn. edu. hstc. providers. bookprovider/book/1/number 这样的 Uri 来访问。

同理，content://cn. edu. hstc. providers. bookprovider/books 访问的就是 book 表的全部数据。

实际上，使用 ContentProvider 所暴露的数据并不是只来源于数据库，有时，XML 文件或网络等其他数据存储方式也能成为数据的来源，此时，假如应用需要通过 Uri 访问一份 XML 文件中的 book 父节点下的 number 子节点，那么该 Uri 内容可以为：content://cn. edu. hstc. providers. bookprovider/books/number/。

至此，相信读者已经对应用开发中所涉及的 Uri 有了一定的认知，接下来将通过对 ContentProviderDemo 中所实现的 BookProvider 类进行详细的介绍，来帮助读者更进一步掌握如何实现自定义的 ContentProvider 组件。

试想一下，现在用户正在实现 ContentProvider 类中的 delete 方法，从业务需求上来说，该方法非常可能会涉及操作删除全部数据或者符合数据库表中对应主键_id 的单条数据，那么同一个方法中是如何判断其他应用所传过来的 Uri 参数是要操作全部数据还是单条数据呢？

为了解决上述问题，Android 提供了 UriMatcher 工具类。该工具类主要提供如下两个方法：

① void addURI(String authority, String path, int code)——该方法实现向 UriMatcher 对象注册 Uri。参数 authority 以及参数 path 组合成为一个 Uri，而 code 参数代表了该 Uri 所对应的标识码。

② intmatch(Uri uri)——该方法返回指定 Uri 所对应的标识码，即 addURI 方法中所传入的 code 参数。如果在已经注册过 Uri 的 UriMatcher 对象中找不到匹配的标识码，则返回－1。

有了 UriMatcher 工具类，程序便可以提前注册好所有可能会出现的 Uri 形式，然后在具体的 CRUD 实现方法中，通过 match(Uri uri)方法的返回值判断出该 Uri 形式，然后实现不同的操作数据的代码。

那么，如何创建一个 UriMatcher 实例呢？带着这个疑问阅读一下 BookProvider 类的代码，会发现如下代码片段：

```
private static UriMatcher matcher = new UriMatcher(UriMatcher.NO_MATCH);
private static final int BOOKS = 1;
private static final int BOOK = 2;
private DBManager dbManager;
static {
    //为 UriMatcher 注册两个 Uri
    matcher.addURI(Constant.AUTHORITY, "books", BOOKS);
    matcher.addURI(Constant.AUTHORITY, "book/#", BOOK);
}
```

第一行代码中，通过 new UriMatcher(UriMatcher. NO_MATCH) 创建了一个 UriMatcher 实例，然后通过静态代码块中的代码，为该 UriMatcher 实例注册了两个 Uri 地址。即 addURI 方法中第一个参数和第二个参数组合成一个 Uri 地址，第三个参数为该 Uri 地址所对应的标识码。那么，可以得出 cn. edu. hstc. providers. bookprovider/books 所对应的标识码为 1，cn. edu. hstc. providers. bookprovider/book/ # 所对应的标识码为 2。

这里需要注意的是，# 是作为通配符存在的，这就意味着，cn. edu. hstc. providers. bookprovider/book/ # 中的 # 可以由任意字符所代替，也就是说，matcher. match(Uri. parse("cn. edu. hstc. providers. bookprovider/book/1"))和 matcher. match(Uri. parse("cn. edu. hstc. providers. bookprovider/book/2"))所返回的标识码同为 2，依此类推。

阅读 BookProvider 类中的 delete 方法代码，发现其正是通过 UriMatcher 对象的 match(Uri uri)方法来获得其他应用所传入的 Uri 地址的返回值，根据该返回值来判断该 Uri 地址的形式并执行不同的代码片段。

细心的读者也许已经发现了，delete 方法中还涉及一个 ContentUris 的工具类。该工具类的静态方法 parseId(Uri uri)返回了一个 long 类型的值。那么这个 ContentUris 工具类又是干什么的呢？

实际上，ContentUris 是一个操作 Uri 字符串的工具类，它提供了如下两个方法：

① Uri withAppendedId(Uri uri, long rowId)——用于为一个 Uri 地址加上 ID 部分。例如 Uri uri = ContentUris. withAppendedId(Uri. parse("cn. edu. hstc. providers. bookprovider/book"),1)所生成的新的 Uri 地址为 cn. edu. hstc. providers. bookprovider/book/1；

② long parseId(Uri uri)——用于从传入的 Uri 参数中解析出指定的 ID 值。例如 long id=ContentUris. parseId(Uri. parse("cn. edu. hstc. providers. bookprovider/book/1"))；所返回的 id 值为 1。

有了这个 ContentUris 工具类，便可以在 delete 方法中通过该工具类解析出其他应用所传入的 Uri 地址所带的 ID 值，然后根据该 ID 值拼接相应的数据库语句，删除指定主键所对应的单条数据。读者可以通过阅读 BookProvider 类中的 delete 方法体会上述流程。

至此，开发自定义 ContentProvider 组件的基本知识已全部介绍完毕，掌握以上知识点，读者便可以迅速开发自己的 ContentProvider 来暴露应用的数据，供其他应用访问。那么，说到其他应用对 ContentProviderDemo 所暴露出来的图书数据的访问，这涉及了 9.1 节中所实现的另一个应用程序——ContentResolverDemo。该应用通过 ContentResolver 来操作 BookProvider 所暴露的数据。接下来将对 ContentResolverDemo 中的源码进行剖析，以帮助用户更好地掌握 ContentResolver 的操作方法。

上面已经讲过，其他应用需要使用 ContentResolver 才能操作 ContentProvider 暴露的数据。阅读 ContentResolverDemo 项目的 MainActivity 类的源码可以发现，创建 ContentResolver 实例是通过 Context 提供的 getContentResolver()方法来获取的。一旦程序获得 ContentResolver 对象之后，便可以调用其 CRUD 四个方法来操作 ContentProvider 暴露的数据。ContentResolver 提供的四个方法如表 9.2 所示。

一般来说，ContentProvider 是单例模式的，当多个应用程序通过 ContentResolver 来操作 ContentProvider 提供的数据时，ContentResolver 调用的数据操作将会委托给同一个 ContentProvider 处理。

表 9.2 ContentResolver 所支持的方法

方 法 名	描 述
insert(Uri uri, ContentValues values)	向 Uri 对应的 ContentProvider 中插入 values 对应的数据
delete(Uri uri, String where, String[]selectionArgs)	删除 Uri 对应的 ContentProvider 中 where 提交匹配的数据
update(Uri uri, ContentValues values , String where, String[]selectionArgs)	更新 Uri 对应的 ContentProvider 中 where 提交匹配的数据
query(uri, projection, selection, selectionArgs, sortOrder)	查询 Uri 对应的 ContentProvider 中 where 提交匹配的数据

可以根据表 9.2 中 ContentResolver 的基本知识，阅读 ContentResolverDemo 的 MainActivity 类的源码，加深对 ContentResolver 操作 ContentProvider 的理解。

9.3 访问通讯录中的联系人和添加联系人

上面已经介绍了如何开发自己的 ContentProvider 以及使用 ContentResolver 操作 ContentProvider 所暴露的数据。实际上，Android 系统自身所带的应用已经暴露了大量的 ContentProvider，允许开发者来操作这些 ContentProvider 所暴露的数据。

回顾上面的介绍，可以知道，使用 ContentResolver 操作数据的步骤如下：

① 调用 Context 的 getContentResolver()获取 ContentResolver 的对象；

② 调用 ContentResolver 的 insert、delete、update、query 方法操作数据。

为了操作系统基本的 ContentProvider，需要了解 ContentProvider 所对应的 Uri。这里，以访问系统通讯录为例，介绍如何使用 ContentResolver 操作系统提供的 ContentProvider。

Android 系统提供了 Contacts 应用程序来管理手机联系人，而且该应用程序还对外提供了 ContentProvider，所以开发者可以开发自己的应用程序，通过 ContentResolver 来管理联系人数据。

Android 系统对联系人管理暴露的 ContentProvider 的 Uri 如下：

- ContactsContract. Contacts. CONTENT_URI——管理联系人的 Uri；
- ContactsContract. CommonDataKinds. Phone. CONTENT_URI——管理联系人的电话的 Uri；
- ContactsContract. CommonDataKinds. Email. CONTENT_URI——管理联系人的 E-mail 的 Uri。

知道了以上手机联系人应用的 ContentProvider 的 Uri 后，就可以在应用程序中使用 ContentResolver 来管理联系人的数据了。下面通过实现一个 APP 来实际操作联系人的数据。

新建一个 Android Application Project，命名为 ContactResolverDemo，填写相关必要的信息，完成创建项目。该应用的需求为：在启动界面告知用户按手机 Menu 键可以查询和添加联系人，然后通过创建 Android OptionsMenu 创建两个菜单项，其中一个用于弹出一个对话框供用户填写手机联系人的姓名、手机号码、邮件地址三个属性，然后单击对话框“确定"按钮完成联系人的添加；另一个菜单项用来供用户单击查询手机联系人列表，然后显示在一个ExpandableListView界面组件中。这两个功能都需要操作手机自带的联系人应用所暴露的ContentProvider。

由于启动界面的布局非常简单，在此不列出源码。当用户单击添加菜单项时，实际上是从启动界面跳转到添加联系人的页面，只不过在全局配置文件中将该页面设置成对话框模式，故面向用户则为一个对话框。该对话框页面的布局文件代码如下：

```
<LinearLayout xmlns:android = "http://schemas.android.com/apk/res/android"
    android:id = "@ + id/layout_dialog"
    android:layout_width = "fill_parent"
    android:layout_height = "fill_parent"
    android:orientation = "vertical"
    android:padding = "10dp" >

    <LinearLayout
        android:layout_width = "wrap_content"
        android:layout_height = "wrap_content"
        android:orientation = "horizontal" >

        <TextView
            android:layout_width = "wrap_content"
            android:layout_height = "wrap_content"
            android:text = "姓名: " />

        <EditText
            android:id = "@ + id/edit_name"
            android:layout_width = "200dp"
            android:layout_height = "wrap_content"
            android:background = "@drawable/contact_edit_edittext_normal"
            android:cursorVisible = "false"
            android:paddingLeft = "2.5dp" />
    </LinearLayout>

    <LinearLayout
        android:layout_width = "wrap_content"
        android:layout_height = "wrap_content"
        android:orientation = "horizontal"
        android:paddingTop = "10dp" >

        <TextView
            android:layout_width = "wrap_content"
            android:layout_height = "wrap_content"
            android:text = "电话: " />

        <EditText
            android:id = "@ + id/edit_phone"
            android:layout_width = "200dp"
```

```
            android:layout_height = "wrap_content"
            android:background = "@drawable/contact_edit_edittext_normal"
            android:cursorVisible = "false"
            android:paddingLeft = "2.5dp"
            android:phoneNumber = "true" />
    </LinearLayout>

    <LinearLayout
        android:layout_width = "wrap_content"
        android:layout_height = "wrap_content"
        android:orientation = "horizontal"
        android:paddingTop = "10dp" >

        <TextView
            android:layout_width = "wrap_content"
            android:layout_height = "wrap_content"
            android:text = "邮件: " />

        <EditText
            android:id = "@ + id/edit_email"
            android:layout_width = "200dp"
            android:layout_height = "wrap_content"
            android:background = "@drawable/contact_edit_edittext_normal"
            android:cursorVisible = "false"
            android:paddingLeft = "2.5dp" />
    </LinearLayout>

    <Button
        android:id = "@ + id/button"
        android:layout_width = "wrap_content"
        android:layout_height = "wrap_content"
        android:layout_gravity = "center_horizontal"
        android:layout_marginTop = "6dp"
        android:background = "@drawable/style_button"
        android:paddingBottom = "3dp"
        android:paddingLeft = "50dp"
        android:paddingRight = "50dp"
        android:paddingTop = "3dp"
        android:text = "添加" />

</LinearLayout>
```

代码文件：codes\09\9.3\ContactResolverDemo\res\layout\dialog_add_contact.xml

上面的布局从上而下放置了三个文本输入框，供用户输入姓名、手机号码、邮箱地址三个手机联系人属性；一个Button按钮供用户单击完成手机联系人的添加。该联系人添加页面对应的核心代码如下：

```
package cn.edu.hstc.contactresolverdemo.activity;
```

```
import android.app.Activity;
import android.content.ContentUris;
import android.content.ContentValues;
import android.net.Uri;
import android.os.Bundle;
import android.provider.ContactsContract.CommonDataKinds.Email;
import android.provider.ContactsContract.CommonDataKinds.Phone;
import android.provider.ContactsContract.CommonDataKinds.StructuredName;
import android.provider.ContactsContract.RawContacts;
import android.provider.ContactsContract.RawContacts.Data;
import android.view.View;
import android.view.Window;
import android.view.WindowManager;
import android.widget.Button;
import android.widget.EditText;
import android.widget.Toast;

public class AddContactActivity extends Activity {
    private EditText nameEdit, phoneEdit, emailEdit;
    private Button addBtn;

    @Override
    protected void onCreate(Bundle savedInstanceState) {
        super.onCreate(savedInstanceState);
        this.getWindow().setFlags(WindowManager.LayoutParams.FLAG_FULLSCREEN,
        WindowManager.LayoutParams.FLAG_FULLSCREEN);
        this.requestWindowFeature(Window.FEATURE_NO_TITLE);
        setContentView(R.layout.dialog_add_contact);
        initView();
    }

    private void initView() {
        nameEdit = (EditText) findViewById(R.id.edit_name);
        phoneEdit = (EditText) findViewById(R.id.edit_phone);
        emailEdit = (EditText) findViewById(R.id.edit_email);
        addBtn = (Button) findViewById(R.id.button);
        addBtn.setOnClickListener(new View.OnClickListener() {
            @Override
            public void onClick(View v) {
                //获取程序界面中的三个文本框
                String nameStr = nameEdit.getText().toString();
                String phoneStr = phoneEdit.getText().toString();
                String emailStr = emailEdit.getText().toString();
                if (nameStr.equals("") || phoneStr.equals("") || emailStr.equals("")) {
                    Toast.makeText(AddContactActivity.this, "请填写完整信息!", 3000).show();
                } else {
                    //创建一个空的 ContentValues
                    ContentValues values = new ContentValues();
```

```
                        //向 RawContacts.CONTENT_URI 执行一个空值插入,
                        //目的是获取系统返回的 rawContactId
                        Uri rawContactUri = getContentResolver().insert(RawContacts.CONTENT_
URI, values);
                        long rawContactId = ContentUris.parseId(rawContactUri);

                        values.clear();
                        values.put(Data.RAW_CONTACT_ID, rawContactId);
                        //设置内容类型
                        values.put(Data.MIMETYPE, StructuredName.CONTENT_ITEM_TYPE);
                        //设置联系人名字
                        values.put(StructuredName.GIVEN_NAME, nameStr);
                        //向联系人 URI 添加联系人名字
                         getContentResolver().insert(android.provider.ContactsContract.Data.
CONTENT_URI, values);

                        values.clear();
                        values.put(Data.RAW_CONTACT_ID, rawContactId);
                        values.put(Data.MIMETYPE, Phone.CONTENT_ITEM_TYPE);
                        //设置联系人的电话号码
                        values.put(Phone.NUMBER, phoneStr);
                        //设置电话类型
                        values.put(Phone.TYPE, Phone.TYPE_MOBILE);
                        //向联系人电话号码 URI 添加电话号码
                         getContentResolver().insert(android.provider.ContactsContract.Data.
CONTENT_URI, values);

                        values.clear();
                        values.put(Data.RAW_CONTACT_ID, rawContactId);
                        values.put(Data.MIMETYPE, Email.CONTENT_ITEM_TYPE);
                        //设置联系人的 E-mail 地址
                        values.put(Email.DATA, emailStr);
                        //设置该电子邮件的类型
                        values.put(Email.TYPE, Email.TYPE_WORK);
                        //向联系人 Email URI 添加 Email 数据
                         getContentResolver().insert(android.provider.ContactsContract.Data.
CONTENT_URI, values);

                        Toast.makeText(AddContactActivity.this, "联系人数据添加成功", 3000).
show();
                        AddContactActivity.this.finish();
                    }
                }
            });
        }
}
```

代码文件：codes\09\9.3\ContactResolverDemo\cn\edu\hstc\activity\AddContactActivity.java

上面的程序实现向手机联系人应用中插入一行空值，然后根据返回的 ID，分三步通过 ContentResolver 插入联系人的姓名、手机号码、邮箱地址三个联系人属性。

接下来，实现查询联系人并显示在 ExpandableListView 组件上，查询操作的核心代码如下：

```
private void initSearch() {
    //定义两个 List 来封装系统的联系人信息、指定联系人的电话号码、E-mail 等详情
    final ArrayList<String> names = new ArrayList<String>();
    final ArrayList<ArrayList<String>> details = new ArrayList<ArrayList<String>>();
    //使用 ContentResolver 查找联系人数据
    Cursor cursor = getContentResolver().query(ContactsContract.Contacts.CONTENT_URI,
null, null, null, null);
    //遍历查询结果，获取系统中所有联系人
    while (cursor.moveToNext()) {
        //获取联系人 ID
        String contactId = cursor.getString(cursor.getColumnIndex(ContactsContract.
Contacts._ID));
        //获取联系人的名字
        String name = cursor.getString(cursor.getColumnIndex(ContactsContract.Contacts.
DISPLAY_NAME));
        names.add(name);
        //使用 ContentResolver 查找联系人的电话号码
        Cursor phones = getContentResolver().query(ContactsContract.CommonDataKinds.Phone.
CONTENT_URI, null,
                ContactsContract.CommonDataKinds.Phone.CONTACT_ID + " = " + contactId,
null, null);
        ArrayList<String> detail = new ArrayList<String>();
        //遍历查询结果，获取该联系人的多个电话号码
        while (phones.moveToNext()) {
            //获取查询结果中电话号码列中数据
            String phoneNumber = phones.getString(phones.getColumnIndex(ContactsContract.
CommonDataKinds.Phone.NUMBER));
            detail.add("电话号码：" + phoneNumber);
        }
        phones.close();
        //使用 ContentResolver 查找联系人的 E-mail 地址
        Cursor emails = getContentResolver().query(ContactsContract.CommonDataKinds.Email.
CONTENT_URI, null,
                ContactsContract.CommonDataKinds.Email.CONTACT_ID + " = " + contactId,
null, null);
        //遍历查询结果，获取该联系人的多个 E-mail 地址
        while (emails.moveToNext()) {
            //获取查询结果中 E-mail 地址列中数据
            String emailAddress = emails.getString(emails.getColumnIndex(ContactsContract.
CommonDataKinds.Email.DATA));
            detail.add("邮件地址：" + emailAddress);
        }
        emails.close();
        details.add(detail);
    }
```

```
    cursor.close();
    //加载 result.xml 界面布局代表的视图
    View resultDialog = getLayoutInflater().inflate(R.layout.listview_result, null);
    //获取 resultDialog 中 ID 为 list 的 ExpandableListView
    ExpandableListView list = (ExpandableListView) resultDialog.findViewById(R.id.list);
    //创建一个 ExpandableListAdapter 对象
    ExpandableListAdapter adapter = new BaseExpandableListAdapter() {
        //获取指定组位置、指定子列表项处的子列表项数据
        @Override
        public Object getChild(int groupPosition, int childPosition) {
            return details.get(groupPosition).get(childPosition);
        }

        @Override
        public long getChildId(int groupPosition, int childPosition) {
            return childPosition;
        }

        @Override
        public int getChildrenCount(int groupPosition) {
            return details.get(groupPosition).size();
        }

        private TextView getTextView() {
            AbsListView.LayoutParams lp = new AbsListView.LayoutParams(ViewGroup.
            LayoutParams.FILL_PARENT, 64);
            TextView textView = new TextView(MainActivity.this);
            textView.setLayoutParams(lp);
            textView.setGravity(Gravity.CENTER_VERTICAL | Gravity.LEFT);
            textView.setPadding(56, 0, 0, 0);
            textView.setHeight(LinearLayout.LayoutParams.WRAP_CONTENT);
            textView.setTextSize(16);
            return textView;
        }

        //该方法决定每个子选项的外观
        @Override
        public View getChildView(int groupPosition, int childPosition, boolean isLastChild,
View convertView, ViewGroup parent) {
            TextView textView = getTextView();
            textView.setText(getChild(groupPosition, childPosition).toString());
            return textView;
        }

        //获取指定组位置处的组数据
        @Override
        public Object getGroup(int groupPosition) {
            return names.get(groupPosition);
        }

        @Override
```

```
        public int getGroupCount() {
            return names.size();
        }

        @Override
        public long getGroupId(int groupPosition) {
            return groupPosition;
        }

        //该方法决定每个组选项的外观
        @Override
         public View getGroupView(int groupPosition, boolean isExpanded, View convertView,
ViewGroup parent) {
            TextView textView = getTextView();
            textView.setText(getGroup(groupPosition).toString());
            return textView;
        }

        @Override
        public boolean isChildSelectable(int groupPosition, int childPosition) {
            return true;
        }

        @Override
        public boolean hasStableIds() {
            return true;
        }
    };
    //为 ExpandableListView 设置 Adapter 对象
    list.setAdapter(adapter);
    //使用对话框来显示查询结果。
    new AlertDialog.Builder(MainActivity.this).setView(resultDialog).setPositiveButton("确
定", null).show();
}
```

上面的程序使用 ContentResolver 匹配 ContactsContract.Contacts.CONTENT_URI 将系统中所有的联系人信息查询出来，接着使用 ContentResolver 匹配 ContactsContract.CommonDataKinds.Phone.CONTENT_URI 查询出指定联系人的电话信息，接着使用 ContentResolver 匹配 ContactsContract.CommonDataKinds.Email.CONTENT_URI 查询出指定联系人的 E-mail 信息。最后，程序通过一个 ExpandableListView 显示了所有的联系人信息。

需要指出的是，本应用涉及读取、添加联系人的操作，因此需要在 AndroidManifest.xml 配置文件中为该应用程序授权，授权代码片段如下：

```
<!-- 授予读联系人 ContentProvider 的权限 -->
<uses-permission android:name="android.permission.READ_CONTACTS" />
<!-- 授予写联系人 ContentProvider 的权限 -->
<uses-permission android:name="android.permission.WRITE_CONTACTS" />
```

本应用中涉及的创建 Menu 菜单项、为菜单项添加选中事件以及使用 ExpandableListView 的知识点并不是本章要介绍的内容，因此此处不再列出这部分的代码，读者可以通过阅读本书所附带的源码进行学习。

为了演示本应用，将程序部署于 Android 模拟器上，部署成功后，先打开 Android 系统自带的联系人应用 Contacts，可以看到如图 9.4 所示的效果。

接着，启动上面所编写的应用，根据界面中的提示按下手机硬键盘上的 Menu 键，可以看到如图 9.5 所示的效果。

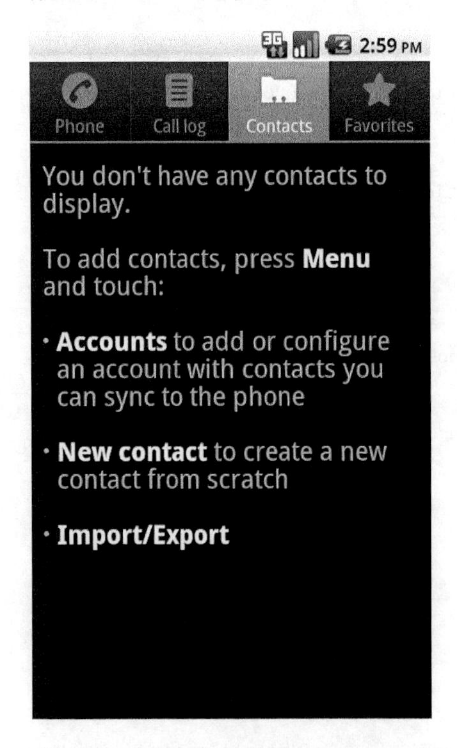

图 9.4　未添加任何联系人的 Contacts

图 9.5　Menu 键功能

单击底部菜单中的“添加”按钮，将会弹出一个新增联系人信息对话框，在该对话框中的三个文本输入框中分别填入姓名、电话、邮件地址信息，如图 9.6 所示。

单击图 9.6 中的“添加”按钮，成功将该联系人信息插入系统联系人应用中，此时，再次打开系统自带的联系人应用 Contacts，可以看到如图 9.7 所示的效果。

单击图 9.7 中的列表项 Tom，可以看到联系人的详细信息，正是图 9.6 中所输入的对应信息，说明刚才通过应用成功将联系人信息插入，其效果如图 9.8 所示。

继续通过应用插入两个联系人信息，然后单击图 9.5 中的第二个菜单项触发查询按钮功能，并展开 ExpandableListView 中的联系人信息，可以看到如图 9.9 所示的效果。

以上通过 ContentResolver 对系统自带的联系人应用所暴露的 ContentProvider 数据进行增、查两个操作。实际上，Android 系统还提供了大量的其他应用的 ContentProvider，例如为音频、视频、图片等多媒体提供的内容提供者 ContentProvider。通过以上介绍的对联系人应用的操作，读者应能够通过查阅 Android 官方文档获得该 ContentProvider 所提供的 Uri，然后进行对应数据管理操作。

图 9.6 填写联系人信息

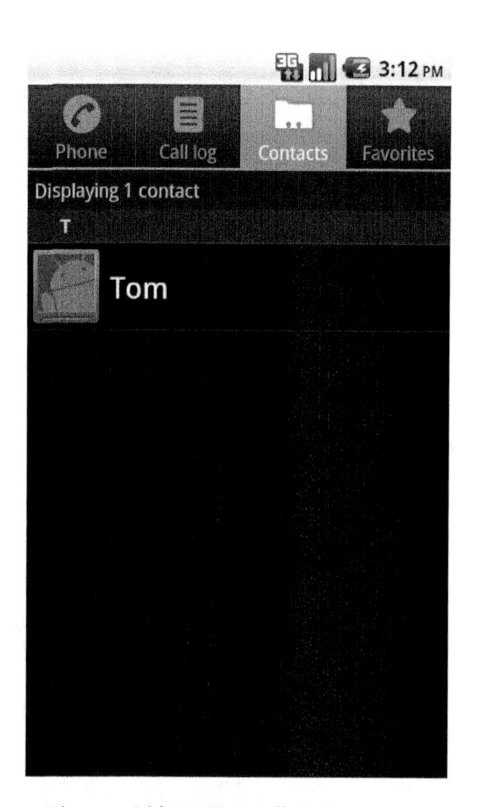

图 9.7 添加了联系人信息的 Contacts

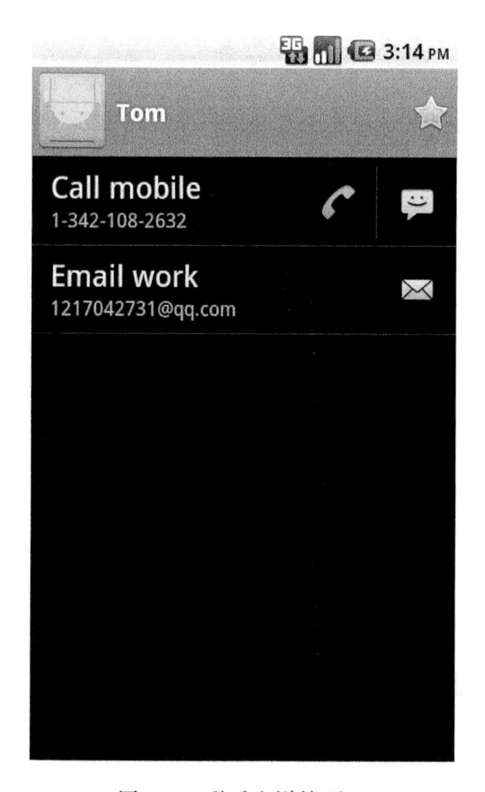

图 9.8 联系人详情页面

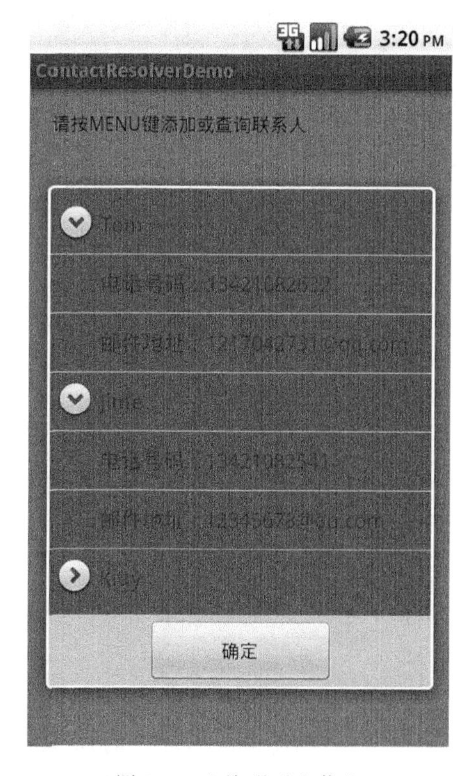

图 9.9 查询联系人信息

9.4 监听 ContentProvider 的数据改变

通过上面的介绍，已经知道，Android 提供了 ContentResolver 供用户操作应用 ContentProvider 所暴露的数据，但这属于用户主动操作 ContentProvider 中的数据，试想：如果应用中的数据变化了，如果此前 ContentResolver 已经查询过 ContentProvider 中的数据了，那么该如何通知 ContentResolver，ContentProvider 中数据的已变化呢？

细心的读者或许已经发现，在实现 9.1 节中的应用 ContentProviderDemo 中的 BookProvider 时，在实现 insert、delete、update 三个方法中，都有如下的一句代码行：getContext().getContentResolver().notifyChange(uri, null)，这行代码的作用正是用来通知注册在该 Uri 上的监听者——该 ContentProvider 所共享的数据发生了改变。只要是 ContentProvider 中的数据发生改变，程序便可以调用该行代码。

为了在应用程序中监听 ContentProvider 中数据的改变，需要利用 Android 提供的 ContentObserver 基类。接下来，将通过一个小程序来介绍如何使用 ContentObserver 来监听 ContentProvider 中数据的变化。

要实现的应用的需求为：通过实现一个 BoradcastReceiver 来拦截 Android 模拟器接收到的短信，并将短信内容显示在页面的 TextView 组件上。由于该应用界面非常简单，只居中放置了一个 TextView 控件，故不在此处给出布局文件源代码。

实现该应用核心功能的思路为自定义一个 ContentObserver 类来监听系统短信应用的 ContentProvider 数据，当监听到 ContentProvider 中有数据变化时，通过 ContentResolver 查询收件箱中数据，如果有未读消息，则将数据以消息形式发送给一个 Handler 类，在该 Handle 类中操作界面组件，显示该条消息。在本应用中，考虑将自定义的 ContentObserver 类作为应用中唯一的 Activity 中的内部类，该应用的 Activity 类实现源码如下：

```
package cn.edu.hstc.contentobserverdemo.activity;

import android.app.Activity;
import android.database.ContentObserver;
import android.database.Cursor;
import android.net.Uri;
import android.os.Bundle;
import android.os.Handler;
import android.widget.TextView;
import cn.edu.hstc.contentobserverdemo.service.R;

public class MonitorSmsActivity extends Activity {
    @Override
    protected void onCreate(Bundle savedInstanceState) {
        super.onCreate(savedInstanceState);
        setContentView(R.layout.activity_monitorsms);
        //为 content://sms 的数据变化注册监听器
```

```
        getContentResolver().registerContentObserver(Uri.parse("content://sms"), true, new
SmsObserver(MyHandler));
    }

    private Handler MyHandler = new Handler() {
        public void handleMessage(android.os.Message msg) {
            switch (msg.what) {
            case 100:
                String body = (String) msg.obj;
                TextView tv = (TextView) findViewById(R.id.txt_show_sms);
                tv.setText(body);              //将短信信息显示在界面上的TextView组件
                break;
            }
        }
    };

    //自定义ContentObserver监听器类
    private final class SmsObserver extends ContentObserver {
        private Handler handler;

        public SmsObserver(Handler handler) {
            super(handler);
            this.handler = handler;
        }

        @Override
        public void onChange(boolean selfChange) {
            //查询收件箱中的短信
            Cursor cursor = getContentResolver().query(Uri.parse("content://sms/inbox"),
null, null, null, "date asc");
            //获取用户最新收到的短信
            cursor.moveToLast();
            StringBuilder sb = new StringBuilder();
            //获取短信的发送地址
            sb.append("短信来自：").append(cursor.getString(cursor.getColumnIndex
            ("address")));
            //获取短信的内容
            sb.append("\r\n短信内容：").append(cursor.getString(cursor.getColumnIndex
("body")));
            //readType表示是否已经读
            int hasRead = cursor.getInt(cursor.getColumnIndex("read"));
            if (hasRead == 0) {                //表示短信未读
                System.out.println("短信未读：" + cursor.getString(cursor.getColumnIndex
("address")));
                handler.obtainMessage(100, sb.toString()).sendToTarget();
```

```
                }
            }
        }
    }
```

代码文件：codes\09\9.4\ContentObserverDemo\cn\edu\hstc\activity\MonitorSmsActivity.java

从上面的程序可以看出，实现监听ContentProvider数据的改变其实非常简单，只需要按照如下步骤即可。

① 自定义一个监听器类继承自ContentObserver基类。

② 重写ContentObserver基类所定义的onChange(booleanselfChange)方法，当所监听的ContentProvider的数据发生改变时，将会触发该方法。

③ 向指定Uri注册ContentObserver监听器，如下面代码所示：

```
//为 content://sms 的数据变化注册监听器
getContentResolver().registerContentObserver(Uri.parse("content://sms"), true, new SmsObserver
(MyHandler));
```

从以上代码行可以看出，注册ContentObserver监听器的方法由ContentResolver提供，实际上，该方法的模型如下：

registerContentObserver(Uri uri, boolean notifyForDescendents, ContentObserver observer)，该方法中的三个参数的说明如表9.3所示。

表9.3 注册ContentObserver监听器的方法的参数说明

参数名	描述
uri	该监听器所监听的ContentProvider的Uri
notifyForDescendents	如果该参数为true，假如注册监听的Uri为content://abc，那么Uri为content://abc/xyz、content://abc/xyz/foo的数据改变时也会触发该监听器；如果该参数为false，假如注册监听的Uri为content://abc，那么只有content://abc的数据发生改变时会触发该监听器
observer	监听器实例

在上面的程序中，当所监听的ContentProvider的数据发生改变时，也就是当系统短信应用的ContentProvider中的数据发生改变时，将会触发上面程序中所实现的ContentObserver基类的onChange(boolean selfChange)方法，因此所实现的逻辑业务代码也将放在该方法中。

将上面所实现的应用部署在Android模拟器5554中，运行效果如图9.10所示。

从图9.10中可以看出，此时的应用界面没有显示任何数据，打开另一个Android模拟器5556并启动其自带的短信应用Messaging，再向模拟器5554中发送一条短信消息“test”，发送成功后，返回模拟器5554，将可以看到如图9.11所示界面。

从图9.11的显示界面可以看出，应用已成功监听到短信应用的ContentProvider中的数据的改变，并将收件箱的Uri中的数据查询出来，如果是未读消息，则显示在界面中。

通过以上应用的实现以及演示，监听ContentProvider中的数据改变的实现流程已全部介绍完毕。

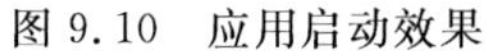

图 9.10　应用启动效果

图 9.11　成功监听短信应用的 ContentProvider

9.5　本章小结

本章主要介绍了 Android 四大组件之一的 ContentProvider 的用法和功能。应用程序可以通过 ContentProvider 暴露自身数据，其他应用通过 ContentResolver 来操作 ContentProvider 所暴露的数据。读者需要掌握的是实现自己的 ContentProvider 以及使用 ContentResolver 操作 Android 系统所提供的大量现有的 ContentProvider 或自定义的 ContentProvider。除此之外，还需要掌握使用另一个工具基类 ContentObserver，实现自定义的 ContentObserver 来监听 ContentProvider 中的数据改变。

第10章 Android的网络编程

手机的确给人们的工作和生活带来不少方便，1995 年问世的第一代(1G)模拟制式手机还只能进行语音通话；1996—1997 年出现的第二代(2G)GSM、CDMA 等数字制式手机增加了接收数据(如电子邮件或网页)等功能；而 2009 年诞生的 3G，其功能更是让人眼花缭乱，比如高速的无线宽带上网、视频通话、无线搜索、手机音乐、手机网游等。无线网络发展的速度也非常迅猛。有了无线网络的支持，人们就不必受时间和空间的限制，可随时随地进行数据交换，浏览 Internet，第一时间获得新闻、资讯。随着人们知识水平的提高，生活圈的扩大，人们更需要网络的帮助来处理一些事务，比如手机炒股、手机证券、手机银行、手机地图等。在 Android 中，掌握了网络通信便可以开发出这些优秀的网络应用。

Android 完全支持 JDK 本身的 TCP、UDP 网络通信协议；也可以使用 ServerSocket、Socket 来建立基于 TCP/IP 的网络通信；还可以使用 DatagramSocket、Datagrampacket、MulticastSocket 来建立基于 UDP 的网络通信。同时，Android 也支持 JDK 提供的 URL、URLConnection 等网络通信 API。

更重要的是，Android 内置的 HttpClient 更是可以非常方便地发送 HTTP 请求，并获取 HTTP 响应，通过 HttpClient，Android 简化了与网站之间的数据交互。接下来将会重点介绍以上提到的 Android 网络编程技术。

10.1 使用 Socket 通信搭建简易聊天室

本节将通过实现一个使用 ServerSocket 搭建的服务端以及一个使用 Socket 进行通信的客户端，形成一个简易的聊天室，向读者介绍基于 TCP 的网络通信的搭建。

需求很简单，服务端不断监听来自客户端的连接，当有一个客户端与其连接成功后，将会启动一个线程为该客户端服务。而客户端只有一个与用户交互的 Activity，在该页面中，放置一个文本输入框供用户输入文本，一个发送按钮供用户单击后将输入文本上传网络中，服务端则循环不断地从客户端中读取发送过来的数据，然后将数据发送到每个已经与服务端形成连接的客户端中，客户端将接收到的数据显示在 Activity 中的聊天记录显示框中。

从以上需求可以得出，至少需要实现一个服务端程序并正常运行。该服务端程序作为普通的 Java Application 存在。新建一个普通的 Java 项目命名为 ServerSocketDemo，新建包，在包下新建一个 Java 类，命名为 TestServerSocket。实现 TestServerSocket 中的程序代码，如下：

```
package cn.edu.hstc.serversocketdemo.util;

import java.io.IOException;
import java.net.ServerSocket;
import java.net.Socket;
import java.util.ArrayList;

public class TestServerSocket {
    //定义保存所有 Socket 的 ArrayList
    public static ArrayList<Socket> socketList = new ArrayList<Socket>();

    public static void main(String[] args) throws IOException {
        ServerSocket serverSocket = new ServerSocket(30000);
        while (true) {
            //此行代码会阻塞,将一直等待别人的连接
            Socket socket = serverSocket.accept();
            socketList.add(socket);
            //每当客户端连接后启动一条 ServerThread 线程为该客户端服务
            new Thread(new ServerThread(socket)).start();
        }
    }
}
```

代码文件：codes\10\10.1\ServerSocketDemo\cn\edu\hstc\util\TestServerSocket.java

上面的程序实现不断监听来自客户端的 Socket 连接，考虑到服务端使用传统的 BufferedReader 的 readLine 方法读取数据时，当该方法成功返回之前，线程被阻塞，程序无法继续执行，一旦连接成功，服务端将会启动一个线程为其服务。程序中代码行 new Thread(new ServerThread(socket)).start 方法就实现了该需求。因此，还需要实现一个线程类，来处理连接后的逻辑业务。该线程为 ServerThread 类所代表，该类源码如下：

```
package cn.edu.hstc.serversocketdemo.util;

import java.io.BufferedReader;
import java.io.IOException;
import java.io.InputStreamReader;
import java.io.OutputStream;
import java.net.Socket;
import java.text.SimpleDateFormat;
import java.util.Date;

public class ServerThread implements Runnable {
    //定义当前线程所处理的 Socket
    Socket socket = null;
    //该线程所处理的 Socket 所对应的输入流
    BufferedReader br = null;

    public ServerThread(Socket socket) throws IOException {
        this.socket = socket;
        //初始化该 Socket 对应的输入流
        br = new BufferedReader(new InputStreamReader(socket.getInputStream(), "utf-8"));
```

```
        }

        public void run() {
            try {
                String content = null;
                //采用循环不断从 Socket 中读取客户端发送过来的数据
                while ((content = readFromClient()) != null) {
                    //遍历 socketList 中的每个 Socket,
                    //将读到的内容向每个 Socket 发送一次
                    for (Socket socket : TestServerSocket.socketList) {
                        OutputStream os = socket.getOutputStream();
                        os.write((packMessage(content) + "\n").getBytes("utf - 8"));
                    }
                }
            } catch (IOException e) {
                e.printStackTrace();
            }
        }

        //定义读取客户端数据的方法
        private String readFromClient() {
            try {
                return br.readLine();
            }
            //如果捕捉到异常,表明该 Socket 对应的客户端已经关闭
            catch (IOException e) {
                //删除该 Socket。
                TestServerSocket.socketList.remove(socket);
            }
            return null;
        }

        private String packMessage(String content) {
            String result = null;
            SimpleDateFormat df = new SimpleDateFormat("HH:mm:ss"); //设置日期格式
            String message = content.substring(content.indexOf(": ") + 1);
                                                            //获取用户发送的真实的信息
            if (content.startsWith("5554")) {
                //获取用户发送的真实的信息
                result = "\n" + "5554 " + df.format(new Date()) + "\n" + message;
            }
            if (content.startsWith("5556")) {
                result = "\n" + "5556  " + df.format(new Date()) + "\n" + message;
            }
            return result;
        }
    }
```

代码文件: codes\10\10.1\ServerSocketDemo\cn\edu\hstc\util\ServerThread.java

上面的程序实现不断从 Socket 中读取来自客户端的数据,当有数据时,则循环将数据写入服务端与各个客户端的 Socket 连接中,并且对数据进行了重新封装,为数据添加了发

送者以及发送时间。

至此,服务端程序的全部代码已全部实现。

服务端程序中的 TestServerSocket 类中有代码行 ServerSocket serverSocket = new ServerSocket(30000),这里实例化了一个 ServerSocket 类。那么该类的作用是什么?又为何在构造器中传入数字 30000 呢?

ServerSocket 类作为 Java 中能够接收其他通信实体连接请求的类存在,ServerSocket 对象用于监听来自客户端的 Socket 连接,如果没有连接,它将一直处于等待状态。表 10.1 简单描述了 ServerSocket 中的监听方法以及各个构造器方法。

表 10.1 ServerSocket 中的监听连接请求的方法以及各个构造器方法

方法名	描述
Socket accept()	如果接收到一个客户端的 Socket 连接请求,该方法将返回一个与连接客户端 Socket 对应的 Socket,否则该方法一直处于等待状态,线程也被阻塞
ServerSocket(int port)	创建一个指定的端口号 port 的 ServerSocket,该端口号 port 的有效取值范围为 0~65 535
ServerSocket(int port, int backlog)	增加一个用来改变连接队列长度的参数 backlog
ServerSocket (int port, int backlog, InteAddress localAddress)	在机器存在多个 IP 地址的情况下,允许通过参数 localAddress 来指定将 ServerSocket 绑定到指定的 IP 地址
close()	当 ServerSocket 使用完毕后,使用该方法来关闭该 ServerSocket

在本实例中的服务端程序 TestServerSocket 类中,利用 while(true){}代码块调用 ServerSocket 的 accept 方法实现服务端不断循环监听来自客户端的连接请求。在此之前,创建了一个端口号为 30000 的 ServerSocket,由于没有指定 IP 地址,则该 ServerSocket 将会绑定到本机默认的 IP 地址。通常推荐使用 1024 以上的端口号,主要是为了避免与其他应用程序的通用端口冲突。

简易聊天室的客户端为常规的 Android 应用。应用涉及聊天,必定涉及与用户的交互,因此,APP 需要实现自定义的交互界面——Activity。该界面布局相对简单,界面顶部分左右两边分别提供一个文本输入框供用户编辑发送文本以及一个 Button 发送按钮供用户单击后将文本输入框中的内容上传网络中,在这两个控件的下方是一个被设置成不可获得焦点以及不可编辑的文本输入框,该输入框用于显示各个客户端之间的聊天记录。布局代码如下:

```
<LinearLayout xmlns:android = "http://schemas.android.com/apk/res/android"
    android:layout_width = "match_parent"
    android:layout_height = "match_parent"
    android:orientation = "vertical" >

    <LinearLayout
        android:layout_width = "fill_parent"
        android:layout_height = "wrap_content"
        android:orientation = "horizontal" >

        <EditText
```

```
            android:id = "@ + id/input"
            android:layout_width = "240px"
            android:layout_height = "wrap_content" />

        <Button
            android:id = "@ + id/send"
            android:layout_width = "fill_parent"
            android:layout_height = "wrap_content"
            android:paddingLeft = "8px"
            android:text = "发送" />
    </LinearLayout>

    <EditText
        android:id = "@ + id/show"
        android:layout_width = "fill_parent"
        android:layout_height = "fill_parent"
        android:cursorVisible = "false"
        android:editable = "false"
        android:gravity = "top" />

</LinearLayout>
```

代码文件：codes\10\10.1\SocketDemo\res\layout\activity_main.xml

编辑完界面布局，需要自定义一个 Activity 类来加载该布局并实现界面中各个控件的业务代码，该 Activity 类 Java 程序代码如下：

```
package cn.edu.hstc.socketdemo.activity;

import java.io.OutputStream;
import java.net.Socket;

import android.app.Activity;
import android.os.Bundle;
import android.os.Handler;
import android.os.Message;
import android.view.View;
import android.view.View.OnClickListener;
import android.widget.Button;
import android.widget.EditText;
import cn.edu.hstc.socketdemo.util.ClientThread;

public class MainActivity extends Activity {
    //定义界面上的两个文本框
    private EditText input, show;
    //定义界面上的一个按钮
    private Button send;
    private OutputStream os;
    private Handler handler;
    private Socket socket;

    @Override
```

```
    protected void onCreate(Bundle savedInstanceState) {
        super.onCreate(savedInstanceState);
        setContentView(R.layout.activity_main);
        input = (EditText) findViewById(R.id.input);
        send = (Button) findViewById(R.id.send);
        show = (EditText) findViewById(R.id.show);

        handler = new Handler() {
            @Override
            public void handleMessage(Message msg) {
                //如果消息来自于子线程
                if (msg.what == 0x123) {
                    //将读取的内容追加显示在文本框中
                    show.append(msg.obj.toString() + "\n");
                }
            }
        };
        try {
            socket = new Socket("192.168.1.102", 30000);
            //客户端启动 ClientThread 线程不断读取来自服务器的数据
            new Thread(new ClientThread(socket, handler)).start();
            os = socket.getOutputStream();
        } catch (Exception e) {
            e.printStackTrace();
        }
        send.setOnClickListener(new OnClickListener() {
            @Override
            public void onClick(View v) {
                try {
                    //将用户在文本框内输入的内容上传网络
                    os.write(("5556: " + input.getText().toString() + "\r\n").getBytes
("utf-8"));
                    //清空 input 文本框
                    input.setText("");
                } catch (Exception e) {
                    e.printStackTrace();
                }
            }
        });
    }
}
```

代码文件：codes\10\10.1\SocketDemo\cn\edu\hstc\socketdemo\activity\MainActivity.java

上面的程序为界面中的发送按钮绑定了事件监听器，实现按钮单击事件，在单击事件方法 onClick 方法中实现将用户在界面文本编辑框中输入的内容写入客户端与服务端连接成功后所返回的 Socket 输出流中并将文本输入框清空。由于该应用并未实现登录操作，故无法确定输出流中的数据来自于哪个客户端，为了在消息记录显示框中显示发送者，这里在写入流中的数据前面加了对应发送者的标识，可用任何字符串类型的文本作区分。比如，将程序中“5556”换成“5554”，则代表输出流中的数据是从 5554 这个模拟器中发送出来的消息。

服务端程序会根据流中内容的前面部分判断消息发送端，这部分逻辑可以通过阅读源码体会理解。

客户端程序使用了 Handler 消息传递机制，以方便地获取来自于子线程中的数据并将数据追加到聊天消息记录框中。这部分逻辑可通过阅读上面程序中实现 Handler 类的代码了解。程序中通过代码行 new Thread(new ClientThread(socket, handler)).start()将所自定义的 Handler 类与子线程关联起来。在该子线程实现类中，主要工作是读取 Socket 输入流中的内容并通过发送 handler 消息通知主界面更新数据。该线程类源码如下：

```
package cn.edu.hstc.socketdemo.util;

import java.io.BufferedReader;
import java.io.IOException;
import java.io.InputStreamReader;
import java.net.Socket;

import android.os.Handler;
import android.os.Message;

public class ClientThread implements Runnable {
    private Handler handler;
    //该线程所处理的 Socket 所对应的输入流
    BufferedReader br = null;

    public ClientThread(Socket socket, Handler handler) throws IOException {
        this.handler = handler;
        br = new BufferedReader(new InputStreamReader(socket.getInputStream()));
    }

    public void run() {
        try {
            String content = null;
            //不断读取 Socket 输入流中的内容
            while ((content = br.readLine()) != null) {
                //每当读到来自服务器的数据之后,发送消息通知程序界面显示该数据
                Message msg = new Message();
                msg.what = 0x123;
                msg.obj = content;
                handler.sendMessage(msg);
            }
        } catch (Exception e) {
            e.printStackTrace();
        }
    }
}
```

代码文件：codes\10\10.1\SocketDemo\cn\edu\hstc\socketdemo\util\ClientThread.java

至此，实现了客户端写入操作以及读取操作，那么，客户端又是如何实现自身与服务端的连接的呢？写入操作以及子线程中的读取操作所对应的该客户端与服务端的 Socket 又是从何而来的呢？实际上，在 MainActivity 类中有如下一行代码：

```
socket = new Socket("192.168.1.102", 30000);
```

客户端正是通过 Socket 的构造器来连接指定的服务端的。Socket 有两个构造器，如表 10.2 所示。

表 10.2　Socket 对应的构造器方法

构造器方法名	描　　述
Socket (InteAddress/String remoteAddress, int port)	创建连接到远程主机、远程端口的 Socket，该构造器默认使用本地主机的默认 IP 地址以及默认使用系统动态指定的端口
Socket (InteAddress/String remoteAddress, int port)	创建连接到远程主机、远程端口的 Socket，并指定本地 IP 地址和端口号，适用于本地主机中有多个 IP 地址的情况

程序通常使用 String 类型的 IP 地址来指定远程服务端。由于服务端程序中实现 ServerSocket 时并没有指定 IP 地址，这里服务端绑定的是作者 PC 的 IP 地址，服务端程序也运行在作者的 PC 中，故 Socket 构造器中的远程 IP 为作者的 PC 的 IP 地址，端口号为创建服务端 Socket 时所开辟的端口号，即 30000。

当程序执行创建 Socket 实例的代码行时，客户端便会连接到指定的服务器，让服务端程序中的 ServerSocket 的 accept 方法向下执行。于是服务端与客户端就产生了一对互相连接的 Socket。之所以说是一对，是因为 Socket 通信是基于 TCP 的，而 TCP 的简单通信示意图如图 10.1 所示。

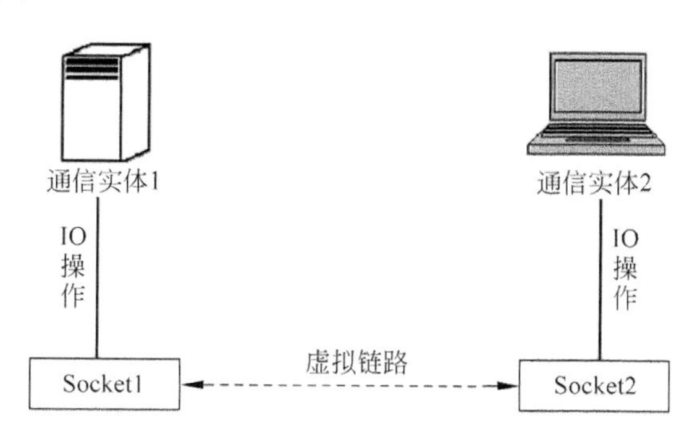

图 10.1　TCP 的通信示意图

从图 10.1 可以看出，TCP 通信的两个通信实体之间并没有服务端、客户端之分，为什么在程序中还需要对其进行区分呢？那是因为，这是两个通信实体已经建立虚拟链路之后的示意图。在两个通信实体没有建立虚拟链路之前，必须有一个通信实体先做出“主动姿态”，主动接收来自其他通信实体的连接请求。通常，将做出主动姿态的一方成为服务端，并将服务端程序运行在 PC 主机中，接收等于或大于 1 的数量的客户端的连接请求。

当客户端与服务端之间产生了对应的 Socket 之后，两者之间的关系就如图 10.1 所示。此时程序不再区分服务器、客户端，而是通过各自的 Socket 进行通信。

细心的读者在阅读了源码后会发现，不管是服务端 Socket 还是客户端 Socket，都可以通过 getInputStream 获得各自 Socket 对象所对应的输入流以及通过 getOutputStream 获得其 Socket 对象所对应的输出流。简单总结如下：

➢ InputStream getInputStream——返回 Socket 对象对应的输入流，程序通过该输入流从 Socket 中读取数据；
➢ OutputStream getOutputStream——返回 Socket 对象对应的输出流，程序通过该输出流向 Socket 中写入数据。

运行服务端，即运行 Java 应用 ServerSocketDemo，然后打开两个模拟器，将客户端程序 SocketDemo 部署在两个模拟器 5554 和 5556 中，然后进行聊天对话，可以看到如图 10.2 和图 10.3 所示效果。

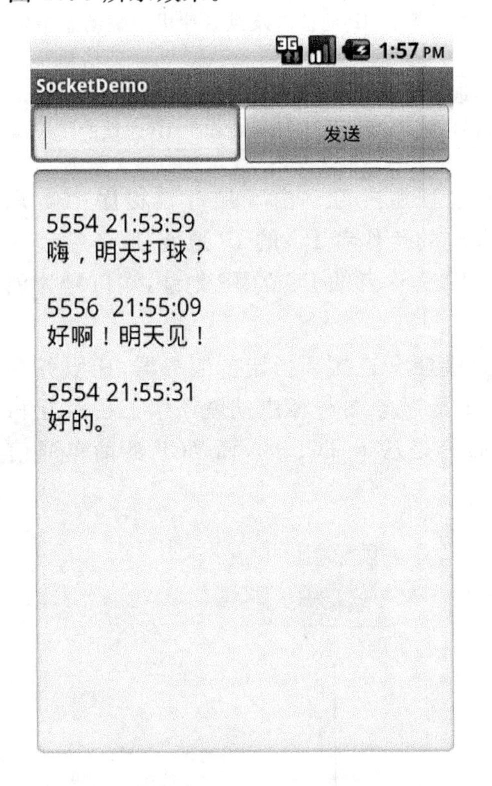

图 10.2　5554 客户端

图 10.3　5556 客户端

10.2 使用 HTTP 访问网络

Android 对 HTTP(超文本传输协议)也提供了很好的支持，在 Android 中，使用 http 访问网络有两种方式：一种是使用 HttpURLConnection；另一种是使用 HttpClient。接下来，先介绍 Android 如何使用 HttpURLConnection 访问网络资源。

10.2.1 使用 HttpURLConnection

使用 HttpURLConnection 能实现简单的 URL 请求、响应功能。HttpURLConnection 继承了 URLConnection，可以发送和接收任何类型和长度的数据，且不用预先知道数据流的长度，可以设置请求方式为 GET 或 POST，可以设置超时时间。

下面通过一个实例来介绍 Android 利用 HttpURLConnection 访问网页、获取网络图片。

首先介绍一下该实例的需求，该实例主要实现两个目的：一个是访问百度首页，获取其返回的 html 字符串；第二个是访问给定的 URL，下载图片并显示出来。该应用只需实现一个 Activity，其对应布局为：界面上方放置一个帧布局，该布局中放置一个 TextView 组件用于显示访问百度首页后所获取的 html 字符串，放置一个 ImageView 组件用于显示从网络上下载的图片，界面底部从左到右放置两个 Button 按钮，其中一个用于触发访问百度首页，另一个用于触发访问下载网络图片。该界面布局比较简单，故在此不进行布局代码的展示。

界面布局所对应的 Activity 程序代码如下所示：

```
package cn.edu.hstc.httpurlconnectiondemo.activity;

import java.io.BufferedReader;
import java.io.IOException;
import java.io.InputStream;
import java.io.InputStreamReader;
import java.net.HttpURLConnection;
import java.net.MalformedURLException;
import java.net.URL;

import android.app.Activity;
import android.graphics.Bitmap;
import android.graphics.BitmapFactory;
import android.os.AsyncTask;
import android.os.Bundle;
import android.view.View;
import android.view.ViewGroup;
import android.widget.Button;
import android.widget.ImageView;
import android.widget.ProgressBar;
import android.widget.RelativeLayout;
import android.widget.TextView;

public class MainActivity extends Activity {
    private Button visitWebBtn;
    private Button downImgBtn;
    private TextView showTextView;
    private ImageView showImageView;
    private String resultString = "";
    private ProgressBar progressBar;
    private ViewGroup viewGroup;

    @Override
    protected void onCreate(Bundle savedInstanceState) {
        super.onCreate(savedInstanceState);
        setContentView(R.layout.activity_main);
        initView();
```

```
        visitWebBtn.setOnClickListener(new View.OnClickListener() {
            @Override
            public void onClick(View v) {
                showImageView.setVisibility(View.GONE);
                showTextView.setVisibility(View.VISIBLE);
                Thread visitBaiduThread = new Thread(new VisitWebRunnable());
                visitBaiduThread.start();
                try {
                    visitBaiduThread.join();
                    if (!resultString.equals("")) {
                        showTextView.setText(resultString);
                    }
                } catch (InterruptedException e) {
                    e.printStackTrace();
                }
            }
        });

        downImgBtn.setOnClickListener(new View.OnClickListener() {
            @Override
            public void onClick(View v) {
                showImageView.setVisibility(View.VISIBLE);
                showTextView.setVisibility(View.GONE);
                String imgUrl = "http://img.shixiu.net/file/news/fjxw/9909e2eb3cc173f10-
7afe2b53a5a2c95.jpg";
                new DownImgAsyncTask().execute(imgUrl);
            }
        });
    }

    private void initView() {
        showTextView = (TextView) findViewById(R.id.textview_show);
        showImageView = (ImageView) findViewById(R.id.imagview_show);
        downImgBtn = (Button) findViewById(R.id.btn_download_img);
        visitWebBtn = (Button) findViewById(R.id.btn_visit_web);
    }

    /**
     * 获取指定 URL 的响应字符串
     */
    private String getURLResponse(String urlString) {
        HttpURLConnection conn = null;                      //连接对象
        InputStream is = null;
        String resultData = "";
        try {
            URL url = new URL(urlString);                   //URL 对象
            conn = (HttpURLConnection) url.openConnection();  //使用 URL 打开一个链接
            conn.setDoInput(true);                          //允许输入流,即允许下载
            conn.setDoOutput(true);                         //允许输出流,即允许上传
            conn.setUseCaches(false);                       //不使用缓冲
```

```
            conn.setRequestMethod("GET");                //使用 get 请求
            is = conn.getInputStream();                  //获取输入流,此时才真正建立链接
            InputStreamReader isr = new InputStreamReader(is);
            BufferedReader bufferReader = new BufferedReader(isr);
            String inputLine = "";
            while ((inputLine = bufferReader.readLine()) != null) {
                resultData += inputLine + "\n";
            }
        } catch (MalformedURLException e) {
            e.printStackTrace();
        } catch (IOException e) {
            e.printStackTrace();
        } finally {
            if (is != null) {
                try {
                    is.close();
                } catch (IOException e) {
                    e.printStackTrace();
                }
            }
            if (conn != null) {
                conn.disconnect();
            }
        }
        return resultData;
    }

    /**
     * 从指定 URL 获取图片
     */
    private Bitmap getImageBitmap(String url) {
        URL imgUrl = null;
        Bitmap bitmap = null;
        try {
            imgUrl = new URL(url);
            HttpURLConnection conn = (HttpURLConnection) imgUrl.openConnection();
            conn.setDoInput(true);
            conn.connect();
            InputStream is = conn.getInputStream();
            bitmap = BitmapFactory.decodeStream(is);
            is.close();
        } catch (MalformedURLException e) {
            e.printStackTrace();
        } catch (IOException e) {
            e.printStackTrace();
        }
        return bitmap;
    }

    class VisitWebRunnable implements Runnable {
        @Override
```

```
        public void run() {
            String data = getURLResponse("http://www.baidu.com/");
            resultString = data;
        }
    }

    class DownImgAsyncTask extends AsyncTask<String, Void, Bitmap> {
        @Override
        protected void onPreExecute() {
            super.onPreExecute();
            showImageView.setImageBitmap(null);
            showProgressBar();                          //显示进度条提示框

        }

        @Override
        protected Bitmap doInBackground(String... params) {
            Bitmap b = getImageBitmap(params[0]);
            return b;
        }

        @Override
        protected void onPostExecute(Bitmap result) {
            super.onPostExecute(result);
            if (result != null) {
                dismissProgressBar();
                showImageView.setImageBitmap(result);
            }
        }
    }

    /**
     * 在母布局中间显示进度条
     */
    private void showProgressBar() {
        progressBar = new ProgressBar(this, null, android.R.attr.progressBarStyleLarge);
        RelativeLayout.LayoutParams params = new RelativeLayout.LayoutParams(
                ViewGroup.LayoutParams.WRAP_CONTENT, ViewGroup.LayoutParams.WRAP_
CONTENT);
        params.addRule(RelativeLayout.CENTER_IN_PARENT, RelativeLayout.TRUE);
        progressBar.setVisibility(View.VISIBLE);
        //Context context = getApplicationContext();
        viewGroup = (ViewGroup) findViewById(R.id.parent_view);
        viewGroup.addView(progressBar, params);
    }

    /**
     * 隐藏进度条
     */
    private void dismissProgressBar() {
        if (progressBar != null) {
```

```
                progressBar.setVisibility(View.GONE);
                viewGroup.removeView(progressBar);
                progressBar = null;
            }
        }
}
```

代码文件：codes\10\10.2\HttpURLConnectionDemo\cn\edu\hstc\activity\MainActivity.java

最后，需要在 AndroidManifest.xml 中添加访问网络的权限：

```
<uses-permission android:name = "android.permission.INTERNET" />
```

从上面的 MainActivity 类的实现代码可以看出，使用 HttpURLConnection 的步骤是先通过 URL 的 openConnection 实例化 HttpURLConnection 对象，然后设置参数，注意此时并没有发生连接。真正发生连接是在获得流时，即代码文件中的 conn.getInputStream 这一代码行，这一点与 TCP Socket 是一样的。

将该应用部署在 Android 模拟器上，分别单击界面中的访问百度的按钮以及下载图片按钮，运行效果如图 10.4 和图 10.5 所示。

```
3:36 PM
HttpURLConnectionDemo
<!DOCTYPE html>
<html><!--STATUS OK--><head><meta http-equiv="Content-Type" content="text/html; charset=utf-8" /><meta http-equiv="Cache-control" content="no-cache" /><meta name="viewport" content="width=device-width,minimum-scale=1.0,maximum-scale=1.0,user-scalable=no"/><style type="text/css">body {margin: 0;text-align: center;font-size: 14px;font-family: Arial, Helvetica,LiHei Pro Medium;color: #262626;}form {position: relative;margin: 12px 15px 91px;height: 41px;}img {border: 0}.wordWrap{margin-right: 85px;}#word {background-color: #FFF;border: 1px solid #6E6E6E;color: #000;font-size: 18px;height: 27px;padding: 6px;width: 100%;-webkit-appearance: none;-webkit-border-radius: 0;border-radius: 0;}.bn {background-color: #F5F5F5;border: 1px solid #787878;font-size: 16px;font-weight: bold;height: 41px;letter-spacing: -1px;line-height: 41px;padding: 0;position: absolute;right: 0;text-align: center; top: 0;width: 72px;-webkit-appearance: none;-webkit-box-sizing: border-box;box-sizing: border-box;-webkit-border-radius: 0;border-
访问百度　　下载图片
```

图 10.4　访问百度首页

图 10.5　下载网络图片

10.2.2　使用 HttpClient 接口

10.2.1 节介绍了通过标准 Java 接口来实现 Android 应用的联网操作。前面只是简单地进行了网络的访问，但是在实际开发中，可能会运用到更复杂的联网操作，例如处理

Session、Cookie 等细节问题，Apache 开源组织提供了 HttpClient 项目，它对 java. net 中的类做了封装和抽象，更适合在 Android 上开发网络应用。简单来说，HttpClient 就是一个增强版的 HttpURLConnection，HttpURLConnection 能做的事情 HttpClient 都可以胜任，并且 HttpClient 提供了 HttpURLConnection 所没有提供的部分功能。Android 已经成功地集成了 HttpClient，这意味着开发人员可以直接在 Android 应用中使用 HttpClient 来访问提交请求、接收响应。要使用 HttpClient，需要了解下面一些接口和类。

1. ClientConnectionManager 接口

ClientConnectionManager 是客户端连接管理器接口，它提供几个抽象方法，如表 10.3 所示。

表 10.3　ClientConnectionManager 的抽象方法

方 法 名	描　述
voidcloseExpiredConnections	关闭所有无效、超时的连接
voidcloseIdleConnections(long idletime, TimeUnit tunit)	关闭空闲的连接
voidreleaseConnection (HttpClientConnection conn, Object newState, long validDuration, TimeUnit timeUnit)	释放一个连接
ConnectionRequest requestConnection(HttpRoute route, Object state)	请求一个新的连接
voidshutdown()	关闭管理器并释放资源

2. DefaultHttpClient

DefaultHttpClient 是默认的一个 HTTP 客户端，可以使用它创建一个 HTTP 连接。代码如下：

```
HttpClient httpClient = new DefaultHttpClient();
```

3. HttpResponse

HttpResponse 是一个 HTTP 连接响应，当执行一个 HTTP 连接后，就会返回一个 HttpResponse，可以通过 HttpResponse 获得一些响应的信息。下面是请求一个 HTTP 连接并获得该请求是否成功的代码。

```
HttpResponse httpResponse = httpClient.execute(httpRequest);
If (httpResponse.getStatusLine().getStatusCode() == HttpStatus.SC_OK) {
    //连接成功
}
```

通过上面几个类和接口的了解，下面运用上面的知识来实现一个从服务端获取通讯录信息的应用，该应用的代码实现中既使用了 Get 的请求方式，也使用了 Post 的请求方式。

首先，为了在项目中更方便地使用 HttpClient 以及使该部分代码可以重用，需要实现一个名为 HttpUtil 的工具类，在该工具类中实现了 Android 使用 HttpClient 访问网络获取数据的核心程序，该工具类源码如下：

```
package cn.edu.hstc.httpclientdemo.util;

import java.io.IOException;
import java.util.List;

import org.apache.http.HttpEntity;
import org.apache.http.HttpResponse;
import org.apache.http.HttpStatus;
import org.apache.http.NameValuePair;
import org.apache.http.client.ClientProtocolException;
import org.apache.http.client.HttpClient;
import org.apache.http.client.entity.UrlEncodedFormEntity;
import org.apache.http.client.methods.HttpGet;
import org.apache.http.client.methods.HttpPost;
import org.apache.http.impl.client.DefaultHttpClient;
import org.apache.http.util.EntityUtils;

public class HttpUtil {
    public static final String BASE_URL = "http://192.168.1.100:8080/ydContacts/servlet/";

    public static String getDataByGet(String url) {
        String resultString = "";
        try {
            //得到HttpClient对象
            HttpClient getClient = new DefaultHttpClient();
            //得到HttpGet对象
            HttpGet request = new HttpGet(url);
            //客户端使用Get方式执行请求,获得服务器端的回应response
            HttpResponse response = getClient.execute(request);
            //判断请求是否成功
            if (response.getStatusLine().getStatusCode() == HttpStatus.SC_OK) {
                resultString = EntityUtils.toString(response.getEntity());
            } else {
                resultString = "请求服务器发生错误!";
            }
        } catch (ClientProtocolException e) {
            System.out.println(e.getMessage().toString());
            resultString = "请求服务器异常!";
        } catch (IOException e) {
            System.out.println(e.getMessage().toString());
            resultString = "请求服务器异常!";
        } catch (Exception e) {
            System.out.println(e.getMessage().toString());
            resultString = "请求服务器异常!";
        }
        return resultString;
    }

    public static String getDataByPost(String url, List<NameValuePair> params) {
        String resultString = "";
        try {
```

```
                HttpClient client = new DefaultHttpClient();
                HttpPost request = new HttpPost(url);
                //实例化 UrlEncodedFormEntity 对象并设置字符集
                HttpEntity entity = new UrlEncodedFormEntity(params, "gb2312");
                //使用 HttpPost 对象来设置 UrlEncodedFormEntity 的 Entity
                request.setEntity(entity);
                HttpResponse response = client.execute(request);
                if (response.getStatusLine().getStatusCode() == HttpStatus.SC_OK) {
                    resultString = EntityUtils.toString(response.getEntity());
                } else {
                    resultString = "请求服务器发生错误!";
                }
            } catch (ClientProtocolException e) {
                System.out.println(e.getMessage().toString());
                resultString = "请求服务器异常!";
            } catch (IOException e) {
                System.out.println(e.getMessage().toString());
                resultString = "请求服务器异常!";
            } catch (Exception e) {
                System.out.println(e.getMessage().toString());
                resultString = "请求服务器异常!";
            }
            return resultString;
        }
    }
```

代码文件：codes\10\10.2\HttpClientDemo\cn\edu\hstc\util\HttpUtil.java

上面的程序实现了使用 HttpClient 发送 Get 请求以及 Post 请求，并且获取响应数据的核心功能，因此以上工具类的代码实现是本节的掌握重点。

由于该应用中获取的通讯录数据是显示在一个 ListView 组件中，并且该 ListView 需要实现下拉刷新以及上拉加载更多的功能，达到 ListView 分页效果，因此需要自定义自己的 ListView 基类，而该基类集成了下拉刷新以及上拉加载更多的功能，在其他 Android 项目中，可以对该基类进行复用，该基类代码如下：

```
package cn.edu.hstc.httpclientdemo.util;

import android.content.Context;
import android.util.AttributeSet;
import android.view.MotionEvent;
import android.view.View;
import android.view.ViewTreeObserver.OnGlobalLayoutListener;
import android.view.animation.DecelerateInterpolator;
import android.widget.AbsListView;
import android.widget.AbsListView.OnScrollListener;
import android.widget.ListAdapter;
import android.widget.ListView;
import android.widget.RelativeLayout;
import android.widget.Scroller;
import android.widget.TextView;
import cn.edu.hstc.httpclientdemo.activity.R;
```

```
public class RefreshListView extends ListView implements OnScrollListener {
    private float mLastY = -1;
    private Scroller mScroller;
    private OnScrollListener mScrollListener;
    private ListViewListener mListViewListener;

    private ListViewHeader mHeaderView;                    //header view
    private RelativeLayout mHeaderViewContent;
    private TextView mHeaderTimeView;
    private int mHeaderViewHeight;                         //header view's height
    private boolean mEnablePullRefresh = true;
    private boolean mPullRefreshing = false;               //is refreashing.

    private ListViewFooter mFooterView;                    //footer view
    private boolean mEnablePullLoad;
    private boolean mPullLoading;
    private boolean mEnableAutoload = true;

    private int mTotalItemCount;

    private int mScrollBack;
    private final static int SCROLLBACK_HEADER = 0;

    private final static int SCROLL_DURATION = 400;       //scroll back duration
    private final static float OFFSET_RADIO = 1.8f;       //support iOS like pull
    private int total;                                     //列表项总数

    public RefreshListView(Context context) {
        super(context);
        initWithContext(context);
    }

    public RefreshListView(Context context, AttributeSet attrs) {
        super(context, attrs);
        initWithContext(context);
    }

    public RefreshListView(Context context, AttributeSet attrs, int defStyle) {
        super(context, attrs, defStyle);
        initWithContext(context);
    }

    private void initWithContext(Context context) {
        mScroller = new Scroller(context, new DecelerateInterpolator());
        super.setOnScrollListener(this);

        mHeaderView = new ListViewHeader(context);
         mHeaderViewContent = (RelativeLayout) mHeaderView.findViewById(R.id.xlistview_
header_content);
        mHeaderTimeView = (TextView) mHeaderView.findViewById(R.id.xlistview_header_time);
```

```
        addHeaderView(mHeaderView);

        mFooterView = new ListViewFooter(context);
        mFooterView.getLoadmore().setOnClickListener(new OnClickListener() {
            @Override
            public void onClick(View v) {
                startLoadMore();
            }
        });

        mHeaderView.getViewTreeObserver().addOnGlobalLayoutListener(new OnGlobalLayoutListener() {
            @Override
            public void onGlobalLayout() {
                mHeaderViewHeight = mHeaderViewContent.getHeight();
                getViewTreeObserver().removeGlobalOnLayoutListener(this);
            }
        });

    }

    public void setTotal(int total) {
        this.total = total;
    }

    public boolean pullRefreshing() {
        return mPullRefreshing;
    }

    public boolean pullLoading() {
        return mPullLoading;
    }

    @Override
    public void setAdapter(ListAdapter adapter) {
        addFooterView(mFooterView);
        super.setAdapter(adapter);
        if (adapter.getCount() == total) {
            removeFooterView(mFooterView);
        }
    }

    public void setPullRefreshEnable(boolean enable) {
        mEnablePullRefresh = enable;
        if (!mEnablePullRefresh) {                    //disable, hide the content
            mHeaderViewContent.setVisibility(View.INVISIBLE);
        } else {
            mHeaderViewContent.setVisibility(View.VISIBLE);
        }
    }

    public void setPullLoadEnable(boolean enable) {
```

```
        mEnablePullLoad = enable;
        mFooterView.setOnClickListener(null);
        if (!mEnablePullLoad) {
            mFooterView.hide();
        } else {
            mPullLoading = false;
            mFooterView.show();
            if (!mEnableAutoload) {
                mFooterView.setState(ListViewFooter.STATE_BUTTON);
            } else {
                mFooterView.setState(ListViewFooter.STATE_NORMAL);
            }
        }
    }

    public void setAutoLoad(boolean enable) {
        mEnableAutoload = enable;
        if (!mPullLoading) {
            if (!mEnableAutoload) {
                mFooterView.setState(ListViewFooter.STATE_BUTTON);
            } else {
                mFooterView.setState(ListViewFooter.STATE_NORMAL);
            }
        }
    }

    public void stopRefresh() {
        if (mPullRefreshing == true) {
            mPullRefreshing = false;
            resetHeaderHeight();
        }
    }

    public void stopLoadMore() {
        if (mPullLoading == true) {
            mPullLoading = false;
            if (!mEnableAutoload) {
                mFooterView.setState(ListViewFooter.STATE_BUTTON);
            } else {
                mFooterView.setState(ListViewFooter.STATE_NORMAL);
            }
        }
    }

    public void setRefreshTime(String time) {
        mHeaderTimeView.setText(time);
    }

    public String getRefreshTime() {
        return mHeaderTimeView.getText().toString();
    }
```

```
private void invokeOnScrolling() {
    if (mScrollListener instanceof OnXScrollListener) {
        OnXScrollListener l = (OnXScrollListener) mScrollListener;
        l.onXScrolling(this);
    }
}

private void updateHeaderHeight(float delta) {
    mHeaderView.setVisiableHeight((int) delta + mHeaderView.getVisiableHeight());
    if (mEnablePullRefresh && !mPullRefreshing) {      //未处于刷新状态,更新箭头
        if (mHeaderView.getVisiableHeight() > mHeaderViewHeight) {
            mHeaderView.setState(ListViewHeader.STATE_READY);
        } else {
            mHeaderView.setState(ListViewHeader.STATE_NORMAL);
        }
    }
    setSelection(0);                                   //scroll to top each time
}

private void resetHeaderHeight() {
    int height = mHeaderView.getVisiableHeight();
    if (height == 0)                                   //not visible.
        return;
    //refreshing and header isn't shown fully. do nothing.
    if (mPullRefreshing && height <= mHeaderViewHeight) {
        return;
    }
    int finalHeight = 0;                          //default: scroll back to dismiss header.
    //is refreshing, just scroll back to show all the header.
    if (mPullRefreshing && height > mHeaderViewHeight) {
        finalHeight = mHeaderViewHeight;
    }
    mScrollBack = SCROLLBACK_HEADER;
    mScroller.startScroll(0, height, 0, finalHeight - height, SCROLL_DURATION);
    //trigger computeScroll
    invalidate();
}

private void startLoadMore() {
    mPullLoading = true;
    mFooterView.setState(ListViewFooter.STATE_LOADING);
    if (mListViewListener != null) {
        mListViewListener.onLoadMore();
    }
}

@Override
public boolean onTouchEvent(MotionEvent ev) {
    if (mLastY == -1) {
        mLastY = ev.getRawY();
```

```
        }

        switch (ev.getAction()) {
        case MotionEvent.ACTION_DOWN:
            mLastY = ev.getRawY();
            break;
        case MotionEvent.ACTION_MOVE:
            final float deltaY = ev.getRawY() - mLastY;
            mLastY = ev.getRawY();
            if (getFirstVisiblePosition() == 0 && (mHeaderView.getVisiableHeight() > 0 ||
deltaY > 0)) {
                //the first item is showing, header has shown or pull down.
                updateHeaderHeight(deltaY / OFFSET_RADIO);
                invokeOnScrolling();
            }
            break;
        default:
            mLastY =-1;                                    //reset
            if (getFirstVisiblePosition() == 0) {
                //invoke refresh
                if (mEnablePullRefresh && mHeaderView.getVisiableHeight() > mHeaderViewHeight) {
                    mHeaderView.setState(ListViewHeader.STATE_REFRESHING);
                    if (mListViewListener != null && !mPullRefreshing) {
                        mPullRefreshing = true;
                        mListViewListener.onRefresh();
                    }
                }
                resetHeaderHeight();
            }
            break;
        }
        return super.onTouchEvent(ev);
    }

    @Override
    public void computeScroll() {
        if (mScroller.computeScrollOffset()) {
            if (mScrollBack == SCROLLBACK_HEADER) {
                mHeaderView.setVisiableHeight(mScroller.getCurrY());
            } else {
                mFooterView.setBottomMargin(mScroller.getCurrY());
            }
            postInvalidate();
            invokeOnScrolling();
        }
        super.computeScroll();
    }

    @Override
    public void setOnScrollListener(OnScrollListener l) {
        mScrollListener = l;
```

```
    }

    @Override
    public void onScrollStateChanged(AbsListView view, int scrollState) {
        if (mScrollListener != null) {
            mScrollListener.onScrollStateChanged(view, scrollState);
        }

        if (scrollState == SCROLL_STATE_TOUCH_SCROLL) {
            int totalCount = 0;                          //内容列表数
            if (getFooterViewsCount() > 0)
                totalCount = mTotalItemCount - 2;
            else
                totalCount = mTotalItemCount - 1;
            if (mTotalItemCount > 0 && totalCount < total) {
                if (getFooterViewsCount() == 0) {
                    addFooterView(mFooterView);
                }
            } else {
                if (getFooterViewsCount() > 0)
                    removeFooterView(mFooterView);
            }
        }
    }

    @Override
    public void onScroll(AbsListView view, int firstVisibleItem, int visibleItemCount, int
totalItemCount) {
        //send to user's listener
        mTotalItemCount = totalItemCount;
        if (mScrollListener != null) {
            mScrollListener.onScroll(view, firstVisibleItem, visibleItemCount, totalItemCount);
        }

          if (getLastVisiblePosition() == mTotalItemCount - 1 && !mPullLoading &&
getFooterViewsCount() > 0) {
            if (mEnableAutoload) {
                startLoadMore();
            }
        }
    }

    public void setListViewListener(ListViewListener l) {
        mListViewListener = l;
    }

    public interface OnXScrollListener extends OnScrollListener {
        public void onXScrolling(View view);
    }

    public interface ListViewListener {
```

```
        public void onRefresh();

        public void onLoadMore();
    }
}
```

代码文件：codes\10\10.2\HttpClientDemo\cn\edu\hstc\util\ RefreshListView.java

上面 RefreshListView 类继承自 ListView 基类，实现 OnScrollListener 接口，实现自定义的下拉刷新以及上拉加载更多的功能。当用户下拉 ListView 列表时，将会在 ListView 的顶部出现"下拉可以刷新"的提示，当用户松手时，将会在顶部出现"正在加载…"的提示；当用户上拉 ListView 列表时，将会在底部提示"正在加载中…"，因此，需要对 ListView 的顶部和底部进行特殊处理，分别新建两个布局类来代表该 ListView 的顶部和底部，顶部的代表类为 ListViewHeader，该类源码如下：

```
package cn.edu.hstc.httpclientdemo.util;

import android.content.Context;
import android.util.AttributeSet;
import android.view.Gravity;
import android.view.LayoutInflater;
import android.view.View;
import android.view.animation.Animation;
import android.view.animation.RotateAnimation;
import android.widget.ImageView;
import android.widget.LinearLayout;
import android.widget.ProgressBar;
import android.widget.TextView;
import cn.edu.hstc.httpclientdemo.activity.R;

public class ListViewHeader extends LinearLayout {
    private LinearLayout mContainer;
    private ImageView mArrowImageView;
    private ProgressBar mProgressBar;
    private TextView mHintTextView;
    private int mState = STATE_NORMAL;

    private Animation mRotateUpAnim;
    private Animation mRotateDownAnim;

    private final int ROTATE_ANIM_DURATION = 180;

    public final static int STATE_NORMAL = 0;
    public final static int STATE_READY = 1;
    public final static int STATE_REFRESHING = 2;

    public ListViewHeader(Context context) {
        super(context);
        initView(context);
    }
```

```
    public ListViewHeader(Context context, AttributeSet attrs) {
        super(context, attrs);
        initView(context);
    }

    private void initView(Context context) {
        //初始情况,设置下拉刷新 view 高度为 0
        LinearLayout.LayoutParams lp = new LinearLayout.LayoutParams(android.view.
ViewGroup.LayoutParams.MATCH_PARENT, 0);
        mContainer = (LinearLayout) LayoutInflater.from(context).inflate(R.layout.
xlistview_header, null);
        addView(mContainer, lp);
        setGravity(Gravity.BOTTOM);

        mArrowImageView = (ImageView) findViewById(R.id.xlistview_header_arrow);
        mHintTextView = (TextView) findViewById(R.id.xlistview_header_hint_textview);
        mProgressBar = (ProgressBar) findViewById(R.id.xlistview_header_progressbar);

        mRotateUpAnim = new RotateAnimation(0.0f, -180.0f, Animation.RELATIVE_TO_SELF,
0.5f, Animation.RELATIVE_TO_SELF, 0.5f);
        mRotateUpAnim.setDuration(ROTATE_ANIM_DURATION);
        mRotateUpAnim.setFillAfter(true);
        mRotateDownAnim = new RotateAnimation(-180.0f, 0.0f, Animation.RELATIVE_TO_SELF,
0.5f, Animation.RELATIVE_TO_SELF, 0.5f);
        mRotateDownAnim.setDuration(ROTATE_ANIM_DURATION);
        mRotateDownAnim.setFillAfter(true);
    }

    public void setState(int state) {
        if (state == mState)
            return;

        if (state == STATE_REFRESHING) {                    //显示进度
            mArrowImageView.clearAnimation();
            mArrowImageView.setVisibility(View.INVISIBLE);
            mProgressBar.setVisibility(View.VISIBLE);
        } else {                                            //显示箭头图片
            mArrowImageView.setVisibility(View.VISIBLE);
            mProgressBar.setVisibility(View.INVISIBLE);
        }

        switch (state) {
        case STATE_NORMAL:
            if (mState == STATE_READY) {
                mArrowImageView.startAnimation(mRotateDownAnim);
            }
            if (mState == STATE_REFRESHING) {
                mArrowImageView.clearAnimation();
            }
            mHintTextView.setText("下拉可以刷新");
            break;
```

```
        case STATE_READY:
            if (mState != STATE_READY) {
                mArrowImageView.clearAnimation();
                mArrowImageView.startAnimation(mRotateUpAnim);
                mHintTextView.setText("松开可以刷新");
            }
            break;
        case STATE_REFRESHING:
            mHintTextView.setText("正在加载...");
            break;
        default:
            break;
        }

        mState = state;
    }

    public void setVisiableHeight(int height) {
        if (height < 0)
            height = 0;
        LinearLayout.LayoutParams lp = (LinearLayout.LayoutParams) mContainer.
        getLayoutParams();
        lp.height = height;
        mContainer.setLayoutParams(lp);
    }

    public int getVisiableHeight() {
        return mContainer.getHeight();
    }
}
```

代码文件：codes\10\10.2\HttpClientDemo\cn\edu\hstc\util\ ListViewHeader.java

接着，需要实现自定义 ListView 的底部代表类，该类命名为 ListViewFooter，该类源码如下：

```
package cn.edu.hstc.httpclientdemo.util;

import android.content.Context;
import android.util.AttributeSet;
import android.view.LayoutInflater;
import android.view.View;
import android.widget.Button;
import android.widget.LinearLayout;
import android.widget.TextView;
import cn.edu.hstc.httpclientdemo.activity.R;

public class ListViewFooter extends LinearLayout {
    public final static int STATE_NORMAL = 0;
    public final static int STATE_LOADING = 1;
    public final static int STATE_BUTTON = 2;
```

```
private Context mContext;

private View mContentView;
private View mProgressBar;
private TextView mHintView;
private Button btn_loadmore;

public ListViewFooter(Context context) {
    super(context);
    initView(context);
}

public ListViewFooter(Context context, AttributeSet attrs) {
    super(context, attrs);
    initView(context);
}

public void setState(int state) {
    if (state == STATE_NORMAL) {
        btn_loadmore.setVisibility(View.GONE);
        mProgressBar.setVisibility(View.GONE);
        mHintView.setVisibility(View.VISIBLE);
        mHintView.setText("查看更多");
    } else if (state == STATE_LOADING) {
        btn_loadmore.setVisibility(View.GONE);
        mProgressBar.setVisibility(View.VISIBLE);
        mHintView.setVisibility(View.VISIBLE);
        mHintView.setText("正在加载中...");
    } else if (state == STATE_BUTTON) {
        mProgressBar.setVisibility(View.GONE);
        mHintView.setVisibility(View.GONE);
        btn_loadmore.setVisibility(View.VISIBLE);
    }
}

public void setBottomMargin(int height) {
    if (height < 0)
        return;
    LinearLayout.LayoutParams lp = (LinearLayout.LayoutParams) mContentView.
    getLayoutParams();
    lp.bottomMargin = height;
    mContentView.setLayoutParams(lp);
}

public int getBottomMargin() {
    LinearLayout.LayoutParams lp = (LinearLayout.LayoutParams) mContentView.
    getLayoutParams();
    return lp.bottomMargin;
}

public void normal() {
```

```
            mHintView.setVisibility(View.VISIBLE);
            mProgressBar.setVisibility(View.GONE);
        }

        public void loading() {
            mHintView.setVisibility(View.GONE);
            mProgressBar.setVisibility(View.VISIBLE);
        }

        public void hide() {
            LinearLayout.LayoutParams lp = (LinearLayout.LayoutParams) mContentView.
            getLayoutParams();
            lp.height = 0;
            mContentView.setLayoutParams(lp);
        }

        public void show() {
            LinearLayout.LayoutParams lp = (LinearLayout.LayoutParams) mContentView.
            getLayoutParams();
            lp.height = android.view.ViewGroup.LayoutParams.WRAP_CONTENT;
            mContentView.setLayoutParams(lp);
        }

        public Button getLoadmore() {
            return btn_loadmore;
        }

        private void initView(Context context) {
            mContext = context;
            LinearLayout moreView = (LinearLayout) LayoutInflater.from(mContext).inflate(R.
    layout.xlistview_footer, null);
            addView(moreView);
            moreView.setLayoutParams(new LinearLayout.LayoutParams(android.view.ViewGroup.
            LayoutParams.MATCH_PARENT, android.view.ViewGroup.LayoutParams.WRAP_CONTENT));

            mContentView = moreView.findViewById(R.id.xlistview_footer_content);
            mProgressBar = moreView.findViewById(R.id.xlistview_footer_progressbar);
            mHintView = (TextView) moreView.findViewById(R.id.xlistview_footer_hint_
            textview);
            btn_loadmore = (Button) moreView.findViewById(R.id.btn_loadmore);
        }
    }
```

代码文件：codes\10\10.2\HttpClientDemo\cn\edu\hstc\util\ ListViewFooter.java

上面所实现的ListView组件的自定义顶部以及底部都有对应的布局代码，首先，实现顶部布局类ListViewHeader所对应的布局文件，其源码如下：

```
<?xml version = "1.0" encoding = "utf - 8"?>
< LinearLayout xmlns:android = "http://schemas.android.com/apk/res/android"
    android:layout_width = "fill_parent"
    android:layout_height = "wrap_content"
```

```
android:background = "@android:color/white"
android:gravity = "bottom" >

<RelativeLayout
    android:id = "@ + id/xlistview_header_content"
    android:layout_width = "fill_parent"
    android:layout_height = "60dp" >

    <LinearLayout
        android:id = "@ + id/xlistview_header_text"
        android:layout_width = "wrap_content"
        android:layout_height = "wrap_content"
        android:layout_centerInParent = "true"
        android:gravity = "center"
        android:orientation = "vertical" >

        <TextView
            android:id = "@ + id/xlistview_header_hint_textview"
            android:layout_width = "wrap_content"
            android:layout_height = "wrap_content"
            android:text = "下拉可以刷新" />

        <LinearLayout
            android:layout_width = "wrap_content"
            android:layout_height = "wrap_content"
            android:layout_marginTop = "3dp" >

            <TextView
                android:layout_width = "wrap_content"
                android:layout_height = "wrap_content"
                android:text = "更新于: "
                android:textSize = "12sp" />

            <TextView
                android:id = "@ + id/xlistview_header_time"
                android:layout_width = "wrap_content"
                android:layout_height = "wrap_content"
                android:textSize = "12sp" />
        </LinearLayout>
    </LinearLayout>

    <ImageView
        android:id = "@ + id/xlistview_header_arrow"
        android:layout_width = "20dp"
        android:layout_height = "wrap_content"
        android:layout_alignLeft = "@id/xlistview_header_text"
        android:layout_centerVertical = "true"
        android:layout_marginLeft = " - 35dp"
        android:src = "@drawable/xlistview_arrow" />

    <ProgressBar
```

```
            android:id = "@ + id/xlistview_header_progressbar"
            android:layout_width = "20dp"
            android:layout_height = "20dp"
            android:layout_alignLeft = "@id/xlistview_header_text"
            android:layout_centerVertical = "true"
            android:layout_marginLeft = " - 40dp"
            android:visibility = "invisible" />
    </RelativeLayout>

</LinearLayout>
```

代码文件：codes\10\10.2\HttpClientDemo\res\layout\ xlistview_header.xml

ListView 自定义底部布局文件如下：

```
<?xml version = "1.0" encoding = "utf - 8"?>
<LinearLayout xmlns:android = "http://schemas.android.com/apk/res/android"
    android:layout_width = "fill_parent"
    android:layout_height = "wrap_content"
    android:background = "@android:color/white">

    <RelativeLayout
        android:id = "@ + id/xlistview_footer_content"
        android:layout_width = "fill_parent"
        android:layout_height = "wrap_content"
        android:padding = "20dp"
        android:gravity = "center">

        <ProgressBar
            android:id = "@ + id/xlistview_footer_progressbar"
            android:layout_width = "20dp"
            android:layout_height = "20dp"
            android:layout_marginRight = "10dp"
            android:visibility = "gone" />

        <TextView
            android:id = "@ + id/xlistview_footer_hint_textview"
            android:layout_width = "wrap_content"
            android:layout_height = "wrap_content"
            android:layout_toRightOf = "@id/xlistview_footer_progressbar"
            android:text = "查看更多"
            android:textColor = "@android:color/black"
            android:textSize = "18sp"
            android:layout_centerVertical = "true"/>

        <Button
            android:id = "@ + id/btn_loadmore"
            android:layout_width = "fill_parent"
            android:layout_height = "wrap_content"
            android:text = "查看更多"
            android:textColor = "@android:color/black"
            android:textSize = "18sp"
```

```
                android:visibility = "gone"/>

        </RelativeLayout >

    </LinearLayout >
```

代码文件: codes\10\10.2\HttpClientDemo\res\layout\ xlistview_footer.xml

当然,既然项目中运用到了 ListView 组件,必定需要为该 ListView 组件的每个列表项定义其布局,在本应用中,ListView 列表项的布局如下:

```
<?xml version = "1.0" encoding = "utf - 8"?>
< RelativeLayout xmlns:android = "http://schemas.android.com/apk/res/android"
    android:layout_width = "match_parent"
    android:layout_height = "match_parent"
    android:paddingBottom = "2dp"
    android:paddingLeft = "20dp"
    android:paddingRight = "20dp"
    android:paddingTop = "2dp" >

    < TextView
        android:id = "@ + id/txt_name"
        android:layout_width = "wrap_content"
        android:layout_height = "wrap_content"
        android:layout_alignParentLeft = "true"
        android:layout_centerVertical = "true" />

    < TextView
        android:id = "@ + id/txt_ge"
        android:layout_width = "wrap_content"
        android:layout_height = "wrap_content"
        android:layout_centerVertical = "true"
        android:layout_toRightOf = "@id/txt_name"
        android:text = "(" />

    < TextView
        android:id = "@ + id/txt_department"
        android:layout_width = "wrap_content"
        android:layout_height = "wrap_content"
        android:layout_centerVertical = "true"
        android:layout_toRightOf = "@id/txt_ge" />

    < TextView
        android:id = "@ + id/txt_ge1"
        android:layout_width = "wrap_content"
        android:layout_height = "wrap_content"
        android:layout_centerVertical = "true"
        android:layout_toRightOf = "@id/txt_department"
        android:text = ")" />

    < ImageView
        android:id = "@ + id/image_icon"
```

```
            android:layout_width = "wrap_content"
            android:layout_height = "wrap_content"
            android:layout_alignParentRight = "true"
            android:layout_centerVertical = "true"
            android:background = "@drawable/right_icon" />

</RelativeLayout>
```

代码文件：codes\10\10.2\HttpClientDemo\res\layout\listview_item.xml

项目中使用了 ListView，必定需要加载数据源，因此，需要自定义 ListView 的数据适配器 MyListViewAdapter，源码如下：

```
package cn.edu.hstc.httpclientdemo.util;

import java.util.ArrayList;
import java.util.List;

import org.json.JSONObject;

import android.content.Context;
import android.view.LayoutInflater;
import android.view.View;
import android.view.ViewGroup;
import android.widget.BaseAdapter;
import android.widget.TextView;
import cn.edu.hstc.httpclientdemo.activity.R;

public class MyListViewAdapter extends BaseAdapter {
    private List<JSONObject> list = new ArrayList<JSONObject>();
    private LayoutInflater inflater;

    //在构造器中注入数据源
    public MyListViewAdapter(Context context, List<JSONObject> list) {
        this.list = list;
        inflater = (LayoutInflater) context.getSystemService(Context.LAYOUT_INFLATER_
SERVICE);
    }

    @Override
    public int getCount() {
        return list.size();
    }

    @Override
    public Object getItem(int position) {
        return list.get(position);
    }

    @Override
    public long getItemId(int position) {
        return position;
```

```
    }

    //自定义列表项
    @Override
    public View getView(int position, View convertView, ViewGroup parent) {
        ViewHolder holder = null;
        if (convertView == null) {
            //加载列表项布局文件
            convertView = inflater.inflate(R.layout.listview_item, null);
            holder = new ViewHolder();
            holder.name = (TextView) convertView.findViewById(R.id.txt_name);
            holder.department = (TextView) convertView.findViewById(R.id.txt_department);
            convertView.setTag(holder);
        } else {
            holder = (ViewHolder) convertView.getTag();
        }
        holder.name.setText(list.get(position).optString("name"));
        holder.department.setText(list.get(position).optString("department"));
        return convertView;
    }

    public static class ViewHolder {
        public TextView name;
        public TextView department;
    }
}
```

代码文件：codes\10\10.2\HttpClientDemo\cn\edu\hstc\\util\ MyListViewAdapter.java

应用一启动，将会向服务器端请求数据，而这需要经过一定的时间段，在该时间内，如果在界面中没有任何提示，将会大大降低应用的用户体验，此时，需要借助一个对话框来提示用户数据正在加载中，该对话框对应的布局文件如下：

```
<?xml version="1.0" encoding="utf-8"?>
<LinearLayout xmlns:android="http://schemas.android.com/apk/res/android"
    android:layout_width="match_parent"
    android:layout_height="match_parent"
    android:gravity="center"
    android:padding="10dp" >

    <ProgressBar
        android:layout_width="wrap_content"
        android:layout_height="wrap_content"
        android:layout_marginLeft="10dp"
        android:layout_marginRight="10dp" />

    <TextView
        android:id="@+id/message"
        android:layout_width="wrap_content"
        android:layout_height="wrap_content"
        android:layout_marginRight="10dp"
        android:textColor="@android:color/white" />
```

```
</LinearLayout>
```

代码文件：codes\10\10.2\HttpClientDemo\res\layout\dialog_prompt.xml

接着，自定义一个对话框工具类来加载上面的布局文件，生成一个 AlertDialog 对话框，该工具类将会在实现软件 Activity 时使用到，当 Activity 生成时，利用该工具类返回一个对话框，以提示用户数据加载中，改进用户体验。该工具类源码如下：

```
package cn.edu.hstc.httpclientdemo.util;

import android.app.Activity;
import android.app.AlertDialog;
import android.widget.LinearLayout;
import android.widget.TextView;
import cn.edu.hstc.httpclientdemo.activity.R;

public class DialogUtil {
    public static AlertDialog prompt(Activity context, String content) {
        AlertDialog.Builder builder = new AlertDialog.Builder(context);
        LinearLayout layout = (LinearLayout) context.getLayoutInflater().inflate(R.layout.
dialog_prompt, null);
        ((TextView) layout.findViewById(R.id.message)).setText(content);
        builder.setView(layout);
        AlertDialog dialog = builder.show();
        dialog.setCanceledOnTouchOutside(false);
        return dialog;
    }
}
```

代码文件：codes\10\10.2\HttpClientDemo\cn\edu\hstc\httpclientdemo\util\DialogUtil.java

完成了以上的一些基础支撑类的创建后，就可以进入应用的界面实现部分了，本应用只有一个用户界面，该界面作为应用的启动界面，软件一启动便加载，然后在该 Activity 中请求后台获取数据，并填充在自定义的 ListView 组件中。该 Activity 所对应的布局代码如下：

```
<LinearLayout xmlns:android="http://schemas.android.com/apk/res/android"
    android:id="@+id/content"
    android:layout_width="match_parent"
    android:layout_height="match_parent"
    android:orientation="vertical" >

    <RelativeLayout
        android:id="@+id/layout_list"
        android:layout_width="fill_parent"
        android:layout_height="wrap_content"
        android:background="@color/gainsboro"
        android:padding="10dp" >

        <TextView
            android:layout_width="wrap_content"
```

```
            android:layout_height = "wrap_content"
            android:layout_centerInParent = "true"
            android:text = "通讯录"
            android:textSize = "22sp" />
    </RelativeLayout>

</LinearLayout>
```

代码文件：codes\10\10.2\HttpClientDemo\res\layout\activity_main.xml

上面提到过，在应用的启动界面中，将实现向特定的后台请求数据并显示在 ListView 组件中，那么，该 Activity 类就实现了使用 HttpClient 访问网络数据的功能。也就是说，在该 Activity 类中，将会调用一开始就已经实现好的 HttpUtil 工具类中的两个方法，使用 HttpClient 发送 Get 请求或 Post 请求并获得 HTTP 响应数据。通常，需要将与服务端交互的实现放在一个线程中去执行，而 Android 提供了封装良好的 AsyncTask 机制，AsyncTask 允许在后台执行一个异步任务，因此可以将耗时的操作放在异步任务当中来执行，并随时将任务执行的结果返回给 UI 线程来更新 UI 控件。因此，只需要在 AsyncTask 实现 HttpClient 访问后台数据并通知 ListView 组件更新数据源即可。

该 Activity 类的实现代码如下：

```
package cn.edu.hstc.httpclientdemo.activity;

import java.text.SimpleDateFormat;
import java.util.ArrayList;
import java.util.Date;
import java.util.List;

import org.apache.http.NameValuePair;
import org.apache.http.message.BasicNameValuePair;
import org.json.JSONArray;
import org.json.JSONObject;

import android.app.Activity;
import android.app.AlertDialog;
import android.content.Context;
import android.content.DialogInterface;
import android.content.DialogInterface.OnCancelListener;
import android.graphics.Color;
import android.os.AsyncTask;
import android.os.Bundle;
import android.util.Log;
import android.view.View;
import android.widget.LinearLayout;
import android.widget.LinearLayout.LayoutParams;
import android.widget.Toast;
import cn.edu.hstc.httpclientdemo.util.DialogUtil;
import cn.edu.hstc.httpclientdemo.util.HttpUtil;
import cn.edu.hstc.httpclientdemo.util.MyListViewAdapter;
import cn.edu.hstc.httpclientdemo.util.RefreshListView;
```

```
public class MainActivity extends Activity {
    private LinearLayout content;
    private CustomListView listview;
    private AlertDialog prompt;
    private MyListViewAdapter myListViewAdapter;
    private AsyncTask<String, Integer, String> access;
    private List<JSONObject> list = new ArrayList<JSONObject>();
    private int index = 1;
    private int count;

    @Override
    protected void onCreate(Bundle savedInstanceState) {
        super.onCreate(savedInstanceState);
        setContentView(R.layout.activity_main);
        content = (LinearLayout) findViewById(R.id.content);
        listview = new CustomListView(this);
        listview.setLayoutParams(new LayoutParams(LayoutParams.MATCH_PARENT, LayoutParams.
MATCH_PARENT));
        listview.setFadingEdgeLength(0);
        listview.setCacheColorHint(Color.TRANSPARENT);
        content.addView(listview);
        myListViewAdapter = new MyListViewAdapter(MainActivity.this, list);
        acquire(true);
    }

    private void acquire(final boolean dialog) {
        access = new AsyncTask<String, Integer, String>() {
            @Override
            protected void onPreExecute() {
                super.onPreExecute();
                if (dialog) {
                    prompt = DialogUtil.prompt(MainActivity.this, "正在加载中...");
                    prompt.setOnCancelListener(new OnCancelListener() {
                        @Override
                        public void onCancel(DialogInterface dialog) {
                            access.cancel(true);
                        }
                    });
                }
            }

            @Override
            protected String doInBackground(String... params) {
                String result = null;
                try {
                    //如果不是加载更多,则获取通讯录总数
                    if (!listview.pullLoading()) {
                        //这里获取通讯录总数用的是 Get 请求方式
                        JSONObject object = new JSONObject(HttpUtil.getDataByGet
                        (HttpUtil.BASE_URL + "GetAddressBookCount"));
                        if (object.has("count")) {
```

```
                count = object.optInt("count");
                Log.i("", "count: " + count);
            } else {
                result = object.optString("error");  //结果为错误信息
            }
        }
        if (count > 0) {
            listview.setTotal(count);
            //下面三行代码用的是 Post 请求方式
            List<NameValuePair> paramList = new ArrayList<NameValuePair>();
            paramList.add(new BasicNameValuePair("index", String.valueOf
            (index)));
            result = HttpUtil.getDataByPost(HttpUtil.BASE_URL +
            "GetAddressBooks", paramList);
            //也可以注释掉上面三行代码,打开下面这一行代码,使用 Get 请求方式
            //获取数据
            /* result = HttpUtil.getDataByGet(HttpUtil.BASE_URL +
            "GetAddressBooks?index=" + index); */
        }
    } catch (Exception e) {
        e.printStackTrace();
    }
    return result;
}

@Override
protected void onPostExecute(String result) {
    try {
        if (result != null) {
            JSONObject object = new JSONObject(result);
            if (object.has("data")) {
                JSONArray array = object.optJSONArray("data");
                if (array != null) {
                    if (array.length() > 0) {
                        if (listview.pullRefreshing() || dialog)
                            list.clear();

                        for (int i = 0; i < array.length(); i++) {
                            list.add(array.optJSONObject(i));
                        }

                        content.setVisibility(View.VISIBLE);

                        if (listview.pullRefreshing() || dialog) {
                            listview.setAdapter(myListViewAdapter);
                        } else {
                            myListViewAdapter.notifyDataSetChanged();
                        }
```

```
                            }
                        }
                    } else if (object.has("error")) {
                        Toast.makeText(MainActivity.this, object.optString("error"),
Toast.LENGTH_LONG).show();
                    }
                } else {
                    list.clear();
                    myListViewAdapter.notifyDataSetChanged();
                    content.setVisibility(View.GONE);
                    Toast.makeText(MainActivity.this, "暂时没有数据!", Toast.LENGTH_
LONG).show();
                }

                if (listview.pullRefreshing()) {
                    listview.stopRefresh();
                    SimpleDateFormat sdf = new SimpleDateFormat("MM-dd HH:mm");
                    listview.setRefreshTime(sdf.format(new Date()));
                } else {
                    listview.stopLoadMore();
                }
            } catch (Exception e) {
                e.printStackTrace();
            }

            if (dialog) {
                prompt.dismiss();
            } else if (listview.pullRefreshing()) {
                listview.stopRefresh();
                SimpleDateFormat sdf = new SimpleDateFormat("MM-dd HH:mm");
                listview.setRefreshTime(sdf.format(new Date()));
            } else {
                listview.stopLoadMore();
            }
        }
    };
    access.execute();
}

    private class CustomListView extends RefreshListView implements RefreshListView.
ListViewListener {
        public CustomListView(Context context) {
            super(context);
            setListViewListener(this);
        }

        @Override
        public void onLoadMore() {
```

```
                stopRefresh();
                access.cancel(true);
                index++;
                acquire(false);
            }

            @Override
            public void onRefresh() {
                stopLoadMore();
                access.cancel(true);
                index = 1;
                acquire(false);
            }
        }
    }
```

代码文件：codes\10\10.2\HttpClientDemo\cn\edu\hstc\httpclientdemo\activity\MainActivity.java

服务端程序并不是本书所学习重点，如果读者有兴趣阅读本应用所访问的后台程序代码，可以在本书所附带源码中获得。启动后台，将该应用部署在Android模拟器上，运行效果如图10.6～图10.10所示。

图10.6 加载数据中

从上面的HttpUtil工具类以及MainActivity类中可以得知，使用HttpClient发送请求、接收响应数据其实很简单，只需要如下几个步骤即可：

① 创建HttpClient对象。

② 如果需要发送Get请求，则创建HttpGet对象；如果需要发送Post请求，则创建HttpPost对象。

③ 如果需要发送请求参数，针对Get请求，可以直接将参数拼接在请求地址中；针对Post请求，可将参数封装在一个List < NameValuePair > params中，然后作为HttpEntity构造器参数创建一个HttpEntity对象，再调用HttpPost对象的setEntity(HttpEntity entity)方法将创建的HttpEntity对象传入，即可传入请求参数；实际上，也可以调用HttpGet、HttpPost共同的setParams(HttpParams params)方法来添加请求参数。

④ 调用HttpClient对象的execute(HttpUriRequest request)发送请求，执行该方法返回一个HttpResponse对象。

⑤ 调用HttpResponse对象的getEntity方法可获取HttpEntity对象，该对象封装了服务器的响应内容，可通过EntityUtils.toString(HttpEntity entity)方法将响应内容转换为字符串形式。

5:37 PM

HttpClientDemo

通讯录

张妙君（技术部）
吴煜城（技术部）
黄桂跃（技术部）
陈新博（技术部）
张立鑫（技术部）
黄舒典（技术部）
李汉杰（技术部）
汤楷洪（技术部）
陈煌婷（技术部）
罗镇武（技术部）
庄桂东（技术部）
罗奕鹏（技术部）
林楚豪（技术部）
陈延武（技术部）
陈丽峰（技术部）
杨居风（技术部）
钟雁云（技术部）

图 10.7　加载完第一页数据

5:38 PM

HttpClientDemo

通讯录

下拉可以刷新
更新于：

张妙君（技术部）
吴煜城（技术部）
黄桂跃（技术部）
陈新博（技术部）
张立鑫（技术部）
黄舒典（技术部）
李汉杰（技术部）
汤楷洪（技术部）
陈煌婷（技术部）
罗镇武（技术部）
庄桂东（技术部）
罗奕鹏（技术部）
林楚豪（技术部）
陈延武（技术部）
陈丽峰（技术部）

图 10.8　下拉 ListView 未松手时

5:39 PM

HttpClientDemo

通讯录

正在加载...
更新于：

张妙君（技术部）
吴煜城（技术部）
黄桂跃（技术部）
陈新博（技术部）
张立鑫（技术部）
黄舒典（技术部）
李汉杰（技术部）
汤楷洪（技术部）
陈煌婷（技术部）
罗镇武（技术部）
庄桂东（技术部）
罗奕鹏（技术部）
林楚豪（技术部）
陈延武（技术部）

图 10.9　下拉松手开始加载

5:42 PM

HttpClientDemo

通讯录

李汉杰（技术部）
汤楷洪（技术部）
陈煌婷（技术部）
罗镇武（技术部）
庄桂东（技术部）
罗奕鹏（技术部）
林楚豪（技术部）
陈延武（技术部）
陈丽峰（技术部）
杨居风（技术部）
钟雁云（技术部）
王煜楠（技术部）
黄宇虹（技术部）
王建辉（技术部）

正在加载中……

图 10.10　上拉加载更多数据

10.3 使用 WebView 视图开发 WebKit 应用

Android 浏览器的内核是 WebKit 引擎，WebKit 的前身是 KDE 小组的 KHTML。Apple 将 KHTML 发扬光大，推出了 KHTML 的改进型的 WebKit 引擎的浏览器 Safari，获得了非常好的反响。WebKit 内核在手机上的应用十分广泛，例如 Google 的手机 Gphone、Apple 的 iPhone 等所使用的 Browser 内核引擎，都是基于 WebKit。随着计算机、手机及联网装置的普及，未来终端运算都会在云端执行，目前云计算技术在网络服务中已经随处可见，例如搜索引擎、网络信箱等，使用者只要输入简单指令即可得到大量信息。未来的手机、GPS 等行动装置都可以通过云计算技术，发展出更多的应用服务。因此人们只要拥有一个功能强大的浏览器，就能满足平时工作生活的需要。本节将通过 Android 中的 WebKit 包来实现这些需求。

10.3.1 WebKit 概述

WebKit 是一个开源浏览器网页排版引擎，与之响应的引擎有 Gecko(Mozila、Firefox 等使用的排版引擎)和 Trident(也称 MSHTML，是 IE 使用的排版引擎)。同时，WebKit 也是苹果 Mac OS-X 系统引擎框架版本的名称，主要用于 Safari、Dashboard、Mail 和其他一些 Mac OS-X 程序。WebKit 所包含的 WebCore 排版引擎和 JSCore 引擎来自于 KDE 的 KHTML 和 KJS，当年苹果比较了 Gecko 和 KHTML 后，仍然选择了后者，就因为它拥有清晰的源码结构、极快的渲染速度。而今 Android 系统也毫不犹豫地选择了 WebKit。它具有各触摸屏、高级图形显示和上网功能，用户能够在手机上查看电子邮件、搜索网址和观看视频节目等。可以看出这是一个非常强大的 Web 应用平台。

WebKit 由三个模块组成：JavaScriptCore、WebCore 和 WebKit。

① WebKit：整个项目的名称。

② JavaScriptCore：JS 虚拟机，相对独立，主要用于操作 DOM，DOM 是 W3C 定义的规范，主要用于定义外部可以操作的浏览器内核的接口，而 WebCore 必须实现 DOM 规范。

③ WebCore：整个项目的核心，用来实现 Render 引擎，解析 Web 页面，生成一个 DOM 树和一个 Render 树。

JavaScriptCore 的主要功能有：

① API——基本 JavaScript 功能。

② Binding——与其他功能绑定的功能，如 DOM、C、JNI。

③ DerviedSource——自动产生的代码。

④ PCRE——Perl-Compatible Regular Expressions(Perd 兼容的规则表达式)。

⑤ KJS——Javascript Kernel(JavaScript 内核)。

WebCore 的主要功能有：

① Loader——加载资源及 Cache 实现(Curl)。

② DOM——HTML 词法分析与语法分析。

③ DOM——DOM 节点与 Render 节点创建，形成 DOM 树。

④ Render——Render 树介绍，RenderBox。

⑤ Layout——排版介绍。

⑥ Css Parser 模块。

⑦ Binding——DOM 与 JavascriptCore 绑定的功能。

对 WebKit 各个模块的功能有了了解，下面看看 WebKit 的解析过程是怎样的。流程如下：

① CURL 获得网站的 Stream。

② 解析划分字符串。

③ 通过 DOM Builder 按合法的 HTML 规范生成 DOM 树。

④ 如果有 JavaScript，JSEngine 就通过 ECMA-262 标准完善 DOM 树。

⑤ 把 DOM 传给 LayoutEngine 进行布局，如果有 CSS 样式，就通过 CSSParser 解析。

⑥ 最后渲染 Rendering 出来。

而 Google 对 WebKit 进行了封装，为开发者提供了丰富的 Java 接口，其中最重要的便是 android. webkit. WebView 控件。下面重点学习 WebView 组件的使用。

10.3.2 使用 WebView 浏览网页

Android 提供了 WebView 组件专门来浏览网页，和其他控件一样，它使用起来非常简单。

WebView 组件本身就是一个浏览器实现，它的内核基于 10.3.1 节介绍的开源 WebKit 引擎。如果对 WebView 进行一些美化、包装，可以非常轻松地开发出自己的浏览器。下面通过一个实例介绍 Android 如何使用 WebView 控件来浏览网页。

该实例需求比较简单，在页面中放置一个 EditText 控件供用户输入要访问的网址，在输入框的右边放置一个 Button(按钮)，在这两个控件的下方放置一个 WebView 组件，当用户输入网址后单击按钮，则 WebView 加载该网址内容，当单击网页中任意链接，需要在当前的 WebView 中打开，当用户按下硬键盘中的返回键，即当后退时，在当前 WebView 中加载上一个网页。应用界面布局如下：

```
<LinearLayout xmlns:android = "http://schemas.android.com/apk/res/android"
    android:layout_width = "match_parent"
    android:layout_height = "match_parent"
    android:orientation = "vertical" >

    <RelativeLayout
        android:layout_width = "fill_parent"
        android:layout_height = "wrap_content"
        android:hint = "请输入您的网址..." >

        <Button
            android:id = "@ + id/btn_go"
            android:layout_width = "wrap_content"
            android:layout_height = "wrap_content"
            android:layout_alignParentRight = "true"
            android:text = "GO" />
```

```
        <EditText
            android:id="@+id/edt_url"
            android:layout_width="fill_parent"
            android:layout_height="wrap_content"
            android:layout_toLeftOf="@id/btn_go"
            android:text="http://www.baidu.com"
            android:hint="请输入您的网址..." />
    </RelativeLayout>

    <WebView
        android:id="@+id/webView"
        android:layout_width="fill_parent"
        android:layout_height="fill_parent" />

</LinearLayout>
```

代码文件：codes\10\10.3\WebViewDemo\res\layout\activity_main.xml

上面布局所对应的 Activity 代码如下：

```
package cn.edu.hstc.webviewdemo.activity;

import android.app.Activity;
import android.graphics.Bitmap;
import android.os.Bundle;
import android.view.KeyEvent;
import android.view.View;
import android.webkit.WebSettings;
import android.webkit.WebView;
import android.webkit.WebViewClient;
import android.widget.Button;
import android.widget.EditText;

public class MainActivity extends Activity {
    private EditText urlEdt;
    private Button goBtn;
    private WebView webView;

    @Override
    protected void onCreate(Bundle savedInstanceState) {
        super.onCreate(savedInstanceState);
        setContentView(R.layout.activity_main);
        initView();
    }

    private void initView() {
        urlEdt = (EditText) findViewById(R.id.edt_url);
        goBtn = (Button) findViewById(R.id.btn_go);
        webView = (WebView) findViewById(R.id.webView);

        goBtn.setOnClickListener(new View.OnClickListener() {
```

```
            @Override
            public void onClick(View v) {
                webView.loadUrl(urlEdt.getText().toString());
            }
        });

        WebSettings webSettings = webView.getSettings();
        //设置支持 JavaScript 脚本
        webSettings.setJavaScriptEnabled(true);
        //设置可以访问文件
        webSettings.setAllowFileAccess(true);
        //设置支持缩放
        webSettings.setBuiltInZoomControls(true);

        webView.setWebViewClient(new WebViewClient() {
            @Override
            public boolean shouldOverrideUrlLoading(WebView view, String url) {
                webView.loadUrl(url);     //当有新的链接时,使用当前的 WebView 来显示
                return true;
            }

            @Override
            public void onPageFinished(WebView view, String url) {
                super.onPageFinished(view, url);
            }

            @Override
            public void onPageStarted(WebView view, String url, Bitmap favicon) {
                super.onPageStarted(view, url, favicon);
            }
        });
    }

    @Override
    public boolean onKeyDown(int keyCode, KeyEvent event) {
        if (keyCode == KeyEvent.KEYCODE_BACK && webView.canGoBack()) {
            webView.goBack();             //后退
            return true;
        }
        return super.onKeyDown(keyCode, event);
    }
}
```

代码文件: codes\10\10.3\WebViewDemo\cn\edu\hstc\webviewdemo\activity\MainActivity.java

最后,不要忘了在 AndroidManifest.xml 中为该应用添加访问网络的权限:

```
<uses-permission android:name="android.permission.INTERNET" />
```

将上面的应用部署在 Android 模拟器中,输入百度首页地址,单击 GO 按钮,可以看到如图 10.11 所示效果图。

单击图 10.11 中的“文库”链接，将可以看到如图 10.12 所示效果。

百度一下

文库 图片 知道 新闻 百科
应用 地图 贴吧 hao123 更多
小说 游戏 下载
下载： 百度客户端 百度应用 地图
完整版

图 10.11 WebView 显示百度首页

3:43 PM
WebViewDemo
http://www.baidu.com GO
Bai du 文库 网页 阅读 知道 百科
搜索文档
最热阅读
[IT/计算机] 常用计算机故障诊断与排除
[医药卫生] 临床病理常见知识问答
[经管营销] 我国企业品牌经营的现状
[信息与通信] 万用表的使用方法和技巧
人气排行榜 | 本周榜 | 本月榜
1.自己动手DIY组装电脑基础教程
2.八大杀毒软件简介
3.十个品牌笔记本介绍
4.Windows 7局域网共享设置的解决方案
5.手诊基础知识-图文
6.哈根达斯 网络营销
7.彩妆品牌前十名_世界国际排行榜
8.年轻人的资产与负债

图 10.12 在当前 WebView 中加载链接

当用户按下硬键盘上的返回键时，将回到百度首页页面，如图 10.11 所示。

阅读 MainActivity 类的源码，可以发现，程序调用 WebView 的 loadUrl(String url)方法加载、显示界面 EditText 输入框中所指定 URL 的网页，该行代码亦是程序的关键代码。实际上，WebView 提供了大量的方法来执行浏览器操作，如表 10.4 所示。

表 10.4 WebView 提供的常用方法

方法名	描述
void goBack	后退
void goForward	前进
void loadUrl(String url)	加载指定 URL 对应的网页
booleanzoomIn	放大网页
boolean zoomOut	缩小网页

继续分析代码可以发现，程序创建了一个 WebSettings 对象，并通过该对象设置了 WebView 的一些属性、状态等。在创建 WebView 时，系统有一个默认的设置，可以通过 WebView. getSettings 来得到这个设置。

WebSettings 和 WebView 在同一个生命周期中存在，当 WebView 被销毁后，如果再使用 WebSettings，则会抛出 IllegalStateException 异常。实际上，除了代码中 WebSettings 的设置外，还提供了如表 10.5 所示的一些常用的属性、状态的设置方法。

表 10.5 WebSettings 提供的常用方法

方法名	描述
setAllowFileAccess	启动或禁止 WebView 访问文件数据
setBlockNetWorkImage	是否显示网络图像
setCacheMode	设置缓冲模式
setDefaultFontSize	设置默认的字体大小
setDefaultTextEncodingName	设置在解码时使用的默认编码
setFixedFontFamily	设置固定使用的字体

WebSettings 所提供的方法太多了，这里只是展示部分常用方法，读者可以通过官方 API 获得更多的方法。

代码中调用了 WebView 对象的 setWebViewClient 方法为 WebView 指定了一个 WebViewClient 对象，并且重写了 WebViewClient 中的 shouldOverrideUrlLoading 方法，正是该方法中的代码行 webView. loadUrl(url)控制了当单击 WebView 中显示网页的内嵌链接时，继续在当前 WebView 中加载。如若不为 WebView 指定 WebViewClient 对象并重写该方法，则默认会在系统自带浏览器中打开新链接。WebViewClient 就是专门辅助 WebView 处理各种通知、请求事件的类。表 10.6 展示了部分 WebViewClient 所提供的方法，可以通过覆盖这些方法来辅助 WebView 浏览网页。

表 10.6 WebViewClient 提供的常用方法

方法名	描述
doUpdateVisitedHistory	更新历史记录
onFormResubmission	应用程序重新请求网页数据
onLoadResource	加载指定网址提供的资源
onPageFinished	网页加载完毕
onPageStarted	网页开始加载
onReceivedError	报告错误信息
onScaleChanged	WebView 发生改变
shouldOverrideUrlLoading	控制新的链接在当前 WebView 中打开

以上便是 WebView 组件开发的简单介绍，读者若有兴趣，可以深入了解 WebView 开发的更多知识，然后完善本实例所开发的浏览器应用，甚至完全可以开发出一个系统自带浏览器的替代产品。

10.3.3 使用 WebView 加载 HTML 代码

在 10.2.1 节所开发的实例中，采用了一个 TextView 来显示从网络上获取到的 HTML 字符串，如图 10.3 所示，可以看到，TextView 并不会对 HTML 标签进行解析，而是把整串标准 HTML 代码字符串直接显示出来，当然这对 EditText 来说，也是一样的效果。那么，Android 中是否有组件可以对 HTML 字符串进行解析，然后显示 HTML 中所包含的页面内容呢？答案是肯定的。10.3.2 节介绍的 WebView 组件便能实现上述功能。

WebView 提供了一个 loadData(String data, String mimeType, String encoding)方法

来加载并显示指定的 HTML 代码，该方法的三个参数含义如下：

- data——指定要加载 HTML 代码。
- mimeType——指定所加载的 HTML 代码的 MIME 类型，对于 HTML 代码可指定为 text/html。
- encoding——指定 HTML 代码编码所用的字符集，比如 GBK 或者 UTF-8。

下面通过一个实例来展示 WebView 加载 HTML 代码，在应用布局界面上放置一个 WebView 组件，用于显示解析 HTML 字符串后的内容，该界面布局代码非常简单，故不在此进行展示，下面直接粘贴 Java 程序代码部分。

```
package cn.edu.hstc.webviewdemo.activity;

import android.app.Activity;
import android.os.Bundle;
import android.webkit.WebView;

public class MainActivity extends Activity {
    private WebView webView;

    @Override
    protected void onCreate(Bundle savedInstanceState) {
        super.onCreate(savedInstanceState);
        setContentView(R.layout.activity_main);
        initView();
    }

    private void initView() {
        webView = (WebView) findViewById(R.id.webView);
        //访问后台,获取响应的 HTML 字符串
        String data = HttpUtil.getDataByGet(HttpUtil.BASE_URL + "TransportationHtml");
        System.out.println(data);
        //使用 loadData 方法加载 HTMl 代码
        webView.loadData(data, "text/html", "utf-8");
    }
}
```

上面的程序实现在应用的启动界面中，访问后台地址，获取返回的 THML 字符串，并通过 WebView 组件进行解析并显示。上面所指定访问的后台与 10.2.2 节的实例中所使用的后台为同一个，程序中所用到的 HttpUtil 工具也是 10.2.2 节的实例中所使用的工具类，也就是说，上面的代码中利用了 HttpClient Get 方式请求了后台并获得了响应内容，只不过该响应内容是一串 HTML 代码，至于代码中所访问的后台地址如何实现返回这一串 HTML 代码并不是我们关心的内容，读者可以通过本书附带源码获取后台的实现源码。

启动服务端，并将上面的应用部署于 Android 模拟器中，可以看到如图 10.13 所示界面。

从图 10.13 可以看到，WebView 所显示的内容出现了乱码，难道后台所返回的 HTML 代码就是如此吗？由于在程序中有对后台的响应内容进行输出，所以可以通过 LogCat 窗口看到此时后台所返回的内容，如图 10.14 所示。

图 10.13　WebView 显示乱码

727	727	cn.edu.hstc.webvi...	System.out	<!DOCTYPE HTML PUBLIC "-//W3C//DTD HTML 4.01 Transitional//EN">
727	727	cn.edu.hstc.webvi...	System.out	<HTML>
727	727	cn.edu.hstc.webvi...	System.out	<HEAD><TITLE>A Servlet</TITLE></HEAD>
727	727	cn.edu.hstc.webvi...	System.out	<BODY>
727	727	cn.edu.hstc.webvi...	System.out	我是HTMLBODY体，我将被加载到WebView中 </BODY>
727	727	cn.edu.hstc.webvi...	System.out	</HTML>

图 10.14　后台响应内容

从图 10.14 中可以看出，WebView 中所显示的内容应该为“我是 HTMLBODY 体，我将被加载到 WebView 中”才对，可是，启动应用后，WebView 组件中所显示的内容为什么会出现乱码呢？怎样才能避免乱码呢？

实际上，即使为上面代码中的 loadData 方法指定 encoding 为“gbk”或“gb2312”，也同样无法解决中文乱码问题。但是程序中代码行 webView.loadData(data, "text/html", "utf-8")却是 Android API 所提供的标准用法，这可以算作 Android API 的一个漏洞吧。幸好，Android WebView 提供了 loadDataWithBaseURL(String baseUrl, String data, String mimeType, String encoding, String historyUrl)方法来加载 HTML 代码。使用该方法，将可以避免中文乱码问题。

将上面的程序代码稍微修改一下，修改后的代码如下所示：

```
package cn.edu.hstc.webviewdemo.activity;

import android.app.Activity;
import android.os.Bundle;
```

```
import android.webkit.WebView;

public class MainActivity extends Activity {
    private WebView webView;

    @Override
    protected void onCreate(Bundle savedInstanceState) {
        super.onCreate(savedInstanceState);
        setContentView(R.layout.activity_main);
        initView();
    }

    private void initView() {
        webView = (WebView) findViewById(R.id.webView);
        //访问后台,获取响应的 HTML 字符串
        String data = HttpUtil.getDataByGet(HttpUtil.BASE_URL + "TransportationHtml");
        System.out.println(data);
        //使用 loadData 方法加载 HTMl 代码,如果解析出来的内容中含有中文,将会造成乱码
        webView.loadData(data, "text/html", "utf-8");
        //使用 loadDataWithBaseURL 方法加载 HTML 代码,避免乱码
        webView.loadDataWithBaseURL(null, data, "text/html", "utf-8", null);
    }
}
```

代码文件: codes\10\10.3\WebViewHTMLDemo\cn\edu\hstc\webviewhtmldemo\activity\MainActivity.java

将应用再次部署于 Android 模拟器中,可以看到如图 10.15 所示的运行界面。

图 10.15 避免中文乱码

通过以上介绍，相信读者已经学会了如何实用 WebView 加载 HTML 代码了。

10.4 本章小结

本章主要介绍了 Android 网络编程的相关知识，通过本章的学习，读者需要掌握如何开发 Android Socket 应用，利用 HttpURLConnection 类使用 HTTP 方式访问网络，由于 Android 还支持 Apache HttpClient，故读者需要掌握使用 HttpClient 接口开发，使用 HttpClient 可以方便地与服务端进行交互，Android 端可以方便地发送请求以及接收响应内容。此外，Android WebView 组件内核基于 WebKit 引擎，可以加载任意网页和解析 HTML 格式的字符串，读者需要学会使用 WebView 视图开发 WebKit 应用。

第11章 二维码应用——QR where

通过前面10个章节的介绍，相信读者已经对开发Android应用的流程有了一定的认知，并且已经掌握了Android的基础知识，掌握了Android四大组件的开发，已经有能力独立开发一个属于自己的Android应用了。在本章将开发一个完整的Android应用，该应用将涉及一些在当今应用市场中比较热门的技术。

11.1 QR where功能需求

本章要实现的Android应用，名叫QR where，通过该名字，也许无从知道该应用是做什么的。事实上，当今社会，到处充斥着各种二维码，扫描二维码可以进行电子支付，可以识别微信公众号或者普通的微信账号，可以跳转到链接等等。而二维码的英文名字也叫QR Code。QR where正是一款对二维码进行解析、对四种特定类型的信息生成二维码的应用。在这四种特定类型中，其中一种是地理位置信息，于是，选取单词QR Code里面的QR，再加上单词where，就构成了应用名称QR where。

QR where提供一个扫描框供用户扫描二维码，应用解析二维码，读取二维码信息，然后跳转到相应的类型的详情页面中。QR where对四种类型的二维码做特别的处理，这四种类型分别是长链接(网址)、手机联系人、地理信息(经纬度)、WiFi信息。这四个类型的详情页面对应着不同的操作界面并提供不同的功能。

对于保存了网址文本的二维码，应用扫描该二维码解析后跳转到的页面显示了解析后的网址文本信息，并在该文本下方生成了一张对应的二维码图片，相当于将扫描到的二维码显示在手机应用中。在页面底部从上而下提供了两个按钮：一个供用户单击后在手机自带浏览器中打开该网址链接；一个供用户单击后将该串链接复制到手机粘贴板中，以便用户进行粘贴文本操作。

对于保存了手机联系人信息的二维码，应用扫描该二维码解析后跳转到的页面显示了该联系人的相关信息，比如姓名、电话、邮箱等。在电话文本的右边提供发送短消息、拨打电话的按钮供用户分别操作这两项功能。在页面中也提供以现有的联系人信息创建新联系人以及将页面中的联系人信息添加到手机现有联系人中的功能。在页面底部还提供了一个按钮供用户单击跳转到另外一张页面显示生成的联系人信息对应的二维码。

对于保存了经纬度文本信息的二维码，应用扫描该二维码解析后跳转到的页面，采用高德地图，定位到该二维码中的经纬度的地点，并在地图中画上标记，在地图的下方显示提供三个文本显示组件分别显示经度、纬度、街道地址。在页面底部提供两个按钮供用户单击根据该经纬度信息生成二维码图片以及通过手机自带浏览器打开地图定位到该经纬度。

对于保存了 WiFi 信息的二维码，应用扫描该二维码解析后跳转到的页面显示了该 Wi-Fi 信息的 SSID、Password、Network Type 三个属性，其实就是直接将二维码中文本信息拆分出来，显示在页面上。在文本信息的下方还提供了一张根据 WiFi 信息所生成的二维码图片，相当于将扫描到的二维码显示在手机中。在二维码图片的下方提供了一个按钮供用户单击复制该串 Wi-Fi 文本信息。

如解析出来的文本并非这四种类型的信息，则跳转到一个显示了包含该文本信息的二维码图片的页面。

在以上所介绍的四种类型以及其他非这四种类型的信息的详情页面中，只要有包含二维码图片的页面，在页面的右上角都会提供一个按钮，单击该按钮，便会在页面底部弹出三个按钮：第一个按钮供用户单击跳转到发送 Email 的应用，并携带该二维码图片，作为 Email 的邮件内容，用户可以将该二维码图片通过邮件发送到其他邮箱中；第二个按钮供用户单击后将该二维码图片保存到手机的系统相册中；第三个按钮则是取消底部弹出菜单的功能。

以上所介绍的是对保存了各种不同信息的二维码进行解析后所做的不同的跳转处理。事实上，我们需要从应用的全局上进行需求分析。QR where 底部有四大菜单，接下来分别介绍这四个菜单中所包含的具体功能需求。

① Scan：第一个底部菜单称为 Scan。大家都知道，扫描的英文单词是 scanning，从字面上理解，该部分的功能是为提供二维码扫描操作。

- 单击底部第一个菜单按钮，将会在页面中提供一个二维码扫描框，用户将该扫描框对准任意二维码，应用将会对该二维码进行解析后跳转到上面所介绍的四种类型的详情页面之一或者弹出一个对话框显示解析出来的文本信息。
- 在页面的左上角则提供了一个相册按钮，单击该按钮，则进入系统相册，选中任意一张图片，会返回应用页面中，应用对选中的图片进行解析，如果该图片不是二维码图片，则提示用户；如果该图片是二维码图片，则将二维码进行解析后，跳转到详情页面。这与通过扫描框解析二维码类似。
- 无论通过扫描框还是通过浏览系统相册的方式，只要是将二维码图片成功解析出来后，则会将该二维码的文本信息保存在 Android 自带的 SQLite 数据库中。
- 在页面的右上角提供了一个按钮供用户打开或关闭手机自带的闪光灯功能。

② History：第二个底部菜单称为 History。History 是历史的英文单词，很容易理解，这部分用于查询应用的 SQLite 数据库中的数据并以列表的形式显示在页面中。

- History 记录分为两种类型：一种是通过底部第一个菜单中扫描或浏览系统相册并解析二维码后所保存的二维码文本信息，另外一种类型则是通过接下来所要介绍的第三个菜单中的生成二维码功能后所保存的二维码文本信息。因此，在该页面中，以 Tab 标签页形式分别显示了 Scan 类型的历史记录列表和 Generate 类型的历史记录列表。
- 单击任意一项列表项，则会跳转到该条历史记录所对应的详情页面，该部分的页面跳转与扫描二维码信息后的页面跳转完全一致。

③ Generator：第三个底部菜单称为Generator，即为生成器，它提供了根据以上所介绍的四种类型的信息生成对应的二维码的功能。因此在该部分的页面上从上至下放置了四个大的按钮供用户单击跳转到不同类型的信息编辑页面中。

- QR/Short URL：单击页面中的QR / Short URL按钮，跳转到网址编辑页面，在页面的网址输入框中输入网址，该输入框提供了对网址的格式验证功能。输入网址信息后，单击页面顶部的生成按钮，则跳转到二维码生成页面，在页面中显示了所生成的二维码，并且同样在页面右上角提供按钮供用户单击显示更多的底部按钮，以供用户操作发送邮件、保存二维码图片到系统相册。
- Contact：单击页面中的Contact按钮，跳转到手机联系人信息编辑页面，在该页面中提供了各种手机联系人的基础信息的输入框。也可单击页面右上角所提供的按钮，应用将会调用系统所自带的手机联系人应用，选中任意一个联系人，则返回本应用并将选中的联系人信息自动打印在页面中的各个对应的编辑框中。然后，用户即可单击页面中的生成按钮，同样会生成一张带有联系人信息的二维码图片。
- Location：单击页面中的Location按钮，跳转到地理位置定位页面，在该页面中集成了高德地图，在地图展示的下方放置了一个检索框供用户输入地点，然后精确搜索出该地理名称的经纬度。该检索框带有自动完成的功能，比如输入潮州两字，将会自动出现潮州宾馆、潮州市人民政府、潮州西湖等地点供用户选择。在搜索出来的地理位置会添加上地图标注，单击该标注，会弹出街道信息，这些都是高德地图开发的内容，在应用开发过程中也将会涉及。单击页面中的生成按钮，生成一张包含所搜索地理位置的经纬度信息的二维码图片。
- WiFi：单击页面中的WiFi按钮，跳转到WiFi信息编辑页面，在该页面中提供了SSID、Password两个文本输入框，提供一个Network Type选择器，该选择器其实就是一个文本显示组件，单击该文本控件，将会在页面底部弹出四个Button按钮，单击前面三个按钮可以选择WEP、WPA/WPA2两中WiFi加密方式，或者选择Noencryption表示该WiFi没有任何加密方式，最后一个Button按钮为取消底部菜单的功能。单击页面中的生成按钮，将会跳转到另一个页面显示包含了WiFi信息的二维码图片。

④ Setting：第四个底部菜单成为Setting，即设置。该部分主要提供两个功能：一个是Clear History（清空数据库），另一个是About（关于）。

以上便是QR where的全部功能需求的介绍。接下来，将会根据QR where底部的四个菜单项，分别介绍开发过程。

11.2 开发启动界面MainActivity

11.1节已经介绍了QR where的功能需求，已经了解了QR where底部有四个大的菜单模块。根据现有的Android知识体系，可以使用Android Fragment来实现底部菜单。

首先，需要为底部菜单定义布局文件，在该文件中实现了底部菜单的布局，QR where底部菜单中的每个菜单按钮为一个图标，因此，只需要在该布局文件中放置一个ImageView组件，布局代码如下所示：

```
<?xml version="1.0" encoding="utf-8"?>
<LinearLayout xmlns:android="http://schemas.android.com/apk/res/android"
    android:layout_width="wrap_content"
    android:layout_height="wrap_content"
    android:gravity="center"
    android:orientation="vertical" >

    <ImageView
        android:layout_width="wrap_content"
        android:layout_height="wrap_content"
        android:focusable="false"
        android:padding="3dp"
        android:src="@drawable/scan_tab_btn"
        android:id="@+id/imageview"/>

</LinearLayout>
```

代码文件：codes\11\11.2\QR where\res\layout\scan_tab_btn.xml

接下来，为 QR where 的启动界面 MainActivity 定义布局文件：

```
<?xml version="1.0" encoding="utf-8"?>
<LinearLayout xmlns:android="http://schemas.android.com/apk/res/android"
    android:layout_width="fill_parent"
    android:layout_height="fill_parent"
    android:orientation="vertical">

    <!-- 放置每个 Fragment 的内容 -->
    <FrameLayout
        android:layout_width="fill_parent"
        android:layout_height="0dip"
        android:layout_weight="1"
        android:id="@+id/realtabcontent"/>

    <!-- 底部菜单 -->
    <android.support.v4.app.FragmentTabHost
        android:layout_width="fill_parent"
        android:layout_height="wrap_content"
        android:background="@drawable/buttom_back"
        android:id="@android:id/tabhost">

        <FrameLayout
            android:layout_width="0dp"
            android:layout_height="0dp"
            android:layout_weight="0"
            android:id="@android:id/tabcontent"/>

    </android.support.v4.app.FragmentTabHost>

</LinearLayout>
```

代码文件：codes\11\11.2\QR where\res\layout\activity_main.xml

有了以上的界面布局基础，便可以实现 QR where 启动界面的 Java 程序了，新建类文件 MainActivity.java。为了在 MainActivity 中加载各个菜单项所对应的 Fragment，需要让 MainActivity 继承 FragmentActivity。在 MainActivity 中，需要实现底部菜单栏，因此，需要在 MainActivity 类中声明以下几个变量，用于存放必要的参数。

```
private FragmentTabHost tabHost;
private LayoutInflater layoutInflater;
//定义数组来存放 Fragment 界面
@SuppressWarnings("rawtypes")
private Class fragmentArr [ ] = { ScanFragment. class, HistoryFragment. class,
GeneratorFragment.class, SettingFragment.class };
//定义数组来存放底部按钮图片
private int tabImageArr[ ] = { R. drawable. scan_tab_btn, R. drawable. history_tab_btn, R.
drawable.generator_tab_btn, R.drawable.setting_tab_btn };
//定义数组来存放底部选项卡文字
//private int tabTextArr[ ] = {R. string. scan, R. string. history, R. string. generator, R.
string.setting};
private String tabTextArr[] = { "scan", "history", "generator", "setting" };
```

在 MainActivity 中，我们主要是靠以下的两个方法来实现底部菜单栏的。

```
/**
 * 实现底部菜单
 */
private void initBottomMenu() {
    //实例化布局对象
    layoutInflater = LayoutInflater.from(this);
    tabHost = (FragmentTabHost) this.findViewById(android.R.id.tabhost);
    tabHost.setup(this, getSupportFragmentManager(), R.id.realtabcontent);
    int count = fragmentArr.length;
    for (int i = 0; i < count; i++) {
        //调用 getTabItemView 方法为每一个 Tab 按钮设置图标、文字
        TabSpec tabSpec = tabHost.newTabSpec(tabTextArr[i]).setIndicator(getTabItemView(i));
        tabHost.addTab(tabSpec, fragmentArr[i], null);
        tabHost.getTabWidget().getChildAt(i).setBackgroundResource(R.drawable.selector_
tab_background);
    }
}

/**
 * 给 Tab 按钮设置图标和文字
 */
private View getTabItemView(int index) {
    //加载底部菜单布局，在该布局中只放置了存放底部菜单图标的 ImageView，并没有在图标下方
放置文本组件，
    //所以在底部菜单的图标下方并没有显示菜单文字
    View view = layoutInflater.inflate(R.layout.tab_item_view, null);
    ImageView imageView = (ImageView) view.findViewById(R.id.imageview);
    imageView.setImageResource(tabImageArr[index]);
    return view;
}
```

上面的程序通过加载 MainActivity 所对应的界面布局 activity_main. xml，加载布局中所对应的 FragmentTabHost 对象，即第一张代码片段的图片中所声明的第一个变量 tabHost。通过循环存放了底部菜单所对应的 Fragment 的数组变量 fragmentArr，调用 getTabItemView 方法为每一个 Tab 按钮设置图标、文字。在该方法中，通过加载底部菜单的样式文件 scan_tab_btn. xml，得到底部每个菜单所对应的图标组件 ImageView，并为其设置对应的 Image 资源，即存放了底部按钮图片的变量 tabImageArr 中所存放的 drawable 资源文件。以 tabImageArr 数组中的第一个元素 R. drawable. scan_tab_btn 为例：

```
<?xml version = "1.0" encoding = "utf - 8"?>
< selector xmlns:android = "http://schemas.android.com/apk/res/android">

    < item android: drawable = " @ drawable/scan _ selected _ item" android: state _ selected =
"true"/>
    < item android:drawable = "@drawable/scan_item"/>

</selector >
```

代码文件：codes\11\11.2\QR where\res\drawable\scan_tab_btn.xml

在该 drawable 资源文件中，定义了选项选中时所对应的样式以及选项未选中时所对应的样式，也就是说，通过以上资源文件，实现当选中底部菜单栏中的第一个菜单时，第一个菜单图标将是图片文件 scan_selected_item，该图片是一张背景颜色为黄色的 png 图片，当选中其他菜单项时，第一个菜单按钮将失去焦点，不再是选中状态，则第一个菜单按钮所对应的图标显示为图片文件 scan_item，该图片是一张背景颜色为黑色的 png 图片，该图片与 scan_selected_item. png 除了图片颜色不一样之外，其他都一样。

因此，需要为底部四个菜单按钮分别准备两张除了颜色不一样，其他完全一样的图片，然后分别定制类似于 scan_tab_btn. xml 这种资源文件，供程序中为底部菜单按钮设置 drawable 资源。实现底部菜单项选中与未选中时显示不同的按钮颜色。其他菜单按钮所对应的 drawable 资源文件，可以通过本书附带源码，去 scan_tab_btn. xml 所在的文件夹目录下寻找获得。

通过以上介绍的 MainActivity 类中的 initBottomMenu()方法以及 getTabItemView (int index)方法，已经实现了 QR where 底部菜单栏以及选中任意菜单项后的跳转。对于各个菜单项所对应的 Fragment，则将在接下来的章节中进行介绍。

根据 QR where 的功能需求，我们知道，在 Scan 菜单所对应的 Fragment 中的左上角有一个相册按钮，单击该按钮，将会调出系统相册，选中任意一张图片，将会返回到 MainActivity 中，在 MainActivity 中对返回的图片进行二维码解析。因此，在 MainActivity 类中需要重写 Activity 的 onActivityResult(int requestCode, int resultCode, Intent data) 方法。下面给出整个 MainActivity 类的源码：

```
package net.takewin.qrwhere.activity;

import java.util.Hashtable;

import net.takewin.qrwhere.R;
import net.takewin.qrwhere.fragment.GeneratorFragment;
```

```
import net.takewin.qrwhere.fragment.HistoryFragment;
import net.takewin.qrwhere.fragment.ScanFragment;
import net.takewin.qrwhere.fragment.SettingFragment;
import net.takewin.qrwhere.util.CommonUtil;
import net.takewin.qrwhere.util.DBManager;
import net.takewin.qrwhere.view.MyScanDialog;
import net.takewin.qrwhere.zxing.decoding.RGBLuminanceSource;
import android.app.Dialog;
import android.app.ProgressDialog;
import android.content.Intent;
import android.database.Cursor;
import android.graphics.Bitmap;
import android.graphics.BitmapFactory;
import android.os.Bundle;
import android.os.Handler;
import android.os.Message;
import android.provider.MediaStore;
import android.support.v4.app.FragmentActivity;
import android.support.v4.app.FragmentTabHost;
import android.text.TextUtils;
import android.util.Log;
import android.view.LayoutInflater;
import android.view.View;
import android.widget.ImageView;
import android.widget.TabHost.TabSpec;

import com.google.zxing.BinaryBitmap;
import com.google.zxing.ChecksumException;
import com.google.zxing.DecodeHintType;
import com.google.zxing.FormatException;
import com.google.zxing.NotFoundException;
import com.google.zxing.Result;
import com.google.zxing.common.HybridBinarizer;
import com.google.zxing.qrcode.QRCodeReader;

public class MainActivity extends FragmentActivity {
    private FragmentTabHost tabHost;
    private LayoutInflater layoutInflater;
    //定义数组来存放四个菜单项所对应的 Fragment
    @SuppressWarnings("rawtypes")
      private Class fragmentArr[] = { ScanFragment.class, HistoryFragment.class,
GeneratorFragment.class, SettingFragment.class };
    //定义数组来存放底部按钮图片
    private int tabImageArr[] = { R.drawable.scan_tab_btn, R.drawable.history_tab_btn, R.
drawable.generator_tab_btn, R.drawable.setting_tab_btn };
    //定义数组来存放底部选项卡文字
    private String tabTextArr[] = { "scan", "history", "generator", "setting" };

    //下面声明打开手机相册利用 ZXing 扫描的相关变量
    private static final int REQUEST_CODE = 100;
    private ProgressDialog mProgress;
```

```
    private Bitmap scanBitmap;
    private String photo_path;
    private static final int PARSE_BARCODE_SUC = 300;
    private static final int PARSE_BARCODE_FAIL = 303;
    //声明 MainActivity.dbManager 对象
    public static DBManager dbManager;

    @Override
    protected void onCreate(Bundle savedInstanceState) {
        super.onCreate(savedInstanceState);
        setContentView(R.layout.activity_main);
        //创建 MainActivity.dbManager 对象,
        //各个 Fragment 将调用该 dbManager 实现数据库操作
        dbManager = new DBManager(MainActivity.this);
        initBottomMenu();                   //实现底部菜单
    }

    @Override
    protected void onActivityResult(int requestCode, int resultCode, Intent data) {
        super.onActivityResult(requestCode, resultCode, data);
        if (resultCode == RESULT_OK) {
            switch (requestCode) {
            case REQUEST_CODE:
                //获取选中图片的路径
                Cursor cursor = MainActivity.this.getContentResolver().query(data.getData(),
null, null, null, null);
                if (cursor.moveToFirst()) {
                        photo_path = cursor.getString(cursor.getColumnIndex(MediaStore.
Images.Media.DATA));
                }
                cursor.close();
                mProgress = new ProgressDialog(MainActivity.this);
                mProgress.setMessage("正在扫描...");
                mProgress.setCancelable(false);
                mProgress.show();

                new Thread(new Runnable() {
                                    //由于在线程中,所以需要利用 Handler 机制实现 UI 操作,
                    @Override                   //例如弹出对话框或页面跳转
                    public void run() {
                        //解析返回的图片,com.google.zxing.Result 对象 result 中包含了二维
                          码文本内容
                        Result result = scanningImage(photo_path);
                        if (result != null) {  //表示解析成功
                            Message m = mHandler.obtainMessage();
                            m.what = PARSE_BARCODE_SUC;
                            m.obj = result.getText();
                            mHandler.sendMessage(m);
                        } else {                    //解析失败
                            Message m = mHandler.obtainMessage();
                            m.what = PARSE_BARCODE_FAIL;
```

```
                        m.obj = "Scan failed!";
                        mHandler.sendMessage(m);
                    }
                }
            }).start();
            break;
        }
    }
}

/**
 * 传入图片路径,对该图片进行解析,
 * 若是二维码图片并且解析成功,则返回对应的 com.google.zxing.Result 对象,
 * 若不是二维码图片或其他原因造成的解析失败,则返回 null
 */
private Result scanningImage(String path) {
    if (TextUtils.isEmpty(path)) {
        return null;
    }
    Hashtable<DecodeHintType, String> hints = new Hashtable<DecodeHintType, String>();
    hints.put(DecodeHintType.CHARACTER_SET, "UTF8");     //设置二维码内容的编码

    BitmapFactory.Options options = new BitmapFactory.Options();
    options.inJustDecodeBounds = true;                   //先获取原大小
    scanBitmap = BitmapFactory.decodeFile(path, options);
    options.inJustDecodeBounds = false;                  //获取新的大小
    int sampleSize = (int) (options.outHeight / (float) 200);
    if (sampleSize <= 0)
        sampleSize = 1;
    options.inSampleSize = sampleSize;
    scanBitmap = BitmapFactory.decodeFile(path, options);
    RGBLuminanceSource source = new RGBLuminanceSource(scanBitmap);
    BinaryBitmap bitmap1 = new BinaryBitmap(new HybridBinarizer(source));
    QRCodeReader reader = new QRCodeReader();
    try {
        return reader.decode(bitmap1, hints);
    } catch (NotFoundException e) {
        e.printStackTrace();
    } catch (ChecksumException e) {
        e.printStackTrace();
    } catch (FormatException e) {
        e.printStackTrace();
    }
    return null;
}

private Handler mHandler = new Handler() {
    @Override
    public void handleMessage(Message msg) {
        super.handleMessage(msg);
        mProgress.dismiss();
```

```
            switch (msg.what) {
            case PARSE_BARCODE_SUC:
                //根据不同的文本内容,保存数据记录到 history 表中并跳转到不同类型的页面
                onResultHandler((String) msg.obj, scanBitmap);
                break;
            case PARSE_BARCODE_FAIL:
                doDialog();                                    //弹出对话框
                break;
            }
        }
    };

    /**
     * 根据解析出来的文本内容,实现保存数据记录到 history 表中并跳转到不同的页面
     */
    private void onResultHandler(String resultString, Bitmap bitmap) {
        if (TextUtils.isEmpty(resultString)) {
            doDialog();
            return;
        }
        Intent intent = null;
        if (CommonUtil.isURL(resultString)) {
            Bundle bundle = new Bundle();
            intent = new Intent(MainActivity.this, ScanResultActivity.class);
            bundle.putString("class", "ScanFragment");
            bundle.putString("result", resultString);
            intent.putExtras(bundle);
        } else if (resultString.contains("BEGIN:VCARD")) {
            String[] content1s = null;
            content1s = resultString.split("\r\n");
            if (content1s.length > 1) {
            } else {
                content1s = resultString.split("\n");
            }
            String nameStr = "", phoneStr = "", emailStr = "", companyStr = "", titleStr =
"", addressStr = "", websiteStr = "";
            for (String content : content1s) {
                if (content.contains("FN:")) {
                    String[] name = content.split(":");
                    nameStr = name[1];
                }
                if (content.contains("TEL:")) {
                    String[] phone = content.split(":");
                    phoneStr = phone[1];
                }
                if (content.contains("EMAIL:")) {
                    String[] email = content.split(":");
                    emailStr = email[1];
                }
                if (content.contains("ORG:")) {
                    String[] company = content.split(":");
```

```
                companyStr = company[1];
            }
            if (content.contains("TITLE:")) {
                String[] title = content.split(":");
                titleStr = title[1];
            }
            if (content.contains("ADR:")) {
                String[] address = content.split(":");
                addressStr = address[1];
            }
            if (content.contains("URL:")) {
                String[] website = content.split(":");
                websiteStr = website[1];
            }
        }
        intent = new Intent(MainActivity.this, ContactDetailActivity.class);
        intent.putExtra("all", resultString);
        intent.putExtra("name", nameStr);
        intent.putExtra("phone", phoneStr);
        intent.putExtra("email", emailStr);
        intent.putExtra("company", companyStr);
        intent.putExtra("title", titleStr);
        intent.putExtra("address", addressStr);
        intent.putExtra("website", websiteStr);
        intent.putExtra("content1", resultString);
    } else if (resultString.contains("geo:")) {
        String[] locations = resultString.split(":");
        locations = locations[1].split(",");
        String lat = locations[0];
        locations = locations[1].split("[?]");
        String lon = locations[0];
        String nameStr = "Geo:" + lat + "," + lon;
        intent = new Intent(MainActivity.this, MapResultActivity.class);
        intent.putExtra("flag", "scan");
        intent.putExtra("title", nameStr);
        intent.putExtra("content1", resultString);
        intent.putExtra("lat", lat);
        intent.putExtra("lon", lon);
        if (locations != null && locations.length > 1) {
            locations = locations[1].split("=");
            String qStr = locations[1];
            intent.putExtra("q", qStr);
        }
    } else if (resultString.contains("WIFI:")) {
        Log.i("wifiStr", resultString);
        intent = new Intent(MainActivity.this, WifiResultActivity.class);
        Bundle bundle = new Bundle();
        bundle.putString("class", "ScanFragment");
        bundle.putString("result", resultString);
        intent.putExtras(bundle);
    } else if (resultString.contains("BEGIN:VEVENT")) {
```

```
            intent = new Intent(MainActivity.this, CalendarDetailActivity.class);
            Bundle bundle = new Bundle();
            bundle.putString("class", "scan");
            bundle.putString("calendar", resultString);
            intent.putExtras(bundle);
        } else {
            intent = new Intent(MainActivity.this, ElseDetailActivity.class);
            Bundle bundle = new Bundle();
            bundle.putString("class", "scan");
            bundle.putString("else", resultString);
            intent.putExtras(bundle);
        }
        MainActivity.this.startActivity(intent);
    }

    /**
     * 弹出对话框显示 No barcode detected.(未检测到的条形码。)
     */
    private void doDialog() {
        Dialog dialog = new MyScanDialog(MainActivity.this, R.style.MyScanDialog);
        dialog.show();
    }

    /**
     * 实现底部菜单
     */
    private void initBottomMenu() {
        //实例化布局对象
        layoutInflater = LayoutInflater.from(this);
        tabHost = (FragmentTabHost) this.findViewById(android.R.id.tabhost);
        tabHost.setup(this, getSupportFragmentManager(), R.id.realtabcontent);
        int count = fragmentArr.length;
        for (int i = 0; i < count; i++) {
            //调用 getTabItemView 方法为每一个 Tab 按钮设置图标、文字
            TabSpec tabSpec = tabHost.newTabSpec(tabTextArr[i]).setIndicator
            (getTabItemView(i));
            tabHost.addTab(tabSpec, fragmentArr[i], null);
            tabHost.getTabWidget().getChildAt(i).setBackgroundResource(R.drawable.
            selector_tab_background);
        }
    }

    /**
     * 给 Tab 按钮设置图标和文字
     */
    private View getTabItemView(int index) {
        //加载底部菜单布局,在该布局中只放置了存放底部菜单图标的 ImageView,并没有在图标
          下方放置文本组件,
        //所以在底部菜单的图标下方并没有显示菜单文字
        View view = layoutInflater.inflate(R.layout.tab_item_view, null);
        ImageView imageView = (ImageView) view.findViewById(R.id.imageview);
```

```
            imageView.setImageResource(tabImageArr[index]);
            return view;
        }
    }
```

代码文件：codes\11\11.2\QR where\cn\edu\hstc\qrwhere\activity\MainActivity.java

上面程序重写了 onActivityResult(int requestCode, int resultCode, Intent data)方法，在该方法中实现获取选中图片的路径，当然，调出系统相册功能则是放在了第一个菜单项所对应的 Fragment 中去实现的。在此，我们先不用关心此处代码实现。获取选中图片的路径后，弹出一个进度条 ProgressDialog 提示用户“正在扫描...”，同时开启一个线程，在该线程中调用 scanningImage(String path)方法，传入选中图片路径，对该图片进行二维码解析。

当前市面上主流的二维码扫描和解析、生成的框架有 ZBar 框架以及谷歌的开源 ZXing 框架。QR where 同时使用了这两个框架。为了使用 ZBar 和 ZXing 框架，读者需要到官方下载 ZBar 和 ZXing 源码库，并将 zbar.jar、zxing.jar 添加到项目 libs 路径下。在本书所附带 QR where 的源码项目中，自然也已经包含了 zbar.jar、zxing.jar，读者可以直接将这两个 jar 包集成到其他需要利用 ZBar 或 ZXing 开发二维码应用的项目中。在 QR where 中，调用手机相册返回图片后，采用的是 ZXing 框架对图片进行二维码解析。使用手机摄像头直接扫描二维码后则是采用 ZBar 框架对其进行解析。本应用中同时使用了 ZBar 以及 ZXing，目的就是为了能让读者通过本应用同时了解市面上这两大主流的二维码和条形码识别工具。

在二维码解析的方法 scanningImage(String path)中，根据传入的图片路径，生成一个 android.graphics.Bitmap 对象 scanBitmap。而二维码解析这部分的真正实现，则是通过继承了 com.google.zxing.LuminanceSource 的 RGBLuminanceSource 类以及 com.google.zxing.qrcode.QRCodeReader 的配合来完成的。RGBLuminanceSource 类用于帮助图像文件到 Android 位图的 RGB 数据的解码，该类的源码在此处不作展示，读者可以通过本书附带源码获得。程序中调用 com.google.zxing.qrcode.QRCodeReader 的 decode 方法返回了一个 com.google.zxing.Result 对象，在该对象中包含了解析出来的二维码文本内容。若图片中不包含二维码信息，或其他原因所造成的解析异常，则方法返回 null。具体的二维码解析操作，请参见程序中的 Result scanningImage(String path)方法。

由于程序使用线程来调用 scanningImage(String path)方法操作二维码的解析，在子线程中不能直接操作 UI 主线程。因此，需要利用 Android Handler 机制，根据解析出来的不同结果，实现不同的操作。在 Handler 类的实现中，停止了进度条使其消失，并根据解析结果的不同进行不同的操作。程序实现当解析异常时，弹出一个对话框提示用户未检测到的条形码。如果解析成功，则会调用 onResultHandler 方法，根据解析出来的不同文本，对文本进行不同的包装处理，然后传入消息传递类 Bundle 的对象中，带着 Bundle 数据跳转到包括 URL、Contact、Location、WiFi 四种不同类型的详情界面中或跳转到一个普通的页面显示文本以及二维码。具体实现请参照程序中的 onResultHandler(String resultString, Bitmap bitmap)方法。至于带着这些文本信息跳转后的详情页面，我们暂且不用关心，后面将会具体介绍。

11.3 开发第一个菜单项所对应的界面 ScanFragment

根据上面功能需求的介绍，我们了解到，在第一个菜单项所对应的界面中，在左上角放置了一个相册按钮用于调出手机系统自带的相册，右上角放置了一个按钮用于打开或关闭手机闪光灯功能。然后，我们还需要根据其业务，在屏幕的中间放置一个手机摄像头预览区，该区域由 SurfaceView 来完成。SurfaceView 可以直接从内存或者 DMA 等硬件接口取得图像数据，是非常重要的绘图容器。也就是说，SurfaceView 可以将手机摄像头所捕捉到的画面绘制进去。还需要在界面中放置一个自定义的 View 组件，该 View 组件称为 FinderView，继承自 View 类，该自定义类主要用于在界面中画出扫描框。ScanFragment 所对应的界面布局代码如下：

```
<LinearLayout xmlns:android = "http://schemas.android.com/apk/res/android"
    android:layout_width = "match_parent"
    android:layout_height = "match_parent"
    android:background = "@color/white"
    android:orientation = "vertical" >

    <!-- 顶部标题 -->
    <RelativeLayout
        android:layout_width = "fill_parent"
        android:layout_height = "wrap_content"
        android:background = "@color/backhost"
        android:padding = "10dip" >

        <ImageView
            android:id = "@ + id/getPhotoImageView"
            android:layout_width = "wrap_content"
            android:layout_height = "wrap_content"
            android:layout_alignParentLeft = "true"
            android:layout_centerVertical = "true"
            android:src = "@drawable/camera" />

        <TextView
            android:id = "@ + id/titleTextView"
            android:layout_width = "wrap_content"
            android:layout_height = "wrap_content"
            android:layout_centerHorizontal = "true"
            android:layout_centerVertical = "true"
            android:paddingTop = "10dp"
            android:paddingBottom = "10dp"
            android:text = "@string/scan"
            android:textColor = "@color/white"
            android:textSize = "20sp" />

        <ImageView
            android:id = "@ + id/openFlashLampImageView"
```

```
            android:layout_width = "wrap_content"
            android:layout_height = "wrap_content"
            android:layout_alignParentRight = "true"
            android:layout_centerVertical = "true"
            android:src = "@drawable/flash" />
    </RelativeLayout>

    <!-- 二维码扫描区 -->
    <FrameLayout
        android:layout_width = "match_parent"
        android:layout_height = "match_parent" >

        <SurfaceView
            android:id = "@ + id/surface_view"
            android:layout_width = "match_parent"
            android:layout_height = "match_parent" />

        <net.takewin.qrwhere.view.FinderView
            android:id = "@ + id/finder_view"
            android:layout_width = "match_parent"
            android:layout_height = "match_parent" />
    </FrameLayout>

</LinearLayout>
```

代码文件：codes\11\11.2\QR where\res\layout\activity_scan.xml

以上布局文件使用了一个自定义的 View 组件 FinderView，该 FinderView 源码如下：

```
package net.takewin.qrwhere.view;

import net.takewin.qrwhere.R;
import android.content.Context;
import android.graphics.Canvas;
import android.graphics.Paint;
import android.graphics.Rect;
import android.graphics.drawable.Drawable;
import android.util.AttributeSet;
import android.view.View;

public class FinderView extends View {
    private static final long ANIMATION_DELAY = 30;
    private Paint finderMaskPaint;
    private int measureedWidth;
    private int measureedHeight;

    public FinderView(Context context) {
        super(context);
        init(context);
    }

    public FinderView(Context context, AttributeSet attrs) {
```

```
        super(context, attrs);
        init(context);
    }

    @Override
    protected void onDraw(Canvas canvas) {
        super.onDraw(canvas);
        //canvas.drawRect(leftRect, finderMaskPaint);
        //canvas.drawRect(topRect, finderMaskPaint);
        //canvas.drawRect(rightRect, finderMaskPaint);
        //canvas.drawRect(bottomRect, finderMaskPaint);
        //画框
        zx_code_kuang.setBounds(middleRect);
        zx_code_kuang.draw(canvas);
        if (lineRect.bottom < middleRect.bottom) {
            zx_code_line.setBounds(lineRect);
            lineRect.top = lineRect.top + lineHeight / 2;
            lineRect.bottom = lineRect.bottom + lineHeight / 2;
        } else {
            lineRect.set(middleRect);
            lineRect.bottom = lineRect.top + lineHeight;
            zx_code_line.setBounds(lineRect);
        }
        //zx_code_line.draw(canvas);
        postInvalidateDelayed(ANIMATION_DELAY, middleRect.left, middleRect.top, middleRect.
right, middleRect.bottom);
    }

    private Rect topRect = new Rect();
    private Rect bottomRect = new Rect();
    private Rect rightRect = new Rect();
    private Rect leftRect = new Rect();
    private Rect middleRect = new Rect();

    private Rect lineRect = new Rect();
    private Drawable zx_code_kuang;
    private Drawable zx_code_line;
    private int lineHeight;

    private void init(Context context) {
        int finder_mask = context.getResources().getColor(R.color.finder_mask);
        finderMaskPaint = new Paint(Paint.ANTI_ALIAS_FLAG);
        finderMaskPaint.setColor(finder_mask);
        zx_code_kuang = context.getResources().getDrawable(R.drawable.scan_view);
        zx_code_kuang.setAlpha(110);
        zx_code_line = context.getResources().getDrawable(R.drawable.zx_code_line);
        lineHeight = 30;
    }

    //////////////新增该方法//////////////////////
    /**
```

```
     * 根据图片 size 求出矩形框在图片所在位置,tip: 相机旋转 90 度以后,拍摄的图片是横着
的,所有传递参数时,做了交换
     * @param w
     * @param h
     * @return
     */
    public Rect getScanImageRect(int w, int h) {
        //先求出实际矩形
        Rect rect = new Rect();
        rect.left = middleRect.left;
        rect.right = middleRect.right;
        float temp = h / (float) measureedHeight;
        rect.top = (int) (middleRect.top * temp);
        rect.bottom = (int) (middleRect.bottom * temp);
        return rect;
    }

    ////////////////////////////////////
    @Override
    protected void onMeasure(int widthMeasureSpec, int heightMeasureSpec) {
        super.onMeasure(widthMeasureSpec, heightMeasureSpec);
        measureedWidth = MeasureSpec.getSize(widthMeasureSpec);
        measureedHeight = MeasureSpec.getSize(heightMeasureSpec);
        int borderWidth = measureedWidth / 2 + 100;
        middleRect.set((measureedWidth - borderWidth) / 2, (measureedHeight - borderWidth) / 2,
(measureedWidth - borderWidth) / 2 + borderWidth, (measureedHeight - borderWidth) / 2 +
borderWidth);
        lineRect.set(middleRect);
        lineRect.bottom = lineRect.top + lineHeight;
        leftRect.set(0, middleRect.top, middleRect.left, middleRect.bottom);
        topRect.set(0, 0, measureedWidth, middleRect.top);
        rightRect.set(middleRect.right, middleRect.top, measureedWidth, middleRect.bottom);
        bottomRect.set(0, middleRect.bottom, measureedWidth, measureedHeight);
    }
}
```

代码文件: codes\11\11.2\QR where\cn\edu\hstc\qrwhere\view\FinderView.java

上面的程序自定义了一个继承自 android.view.View 的类 FinderView,该类重写了 onDraw(Canvas canvas)方法,画出了一个取景框,二维码扫描框就是依靠这个类画出来的。

接下来,直接看看第一个菜单项所对应的 Fragment 的实现源码:

```
package net.takewin.qrwhere.fragment;

import net.sourceforge.zbar.Config;
import net.sourceforge.zbar.Image;
import net.sourceforge.zbar.ImageScanner;
import net.sourceforge.zbar.Symbol;
import net.sourceforge.zbar.SymbolSet;
import net.takewin.qrwhere.R;
import net.takewin.qrwhere.activity.CalendarDetailActivity;
import net.takewin.qrwhere.activity.ContactDetailActivity;
```

```
import net.takewin.qrwhere.activity.ElseDetailActivity;
import net.takewin.qrwhere.activity.MapResultActivity;
import net.takewin.qrwhere.activity.ScanResultActivity;
import net.takewin.qrwhere.activity.WifiResultActivity;
import net.takewin.qrwhere.util.CommonUtil;
import net.takewin.qrwhere.view.FinderView;
import android.content.Intent;
import android.content.pm.FeatureInfo;
import android.content.pm.PackageManager;
import android.graphics.Rect;
import android.hardware.Camera;
import android.hardware.Camera.AutoFocusCallback;
import android.hardware.Camera.Parameters;
import android.hardware.Camera.PreviewCallback;
import android.hardware.Camera.Size;
import android.os.AsyncTask;
import android.os.Bundle;
import android.os.Handler;
import android.support.v4.app.Fragment;
import android.util.Log;
import android.view.LayoutInflater;
import android.view.SurfaceHolder;
import android.view.SurfaceView;
import android.view.View;
import android.view.ViewGroup;
import android.widget.ImageView;
import android.widget.Toast;

public class ScanFragment extends Fragment implements SurfaceHolder.Callback {
    private static String TAG = "xiaoqiang";
    private Camera mCamera;
    private SurfaceHolder mHolder;
    private SurfaceView surface_view;
    private ImageScanner scanner;
    private Handler autoFocusHandler;
    private AsyncDecode asyncDecode;
    private FinderView finder_view;

    private ImageView getPhotoImageView;                    //打开手机相册按钮
    private ImageView getFlashImageView;                    //打开手机闪光灯按钮
    private Parameters parameters = null;
    private static final int REQUEST_CODE = 100;
    private boolean flag, b;

    static {
        System.loadLibrary("iconv");
    }

    @Override
    public View onCreateView(LayoutInflater inflater, ViewGroup container, Bundle savedInstanceState)
{
```

```
        return inflater.inflate(R.layout.activity_scan, null);
    }

    @Override
    public void onActivityCreated(Bundle savedInstanceState) {
        super.onActivityCreated(savedInstanceState);
        initView();
    }

    @Override
    public void onResume() {
        super.onResume();
        flag = true;
        b = true;
    }

    private void initView() {
        surface_view = (SurfaceView) getActivity().findViewById(R.id.surface_view);
        finder_view = (FinderView) getActivity().findViewById(R.id.finder_view);
        mHolder = surface_view.getHolder();
        mHolder.setType(SurfaceHolder.SURFACE_TYPE_PUSH_BUFFERS);
        mHolder.addCallback(this);
        scanner = new ImageScanner();
        scanner.setConfig(0, Config.X_DENSITY, 3);
        scanner.setConfig(0, Config.Y_DENSITY, 3);
        autoFocusHandler = new Handler();
        asyncDecode = new AsyncDecode();
        initGetPhotoImageView();
        initGetFlashImageView();
    }

    private void initGetPhotoImageView() {
        getPhotoImageView = (ImageView) getActivity().findViewById(R.id.
        getPhotoImageView);
        getPhotoImageView.setOnClickListener(new View.OnClickListener() {
            @Override
            public void onClick(View v) {
                //打开手机中的相册
                Intent innerIntent = new Intent(Intent.ACTION_GET_CONTENT); //"android.
intent.action.GET_CONTENT"
                innerIntent.setType("image/*");
                Intent wrapperIntent = Intent.createChooser(innerIntent, "选择二维码图片");
                getActivity().startActivityForResult(wrapperIntent, REQUEST_CODE);
            }
        });
    }

    private void initGetFlashImageView() {
        getFlashImageView = (ImageView) getActivity().findViewById(R.id.
        openFlashLampImageView);
        getFlashImageView.setOnClickListener(new View.OnClickListener() {
```

```
            @Override
            public void onClick(View v) {
                PackageManager pm = ScanFragment.this.getActivity().getPackageManager();
                FeatureInfo[] features = pm.getSystemAvailableFeatures();
                for (FeatureInfo f : features) {
                    if (PackageManager.FEATURE_CAMERA_FLASH.equals(f.name)) {
                        if (mCamera != null) {
                            mCamera.startPreview();
                            parameters = mCamera.getParameters();
                             if (parameters.getFlashMode().equals(Parameters.FLASH_MODE_
OFF)) {

                                parameters.setFlashMode(Parameters.FLASH_MODE_TORCH);
                                mCamera.setParameters(parameters);
                            } else {
                                parameters.setFlashMode(Parameters.FLASH_MODE_OFF);
                                mCamera.setParameters(parameters);
                            }
                        } else {
                            Toast.makeText(getActivity(), "Fail to open the flash", 3000).
show();
                        }
                    }
                }
            }
        });
    }

    @Override
    public void surfaceChanged(SurfaceHolder holder, int format, int width, int height) {
        if (mHolder.getSurface() == null) {
            return;
        }
        try {
            if (mCamera != null) {
                mCamera.stopPreview();
            }
        } catch (Exception e) {
        }
        try {
            mCamera.setDisplayOrientation(90);
            mCamera.setPreviewDisplay(mHolder);
            mCamera.setPreviewCallback(previewCallback);
            mCamera.startPreview();
            mCamera.autoFocus(autoFocusCallback);
        } catch (Exception e) {
            Log.d("DBG", "Error starting camera preview: " + e.getMessage());
        }
    }

    /**
     * 预览数据
```

```
    */
    PreviewCallback previewCallback = new PreviewCallback() {
        public void onPreviewFrame(byte[] data, Camera camera) {
            if (asyncDecode.isStoped()) {
                Camera.Parameters parameters = camera.getParameters();
                Size size = parameters.getPreviewSize();
                //图片是被旋转了 90 度的
                Image source = new Image(size.width, size.height, "Y800");
                Rect scanImageRect = finder_view.getScanImageRect(size.height, size.width);
                //图片旋转了 90 度,将扫描框的 TOP 作为 left 裁剪
                source.setCrop(scanImageRect.top, scanImageRect.left, scanImageRect.
                bottom, scanImageRect.right);
                source.setData(data);
                asyncDecode = new AsyncDecode();
                asyncDecode.execute(source);
            }
        }
    };

    private class AsyncDecode extends AsyncTask<Image, Void, Void> {
        private boolean stoped = true;
        private String str = "";

        @Override
        protected Void doInBackground(Image... params) {
            if (flag) {
                stoped = false;
                StringBuilder sb = new StringBuilder();
                Image barcode = params[0];
                int result = scanner.scanImage(barcode);
                if (result != 0) {
                    SymbolSet syms = scanner.getResults();
                    for (Symbol sym : syms) {
                        switch (sym.getType()) {
                        case Symbol.CODABAR:
                            Log.d(TAG, "条形码  " + sym.getData());
                            //条形码
                            sb.append(sym.getData() + "\n");
                            break;
                        case Symbol.CODE128:
                            //128 编码格式二维码
                            Log.d(TAG, "128 编码格式二维码:  " + sym.getData());
                            sb.append(sym.getData() + "\n");
                            break;
                        case Symbol.QRCODE:
                            //QR 码二维码
                            Log.d(TAG, "QR 码二维码  :" + sym.getData());
                            sb.append(sym.getData() + "\n");
                            break;
                        case Symbol.ISBN10:
                            //ISBN10 图书查询
```

```
                Log.d(TAG, "ISBN10 图书查询   :   " + sym.getData());
                sb.append(sym.getData() + "\n");
                break;
            case Symbol.ISBN13:
                //ISBN13 图书查询
                Log.d(TAG, "ISBN13 图书查询   : " + sym.getData());
                sb.append(sym.getData() + "\n");
                break;
            case Symbol.NONE:
                Log.d(TAG, "未知   : " + sym.getData());
                sb.append(sym.getData() + "\n");
                break;
            default:
                Log.d(TAG, "其他:   " + sym.getData());
                sb.append(sym.getData() + "\n");
                break;
            }
        }
    }
    str = sb.toString();
    }
    return null;
}

@Override
protected void onPostExecute(Void result) {
    super.onPostExecute(result);
    stoped = true;
    if (null == str || str.equals("")) {
    } else {
        if (b) {
            b = false;
            Intent intent = null;
            Bundle bundle = new Bundle();
            if (CommonUtil.isURL(str) == true) {
                intent = new Intent(getActivity(), ScanResultActivity.class);
                bundle.putString("class", "ScanFragment");
                bundle.putString("result", str);
                intent.putExtras(bundle);
            } else if (str.contains("BEGIN:VCARD")) {
                String[] content1s = null;
                content1s = str.split("\r\n");
                if (content1s.length > 1) {
                } else {
                    content1s = str.split("\n");
                }
                String nameStr = "", phoneStr = "", emailStr = "", companyStr =
"", titleStr = "", addressStr = "", websiteStr = "";
                for (String content : content1s) {
                    if (content.contains("FN:")) {
                        String[] name = content.split(":");
```

```
                    nameStr = name[1];
                }
                if (content.contains("TEL:")) {
                    String[] phone = content.split(":");
                    phoneStr = phone[1];
                }
                if (content.contains("EMAIL:")) {
                    String[] email = content.split(":");
                    emailStr = email[1];
                }
                if (content.contains("ORG:")) {
                    String[] company = content.split(":");
                    companyStr = company[1];
                }
                if (content.contains("TITLE:")) {
                    String[] title = content.split(":");
                    titleStr = title[1];
                }
                if (content.contains("ADR:")) {
                    String[] address = content.split(":");
                    addressStr = address[1];
                }
                if (content.contains("URL:")) {
                    String[] website = content.split(":");
                    websiteStr = website[1];
                }
            }
            intent = new Intent(getActivity(), ContactDetailActivity.class);
            intent.putExtra("all", str);
            intent.putExtra("name", nameStr);
            intent.putExtra("phone", phoneStr);
            intent.putExtra("email", emailStr);
            intent.putExtra("company", companyStr);
            intent.putExtra("title", titleStr);
            intent.putExtra("address", addressStr);
            intent.putExtra("website", websiteStr);
            intent.putExtra("content1", str);
        } else if (str.contains("geo:")) {
            Log.i("location", str);
            String[] locations = str.split(":");
            locations = locations[1].split(",");
            String lat = locations[0];
            locations = locations[1].split("[?]");
            String lon = locations[0];
            String nameStr = "Geo:" + lat + "," + lon;
            intent = new Intent(getActivity(), MapResultActivity.class);
            intent.putExtra("flag", "scan");
            intent.putExtra("title", nameStr);
            intent.putExtra("content1", str);
            intent.putExtra("lat", lat);
            intent.putExtra("lon", lon);
```

```
                    if (locations != null && locations.length > 1) {
                        locations = locations[1].split("=");
                        String qStr = locations[1];
                        intent.putExtra("q", qStr);
                    }
                } else if (str.contains("WIFI:")) {
                    intent = new Intent(getActivity(), WifiResultActivity.class);
                    bundle.putString("class", "ScanFragment");
                    bundle.putString("result", str);
                    intent.putExtras(bundle);
                } else if (str.contains("BEGIN:VEVENT")) {
                    intent = new Intent(ScanFragment.this.getActivity(),
                    CalendarDetailActivity.class);
                    bundle = new Bundle();
                    bundle.putString("class", "scan");
                    bundle.putString("calendar", str);
                    intent.putExtras(bundle);
                } else {
                    intent = new Intent(ScanFragment.this.getActivity(),
                    ElseDetailActivity.class);
                    bundle = new Bundle();
                    bundle.putString("class", "scan");
                    bundle.putString("else", str);
                    intent.putExtras(bundle);
                }
                getActivity().startActivity(intent);
                flag = false;
            }
        }
    }

    public boolean isStoped() {
        return stoped;
    }
}

/**
 * 自动对焦回调
 */
AutoFocusCallback autoFocusCallback = new AutoFocusCallback() {
    public void onAutoFocus(boolean success, Camera camera) {
        autoFocusHandler.postDelayed(doAutoFocus, 1000);
    }
};

//自动对焦
private Runnable doAutoFocus = new Runnable() {
    public void run() {
        if (null == mCamera || null == autoFocusCallback) {
            return;
        }
```

```
                mCamera.autoFocus(autoFocusCallback);
            }
        };

        @Override
        public void surfaceCreated(SurfaceHolder holder) {
            try {
                if (mCamera == null) {
                    mCamera = Camera.open();
                }
            } catch (Exception e) {
                mCamera = null;
            }
        }

        @Override
        public void surfaceDestroyed(SurfaceHolder holder) {
            if (mCamera != null) {
                mCamera.setPreviewCallback(null);
                mCamera.release();
                mCamera = null;
            }
        }
    }
```

代码文件：codes\11\11.2\QR where\cn\edu\hstc\qrwhere\fragment\ScanFragment.java

上面的程序实现了第一个菜单项所对应的界面。ScanFragment 类需要继承 Fragment 并且实现 android. view. SurfaceHolder. Callback。由于这里实现的是一个 Fragment，因此需要重写 onCreateView 方法，并在该方法中加载界面布局，如上面程序中的代码片段：

```
@Override
public View onCreateView (LayoutInflater inflater, ViewGroup container, Bundle savedInstanceState) {
    return inflater.inflate(R.layout.activity_scan, null);
}
```

上面的代码片段为固定写法。接着，程序重写 onActivityCreated 方法，在该方法中初始化所有界面中的 View 组件，包括二维码扫描解析操作、打开系统图库的按钮的实现、打开或关闭闪关灯的按钮的实现。

程序调用自定义方法 initGetPhotoImageView()，实现加载界面左上角的打开系统图库的按钮并为其添加单击事件监听器。程序直接调用 Android API 实现打开系统图库的操作。这里代码行 getActivity ( ). startActivityForResult (wrapperIntent, REQUEST_CODE)中所传入的 REQUEST_CODE 需要与 MainActivity 类中的 onActivityResult(int requestCode, int resultCode, Intent data)方法中的 REQUEST_CODE 相等。因为在这个方法中，程序将会检查 REQUEST_CODE 是否是之前传给 startActivityForResult()方法的 REQUEST_CODE，如是，则做出相应的处理。一旦用户选择了系统图库中的图片，MainActivity 类中的 onActivityResult 方法将会被调用，然后处理所得到的数据。

这里为什么可以在 ScanFragment 中调用 startActivityForResult 方法，而在 MainActivity 中的 onActivityResult 可以响应呢？那是因为 ScanFragment 始终是存在于 MainActivity 中的。注意到 startActivityForResult（wrapperIntent，REQUEST_CODE）是通过 getActivity()获取父容器，即 MainActivity 后才能调用的。

程序调用自定义方法 initGetFlashImageView()，实现加载界面右上角的按钮并为其添加单击事件监听器。程序也是通过直接调用相关 Android API 操作手机硬件，实现打开或关闭手机闪关灯功能。

程序中还实现了摄像头预览、摄像头自动对焦、自动对焦回调等 Android 多媒体功能。读者可以通过阅读源码学习了解。

程序中通过异步任务，实现二维码扫描解析操作。这样可以避免摄像头预览界面卡死。在预览数据时，生成了 net. sourceforge. zbar. Image 对象，将该对象作为源传入异步任务的 doInBackground 方法中，在该方法中调用 net. sourceforge. zbar. ImageScanner 对象的 scanImage(Image image)方法，该方法的 image 参数正是预览时生成的 net. sourceforge. zbar. Image 对象，该方法正是 ZBar 框架对二维码扫描解析的操作。从代码中可以看出，当 scanImage 方法所返回的整型返回值不为零时，代表了成功解析，则调用 net. sourceforge. zbar. ImageScanner 对象的 getResults()方法返回结果集 net. sourceforge. zbar. SymbolSet，强制循环该结果集，获取结果集中的每个 net. sourceforge. zbar. Symbol 对象，判断其类型，包括条形码、128 编码格式二维码、QR 码二维码、ISBN10 图书查询、ISBN13 图书查询等类型；然后通过 Symbol 对象的 getData()方法获取包括二维码的各种类型的格式码中所包含的文本信息，保存在一个 StringBuilder 对象中。以上正是通过 ZBar 开源库解析二维码的实现过程。

程序重写异步任务的 onPostExecute 方法，判断所解析出来的二维码文本，根据文本内容的不同，将文本内容放入 Bundle 对象中，然后跳转到不同的详情页面中。这部分与解析系统图库中的图片后所执行的跳转操作相同。

读者需要结合以上讲解，阅读 ScanFragment 类的源码加以体会，并在其他需要实现二维码扫描操作中对以上的代码加以复用。

11.4 开发第二个菜单项所对应的界面 HistoryFragment

HistoryFragment 对应第二个菜单项的应用界面。该页面主要用于查询数据库中的历史记录，并拼接在各自的 Tab 标签页中。根据之前介绍 Android SQLite 数据库的章节，我们知道，为了操作数据库的每一行数据，需要为数据库中的每一张表建立对应的实体类，将数据表中的行记录对象化。接下来看看 QR where 应用的 SQLite 数据库中唯一一张表 history 所对应的实体类 History 的源码，以了解该表的结构设计。

```
package net.takewin.qrwhere.entity;

import java.io.Serializable;

public class History implements Serializable {
```

```
    private static final long serialVersionUID = 1L;
    public int _id;             //主键 ID
    public String createTime; //行记录插入时间
    public String title;        //标题,显示在历史记录列表页面每个列表项
    public String content1;     //二维码中的文本内容
    public String content2;     //该属性对应表中的字段 content2,该字段暂时不存放任何内容(废弃)
    public String type1; // 类型,代表历史记录是解析二维码图片或是用户输入文本后生成二维码
    public String type2;        //类型,代表历史记录中所保存的信息是四大类型之一或者其他

    public History() {
    }

    public History(String createTime, String title, String content1, String content2, String
type1, String type2) {
        this.createTime = createTime;
        this.title = title;
        this.content1 = content1;
        this.content2 = content2;
        this.type1 = type1;
        this.type2 = type2;
    }

    public History(int _id, String createTime, String title, String content1, String content2,
String type1, String type2) {
        this._id = _id;
        this.createTime = createTime;
        this.title = title;
        this.content1 = content1;
        this.content2 = content2;
        this.type1 = type1;
        this.type2 = type2;
    }
}
```

代码文件: codes\11\11.2\QR where\cn\edu\hstc\qrwhere\entity\History.java

以上类定义了表 history 所对应的实体类。对 type1 属性需要进一步说明,该属性对应表中的 type1 字段,它的值只有两个:要么是 scan,要么是 generate。如果值为 scan,就代表这条记录中的 content1 字段保存的是解析二维码图片后所得到的字符串文本。如果用户单击应用底部菜单中的第三个菜单按钮,然后单击页面四个按钮中的任意一个,则跳转到对应的文本信息输入页面,输入对应的文本内容,然后单击生成按钮生成二维码图片,此时会将这些文本信息有规律地拼接在一起组成一串新的字符串,然后保存到数据表中的 content1 字段,那么此时这条数据行中的 type1 字段保存为 generate。

对实体类中的 type2 属性也需要作进一步说明。该属性对应表中的 type2 字段,它的值为"url""contact""location""wifi""else"中的任意一个。前面四个代表用户生成二维码的类型,例如,如果 type2 属性的值为"location",则代表该条记录是用户单击第三个菜单项所对应的界面中的第三个按钮后,在高德地图中定位了一个地址后单击生成按钮所保存的记录,该条记录中的 content1 字段所保存的信息是地理位置的经纬度信息。只有当记录是由

应用解析二维码图片后所保存的记录时，该条记录的type2字段的值才有可能为“else”。在这种情况下，代表应用所解析的二维码图片中所包含的文本信息不属于四大类型中的任何一种，所以归为其他类型。

通过阅读以上History类的源码以及所附加的简单说明，相信读者此时已经对数据表的结构以及各个字段的含义有所了解。在HistoryFragment中，将会分成Scan以及Generate两个Tab标签页。程序将查询应用的history表中的数据，然后根据数据的type1的值，将该条数据“归类”到对应的Tab标签页下显示。

每个Tab页的内容为一行一行的列表项拼接而成的列表。每个列表项分为左、中、右三个部分。根据查询出来的数据的type2字段中所保存的值，将会在“左”部分显示对应的图标，例如，如果该列表项的数据为联系人信息，该列表项的“左”部分将显示图标。“中”部分则显示数据的title字段所保存的值，“右”部分则为固定的图标>。

下面的XML代码为HistoryFragment所对应的界面布局源码：

```
<LinearLayout xmlns:android = "http://schemas.android.com/apk/res/android"
    android:layout_width = "match_parent"
    android:layout_height = "match_parent"
    android:background = "@color/white"
    android:orientation = "vertical" >

    <!-- 顶部标题 -->
    <RelativeLayout
        android:layout_width = "fill_parent"
        android:layout_height = "wrap_content"
        android:background = "@color/backhost"
        android:padding = "10dip" >

        <TextView
            android:id = "@ + id/titleTextView"
            android:layout_width = "wrap_content"
            android:layout_height = "wrap_content"
            android:layout_centerHorizontal = "true"
            android:layout_centerVertical = "true"
            android:paddingTop = "10dp"
            android:paddingBottom = "10dp"
            android:text = "@string/history"
            android:textColor = "@color/white"
            android:textSize = "20sp" />
    </RelativeLayout>

    <LinearLayout
        android:layout_width = "fill_parent"
        android:layout_height = "fill_parent" >

        <TabHost
            android:id = "@ + id/tabHost"
            android:layout_width = "fill_parent"
            android:layout_height = "fill_parent" >
```

```
                <RelativeLayout
                    android:layout_width="fill_parent"
                    android:layout_height="fill_parent">

                    <TabWidget
                        android:id="@android:id/tabs"
                        android:layout_width="fill_parent"
                        android:layout_height="wrap_content"
                        android:tabStripEnabled="false" />

                    <FrameLayout
                        android:id="@android:id/tabcontent"
                        android:layout_width="fill_parent"
                        android:layout_height="wrap_content"
                        android:layout_below="@android:id/tabs">

                        <ScrollView
                            android:layout_width="fill_parent"
                            android:layout_height="wrap_content">

                            <LinearLayout
                                android:id="@+id/tab1"
                                android:layout_width="fill_parent"
                                android:layout_height="wrap_content"
                                android:orientation="vertical">
                            </LinearLayout>
                        </ScrollView>

                        <ScrollView
                            android:layout_width="fill_parent"
                            android:layout_height="wrap_content">

                            <LinearLayout
                                android:id="@+id/tab2"
                                android:layout_width="fill_parent"
                                android:layout_height="wrap_content"
                                android:orientation="vertical">
                            </LinearLayout>
                        </ScrollView>
                    </FrameLayout>
                </RelativeLayout>
            </TabHost>
        </LinearLayout>

    </LinearLayout>
```

代码文件：codes\11\QR where\res\layout\activity_history.xml

上面布局在 TabHost 元素下的 FrameLayout 元素下放置了两个 ScrollView，代表了界面中的两个 Tab 标签页的内容。在这两个 ScrollView 中都包含了一个 LinearLayout 布局，在 HistoryFragment 中会将查询出来的一条一条的数据显示在一个动态创建的 View

组件中,然后将创建的 View 添加到 LinearLayout 布局中。

下面给出 HistoryFragment 的实现源码供读者阅读体会:

```
package net.takewin.qrwhere.fragment;

import java.util.ArrayList;
import java.util.List;

import net.takewin.qrwhere.R;
import net.takewin.qrwhere.activity.ContactDetailActivity;
import net.takewin.qrwhere.activity.ElseDetailActivity;
import net.takewin.qrwhere.activity.MainActivity;
import net.takewin.qrwhere.activity.MapResultActivity;
import net.takewin.qrwhere.activity.ScanResultActivity;
import net.takewin.qrwhere.activity.WifiResultActivity;
import net.takewin.qrwhere.entity.History;
import net.takewin.qrwhere.util.CommonUtil;
import android.content.Intent;
import android.database.Cursor;
import android.os.Bundle;
import android.support.v4.app.Fragment;
import android.view.LayoutInflater;
import android.view.View;
import android.view.ViewGroup;
import android.widget.ImageView;
import android.widget.LinearLayout;
import android.widget.LinearLayout.LayoutParams;
import android.widget.RelativeLayout;
import android.widget.TabHost;
import android.widget.TabHost.OnTabChangeListener;
import android.widget.TabHost.TabSpec;
import android.widget.TextView;

public class HistoryFragment extends Fragment {
    private TabHost tabHost;

    @Override
    public View onCreateView(LayoutInflater inflater, ViewGroup container, Bundle
    savedInstanceState) {
        return inflater.inflate(R.layout.activity_history, null);
    }

    @Override
    public void onActivityCreated(Bundle savedInstanceState) {
        super.onActivityCreated(savedInstanceState);
        initTabHost();
        initScanLinear();
        initGenerateLinear();
    }
```

```
/**
 * 初始化 TabHost
 */
private void initTabHost() {
    tabHost = (TabHost) getActivity().findViewById(R.id.tabHost);
    tabHost.setup();

    //新建 Tab 页,显示经过第一个菜单项,
    //对二维码图片进行解析后所保存的历史记录
    TabSpec spec1 = tabHost.newTabSpec("Scan");
    spec1.setIndicator("Scan");
    spec1.setContent(R.id.tab1);

    //新建 Tab 页,显示经过第三个菜单项,
    //所有所保存的历史记录都是经过输入文本后生成二维码的
    TabSpec spec2 = tabHost.newTabSpec("Generate");
    spec2.setIndicator("Generate");
    spec2.setContent(R.id.tab2);

    tabHost.addTab(spec1);
    tabHost.addTab(spec2);
    updateTab(tabHost);
    tabHost.setOnTabChangedListener(new OnTabChangeListener() {
        @Override
        public void onTabChanged(String tabId) {
            updateTab(tabHost);
        }
    });
}

/**
 * 控制 TabHost 的标签样式
 * @param tabHost
 */
private void updateTab(final TabHost tabHost) {
    for (int i = 0; i < tabHost.getTabWidget().getChildCount(); i++) {
        tabHost.getTabWidget().getChildAt(i).getLayoutParams().height = 80;
        View view = tabHost.getTabWidget().getChildAt(i);
        TextView tv = (TextView) tabHost.getTabWidget().getChildAt(i).findViewById
(android.R.id.title);
        tv.setTextSize(20);
        RelativeLayout.LayoutParams params = (RelativeLayout.LayoutParams) tv.
        getLayoutParams();
        params.addRule(RelativeLayout.ALIGN_PARENT_BOTTOM, 0); //取消文字底边对齐
        params.addRule(RelativeLayout.CENTER_IN_PARENT, RelativeLayout.TRUE);
        //tv.setTypeface(Typeface.SERIF, 2);                   //设置字体和风格
        if (tabHost.getCurrentTab() == i) {                     //选中
            view.setBackgroundDrawable(getResources().getDrawable(R.drawable.
            segmented_selected_bg));                            //选中后的背景
            tv.setTextColor(this.getResources().getColorStateList(R.color.backhost));
        } else {                                                //不选中
```

```
                view.setBackgroundDrawable(getResources().getDrawable(R.drawable.
                segmented_bg));                                       //非选择的背景
                 tv.setTextColor(this.getResources().getColorStateList(android.R.color.
white));
            }
        }
    }

    /**
     * Scan类型的历史记录
     */
    private void initScanLinear() {
        final LinearLayout objScanLinear = (LinearLayout) getActivity().findViewById(R.id.
tab1);
        //拼接列表
        initTabLinear(objScanLinear, "scan");
    }

    /**
     * Generate类型的历史记录
     */
    private void initGenerateLinear() {
        final LinearLayout objScanLinear = (LinearLayout) getActivity().findViewById(R.id.
tab2);
        //拼接列表
        initTabLinear(objScanLinear, "generate");
    }

    /**
     * 查询数据库,以创建时间为分类,拼接列表
     */
    private void initTabLinear(final LinearLayout objScanLinear, String type) {
        String[] typeRaw = {type};
        List<String> listCreateTime = getTime(typeRaw);
        for (String createTime : listCreateTime) {
            final List<History> temp = new ArrayList<History>();
            String[] types = new String[2];
            types[0] = type;
            types[1] = createTime;
            final List<History> listHistories = handleDB(types);
            final TextView objTitleText = new TextView(getActivity());
            objTitleText.setLayoutParams(new LayoutParams(LayoutParams.MATCH_PARENT,
            LayoutParams.WRAP_CONTENT));
            objTitleText.setBackgroundColor(R.color.backtext);
            objTitleText.setPadding(25, 5, 0, 5);
            objTitleText.setTextColor(R.color.white);
            objTitleText.setText(createTime);
            objScanLinear.addView(objTitleText);
            for (int i = 0;i < listHistories.size();i++) {
                final int item = i;
                View view = LayoutInflater.from(getActivity()).inflate(R.layout.listview_
```

```
item, null);
                RelativeLayout objRelative = (RelativeLayout) view.findViewById(R.id.
linearLayout_item_scan);
                ImageView objLeftImage = (ImageView) view.findViewById(R.id.imageView_item_
icon_left_scan);
                TextView objText = (TextView) view.findViewById(R.id.textView_item_name_
scan);
                if (listHistories.get(i).type2.equals("url")) {
                    objLeftImage.setImageResource(R.drawable.link);
                } else if (listHistories.get(i).type2.equals("contact")) {
                    objLeftImage.setImageResource(R.drawable.business_card);
                } else if (listHistories.get(i).type2.equals("location")) {
                    objLeftImage.setImageResource(R.drawable.map_pin);
                } else if (listHistories.get(i).type2.equals("wifi") && listHistories.get
(i).title.contains("wifi:")) {
                    objLeftImage.setImageResource(R.drawable.wifi);
                } else if (listHistories.get(i).type2.equals("wifi") && listHistories.get
(i).title.contains("WIFI:")) {
                    objLeftImage.setImageResource(R.drawable.wifi_2);
                } else if (listHistories.get(i).type2.equals("calendar")) {
                    objLeftImage.setImageResource(R.drawable.text);
                } else if (listHistories.get(i).type2.equals("else")) {
                    objLeftImage.setImageResource(R.drawable.text);
                }
                objText.setText(listHistories.get(i).title);
                objRelative.setOnClickListener(new View.OnClickListener() {
                    @Override
                    public void onClick(View v) {
                        initItemOnClick(listHistories, item);
                    }
                });
                objRelative.setOnLongClickListener(new View.OnLongClickListener() {
                    @Override
                    public boolean onLongClick(View v) {
                        int theId = listHistories.get(item)._id;
                        MainActivity.dbManager.delete(theId);
                        //listHistories.remove(item);
                        temp.add(listHistories.get(item));
                        objScanLinear.removeView(v);
                        if (listHistories.size() == temp.size()) {
                            objScanLinear.removeView(objTitleText);
                            temp.clear();
                            listHistories.clear();
                        }
                        return true;
                    }
                });
                objScanLinear.addView(objRelative);
            }
        }
    }
```

```
/**
 * 根据文本字符串判断出类型,单击列表项跳转到不同页面
 */
private void initItemOnClick(List<History> listHistories, int item) {
    Intent intent = null;
    History history = listHistories.get(item);
    if (CommonUtil.isURL(history.content1)) {                    //若是 URL
        intent = new Intent(HistoryFragment.this.getActivity(), ScanResultActivity.
class);
        Bundle bundle = new Bundle();
        bundle.putString("class", "HistoryFragment");
        bundle.putSerializable("history", history);
        intent.putExtras(bundle);
    } else if (history.content1.contains("BEGIN:VCARD")) {       //若是联系人
        String content1 = history.content1;
        String[] content1s = content1.split("\r\n");
        if (content1s.length>1) {
        } else {
            content1s = content1.split("\n");
        }
        String nameStr = "", phoneStr = "", emailStr = "", companyStr = "", titleStr =
"", addressStr = "", websiteStr = "";
        for (String content : content1s) {
            if (content.contains("FN:")) {
                String[] name = content.split(":");
                nameStr = name[1];
            }
            if (content.contains("TEL:")) {
                String[] phone = content.split(":");
                phoneStr = phone[1];
            }
            if (content.contains("EMAIL:")) {
                String[] email = content.split(":");
                emailStr = email[1];
            }
            if (content.contains("ORG:")) {
                String[] company = content.split(":");
                companyStr = company[1];
            }
            if (content.contains("TITLE:")) {
                String[] title = content.split(":");
                titleStr = title[1];
            }
            if (content.contains("ADR:")) {
                String[] address = content.split(":");
                addressStr = address[1];
            }
            if (content.contains("URL:")) {
                String[] website = content.split(":");
                websiteStr = website[1];
```

```
                }
            }
            intent = new Intent(HistoryFragment.this.getActivity(), ContactDetailActivity.
class);
            intent.putExtra("all", "");
            intent.putExtra("name", nameStr);
            intent.putExtra("phone", phoneStr);
            intent.putExtra("email", emailStr);
            intent.putExtra("company", companyStr);
            intent.putExtra("title", titleStr);
            intent.putExtra("address", addressStr);
            intent.putExtra("website", websiteStr);
            intent.putExtra("content1", history.content1);
        } else if (history.content1.contains("geo")) {          //若是经纬度
            intent = new Intent(getActivity(), MapResultActivity.class);
            String[] locations = history.content1.split(":");
            locations = locations[1].split(",");
            intent.putExtra("lat", locations[0]);
            locations = locations[1].split("[?]");
            intent.putExtra("lon", locations[0]);
            if (locations!= null && locations.length>1) {
                locations = locations[1].split("=");
                intent.putExtra("q", locations[1]);
            }
            intent.putExtra("title", history.title);
            intent.putExtra("content1", history.content1);
            intent.putExtra("flag", "history");
        } else if (history.content1.contains("WIFI:")) {       //若是 WIFI
            intent = new Intent(HistoryFragment.this.getActivity(), WifiResultActivity.
class);
            Bundle bundle = new Bundle();
            bundle.putString("class", "HistoryFragment");
            bundle.putSerializable("history", history);
            intent.putExtras(bundle);
        } else {                                                //其他
            intent = new Intent(HistoryFragment.this.getActivity(), ElseDetailActivity.
class);
            Bundle bundle = new Bundle();
            bundle.putString("class", "history");
            bundle.putString("else", history.content1);
            intent.putExtras(bundle);
        }
        startActivity(intent);
    }

    /**
     * 操作数据库,查询表中所有不重复的时间字段
     */
    private List<String> getTime(String[] typeRaw) {
        List<String> listCreateTime = new ArrayList<String>();
        Cursor cursor = MainActivity.dbManager.queryTheCreateTime(typeRaw);
```

```
        getActivity().startManagingCursor(cursor);
        while(cursor.moveToNext()) {
            String createTime = cursor.getString(0);
            listCreateTime.add(createTime);
        }
        return listCreateTime;
    }

    /**
     * 操作数据库,查询表
     */
    private List<History> handleDB(String[] typeRaw) {
        List<History> listHistories = new ArrayList<History>();
        Cursor cursor = null;
        if (typeRaw[0].equals("scan")) {
            cursor = MainActivity.dbManager.queryTheScan(typeRaw);
        } else if (typeRaw[0].equals("generate")) {
            cursor = MainActivity.dbManager.queryTheGenerate(typeRaw);
        }
        getActivity().startManagingCursor(cursor); //托付给 activity 根据自己的生命周期去
管理 Cursor 的生命周期
        while(cursor.moveToNext()) {
            int _id = cursor.getInt(0);
            String createTime = cursor.getString(1);
            String title = cursor.getString(2);
            String content1 = cursor.getString(3);
            String content2 = cursor.getString(4);
            String type1 = cursor.getString(5);
            String type2 = cursor.getString(6);
            History history = new History(_id, createTime, title, content1, content2, type1,
type2);
            listHistories.add(history);
        }
        return listHistories;
    }
}
```

代码文件：codes\11\QR where\cn\edu\hstc\qrwhere\fragment\HistoryFragment.java

在上面的程序中，在拼接列表时，程序调用自定义方法 getTime(String[] typeRaw)，该方法调用 MainActivity.dbManager 的 queryTheCreateTime(String[] type)方法并根据 type1 字段查询出该类型下所有数据的 createTime 字段，该字段对应数据的插入时间，即创建时间。经过 queryTheCreateTime 的查询 SQL 语句中的“group by”关键字，过滤掉重复的时间字段的数据，剩下不重复的。这部分内容所对应的知识属于 Android SQLite 部分的知识，此处不再赘述，读者可以阅读该部分的章节进行巩固。

程序查询出不重复的时间字段后，放入一个 List 数组中，然后在拼接列表项的方法 initTabLinear(final LinearLayout objScanLinear, String type)中遍历该数组，每遍历出一个时间字段的值，则动态创建一个 TextView。在该 TextView 中显示了该时间字段的值，然后根据该时间字段以及 type1 字段查询出历史记录数据，存放在数组中，然后遍历该数组，根据数据中 type2 的值以及 title 值，拼接出对应的列表项。这样，经过双重的循环遍历，则可以将相同时间的数据集中在一起。

程序中还为每个列表项添加了单击事件监听器以及长按事件监听器。单击事件监听器实现了单击列表项跳转到对应的详情页面中，这部分与 ScanFragment 中扫描二维码后的跳转类似，不同之处在于在每个跳转时 Bundle 或 Intent 所带的数据中会加入不同的标识字段数据，比如，跳转到联系人信息详情页面时，如果是从 ScanFragment 页面跳转的，则会在 Intent 中带入"all"数据，该数据不为空；而如果是从 HistoryFragment 页面跳转的，则 Intent 中所带该数据为空。这样，在详情页面中获取该数据，如果不为空，则表示不是从历史记录页面中跳转过来的，那么就需要将该数据插入历史表中；如果为空，则代表不再需要将数据插入历史表中。长按事件监听器则实现长按列表项删除历史表中对应的行记录以及将该列表项从列表页中删除。

读者应仔细阅读 HistoryFragment 类的源码，结合以上的功能分解，加深体会。

11.5 开发第三个菜单项所对应的界面 GeneratorFragment

第三个菜单项所对应的界面布局比较简单，从上而下放置四个按钮供用户单击跳转到对应类型的文本编辑页面，包括 URL、Contact、Location、WiFi 四个类型。下面将界面布局的代码贴出：

```
<LinearLayout xmlns:android="http://schemas.android.com/apk/res/android"
    android:layout_width="match_parent"
    android:layout_height="match_parent"
    android:background="@drawable/back_generator"
    android:orientation="vertical" >

    <!-- 顶部标题 -->

    <RelativeLayout
        android:layout_width="fill_parent"
        android:layout_height="wrap_content"
        android:background="@color/backhost"
        android:padding="10dip" >

        <TextView
            android:id="@+id/titleTextView"
            android:layout_width="wrap_content"
            android:layout_height="wrap_content"
            android:layout_centerHorizontal="true"
            android:layout_centerVertical="true"
            android:paddingTop="10dp"
            android:paddingBottom="10dp"
            android:text="@string/generator"
            android:textColor="@color/white"
            android:textSize="20sp" />
    </RelativeLayout>
```

```
<TableLayout
    android:layout_width = "fill_parent"
    android:layout_height = "fill_parent"
    android:stretchColumns = "0,3" >

    <TableRow
        android:id = "@+id/row_shortURL"
        android:layout_width = "fill_parent"
        android:layout_height = "wrap_content"
        android:layout_weight = "1" >

        <TextView
            android:layout_width = "fill_parent"
            android:layout_height = "wrap_content" />

        <ImageView
            android:layout_width = "wrap_content"
            android:layout_height = "wrap_content"
            android:layout_gravity = "center_vertical"
            android:src = "@drawable/website_icon" />

        <TextView
            android:layout_width = "wrap_content"
            android:layout_height = "wrap_content"
            android:layout_gravity = "center_vertical"
            android:paddingLeft = "20dp"
            android:text = "QR / Short URL"
            android:textColor = "@color/white"
            android:textSize = "20sp" />

        <TextView
            android:layout_width = "fill_parent"
            android:layout_height = "wrap_content" />
    </TableRow>

    <TableRow
        android:layout_width = "fill_parent"
        android:layout_height = "wrap_content" >

        <TextView
            android:layout_width = "fill_parent"
            android:layout_height = "0.5dp"
            android:layout_span = "4"
            android:background = "@color/gainsboro" />
    </TableRow>

    <TableRow
        android:id = "@+id/row_contact"
        android:layout_width = "fill_parent"
        android:layout_height = "wrap_content"
        android:layout_weight = "1" >
```

```
        <TextView
            android:layout_width = "fill_parent"
            android:layout_height = "wrap_content" />

        <ImageView
            android:layout_width = "wrap_content"
            android:layout_height = "wrap_content"
            android:layout_gravity = "center_vertical"
            android:src = "@drawable/contact_icon" />

        <TextView
            android:layout_width = "wrap_content"
            android:layout_height = "wrap_content"
            android:layout_gravity = "center_vertical"
            android:paddingLeft = "20dp"
            android:text = "Contact"
            android:textColor = "@color/white"
            android:textSize = "20sp" />

        <TextView
            android:layout_width = "fill_parent"
            android:layout_height = "wrap_content" />
    </TableRow>

    <TableRow
        android:layout_width = "fill_parent"
        android:layout_height = "wrap_content" >

        <TextView
            android:layout_width = "fill_parent"
            android:layout_height = "0.5dp"
            android:layout_span = "4"
            android:background = "@color/gainsboro" />
    </TableRow>

    <TableRow
        android:id = "@ + id/row_location"
        android:layout_width = "fill_parent"
        android:layout_height = "wrap_content"
        android:layout_weight = "1" >

        <TextView
            android:layout_width = "fill_parent"
            android:layout_height = "wrap_content" />

        <ImageView
            android:layout_width = "wrap_content"
            android:layout_height = "wrap_content"
            android:layout_gravity = "center_vertical"
            android:src = "@drawable/location_icon" />
```

```
    <TextView
        android:layout_width = "wrap_content"
        android:layout_height = "wrap_content"
        android:layout_gravity = "center_vertical"
        android:paddingLeft = "20dp"
        android:text = "Location"
        android:textColor = "@color/white"
        android:textSize = "20sp" />

    <TextView
        android:layout_width = "fill_parent"
        android:layout_height = "wrap_content" />
</TableRow>

<TableRow
    android:layout_width = "fill_parent"
    android:layout_height = "wrap_content" >

    <TextView
        android:layout_width = "fill_parent"
        android:layout_height = "0.5dp"
        android:layout_span = "4"
        android:background = "@color/gainsboro" />
</TableRow>

<TableRow
    android:id = "@ + id/row_wifi"
    android:layout_width = "fill_parent"
    android:layout_height = "wrap_content"
    android:layout_weight = "1" >

    <TextView
        android:layout_width = "fill_parent"
        android:layout_height = "wrap_content" />

    < ImageView
        android:layout_width = "wrap_content"
        android:layout_height = "wrap_content"
        android:layout_gravity = "center_vertical"
        android:src = "@drawable/wifi_icon" />

    <TextView
        android:layout_width = "wrap_content"
        android:layout_height = "wrap_content"
        android:layout_gravity = "center_vertical"
        android:paddingLeft = "20dp"
        android:text = "Wifi"
        android:textColor = "@color/white"
        android:textSize = "20sp" />
```

```
            <TextView
                android:layout_width="fill_parent"
                android:layout_height="wrap_content" />
        </TableRow>
    </TableLayout>

</LinearLayout>
```

代码文件：codes\11\QR where\res\layout\activity_generator.xml

注意，上面的布局使用了 TableLayout 布局，每一种类型占据一个 TableRow，这样可以在该布局所对应的 Java 程序中为整个 TableRow 添加单击事件监听器，保证用户单击一行中的任意位置都可以触发跳转页面事件，而无须对准图标或者图标右边的文字单击。

GeneratorFragment 界面的每个 TableRow 都添加了单击事件监听器，而该单击事件监听器只是实现页面的跳转，也就是从当前页面 MainActivity 跳转到对应类型的生成二维码的页面。因此，GeneratorFragment 只需要重写 onCreateView 加载对应的布局文件，然后在 onActivityCreated 方法中为各个 TableRow 添加单击事件监听器。GeneratorFragment 界面的 Java 程序如下：

```
package net.takewin.qrwhere.fragment;

import net.takewin.qrwhere.R;
import net.takewin.qrwhere.activity.GenerateContactActivity;
import net.takewin.qrwhere.activity.GenerateLocationActivity;
import net.takewin.qrwhere.activity.GenerateURLActivity;
import net.takewin.qrwhere.activity.GenerateWifiActivity;
import android.content.Intent;
import android.os.Bundle;
import android.support.v4.app.Fragment;
import android.view.LayoutInflater;
import android.view.View;
import android.view.ViewGroup;
import android.widget.TableRow;

public class GeneratorFragment extends Fragment {
    private TableRow shortURLRow, contactRow, locationRow, wifiRow;

    @Override
    public View onCreateView(LayoutInflater inflater, ViewGroup container, Bundle
    savedInstanceState) {
        return inflater.inflate(R.layout.activity_generator, null);
    }

    @Override
    public void onActivityCreated(Bundle savedInstanceState) {
        super.onActivityCreated(savedInstanceState);
        initShortURLRow();
        initContactRow();
        initLocationRow();
        initWifiRow();
```

```
    }

    private void initShortURLRow() {
        shortURLRow = (TableRow) getActivity().findViewById(R.id.row_shortURL);
        shortURLRow.setOnClickListener(new View.OnClickListener() {
            @Override
            public void onClick(View v) {
                Intent intent = new Intent(getActivity(), GenerateURLActivity.class);
                startActivity(intent);
            }
        });
    }

    private void initContactRow() {
        contactRow = (TableRow) getActivity().findViewById(R.id.row_contact);
        contactRow.setOnClickListener(new View.OnClickListener() {
            @Override
            public void onClick(View v) {
                Intent intent = new Intent(getActivity(), GenerateContactActivity.class);
                startActivity(intent);
            }
        });
    }

    private void initLocationRow() {
        locationRow = (TableRow) getActivity().findViewById(R.id.row_location);
        locationRow.setOnClickListener(new View.OnClickListener() {
            @Override
            public void onClick(View v) {
                Intent intent = new Intent(getActivity(), GenerateLocationActivity.class);
                startActivity(intent);
            }
        });
    }

    private void initWifiRow() {
        wifiRow = (TableRow) getActivity().findViewById(R.id.row_wifi);
        wifiRow.setOnClickListener(new View.OnClickListener() {
            @Override
            public void onClick(View v) {
                Intent intent = new Intent(getActivity(), GenerateWifiActivity.class);
                startActivity(intent);
            }
        });
    }
}
```

代码文件：codes\11\QR where\cn\edu\hstc\qrwhere\fragment\GeneratorFragment.java

11.5.1 开发 URL 编辑页面 GenerateURLActivity

当用户单击 GeneratorFragment 界面中的第一个 TableRow 时，将跳转到编辑 URL 地址的页面，即 GenerateURLActivity。该页面布局十分简单，左上角放置了一个返回按钮供用户单击结束该页面；在返回按钮的右边放置了一个 TextView 组件用于显示页面标题"OR / Short URL"；在页面标题的正下方放置了一个 ImageView 组件，该组件加载了一张事先设计好的 png 格式的图片，整个 ImageView 作为一个生成二维码的按钮供用户单击跳转到另一个页面，根据输入的 URL 地址生成二维码图片并显示出来。在生成按钮的下方放置了一个 EditText 输入框供用户编辑 URL 地址。该界面所对应的 XML 布局源码比较简单，在此不作粘贴展示，读者可以通过本书附带源码获得。下面给出实现该布局的 Java 程序，即 GenerateURLActivity. java。

```
package net.takewin.qrwhere.activity;

import net.takewin.qrwhere.R;
import net.takewin.qrwhere.util.CommonUtil;
import net.takewin.qrwhere.view.MyGenerateUrlDialog;
import android.app.Activity;
import android.app.Dialog;
import android.content.Context;
import android.content.Intent;
import android.os.Bundle;
import android.text.Editable;
import android.text.TextWatcher;
import android.view.Gravity;
import android.view.View;
import android.view.ViewTreeObserver.OnGlobalLayoutListener;
import android.view.inputmethod.InputMethodManager;
import android.widget.EditText;
import android.widget.ImageView;

public class GenerateURLActivity extends Activity {
    private ImageView backToGenerator, generateUrlImageView;
    private EditText longURLEdt;

    @Override
    protected void onCreate(Bundle savedInstanceState) {
        super.onCreate(savedInstanceState);
        setContentView(R.layout.activity_generate_url);
        initLongURLEdt();
        initBackToGenerator();
        initGenerateUrlBtn();
    }

    /**
     * 为页面左上角的返回按钮添加单击事件监听器
     * 实现当单击返回按钮时，检查软键盘是否弹出，
     * 如果弹出则先隐藏软键盘，然后再结束页面，
     * 防止因为没有先将弹出的键盘隐藏而导致的整个页面的震动
     */
    private void initBackToGenerator() {
        backToGenerator = (ImageView) findViewById(R.id.back_to_generator);
        backToGenerator.setOnClickListener(new View.OnClickListener() {
```

```
            @Override
            public void onClick(View v) {
                 InputMethodManager imm = (InputMethodManager) getSystemService(Context.
INPUT_METHOD_SERVICE);
                boolean isOpen = imm.isActive();
                if (isOpen) {
                    imm.hideSoftInputFromWindow(longURLEdt.getWindowToken(), 0);
                }
                GenerateURLActivity.this.finish();
            }
        });
    }

    /**
     * 为页面中的生成按钮添加单击事件监听器
     */
    private void initGenerateUrlBtn() {
        //加载布局中的生成按钮
        generateUrlImageView = (ImageView) findViewById(R.id.imageview_url_btn);
        //为生成按钮添加事件监听器
        generateUrlImageView.setOnClickListener(new View.OnClickListener() {
            @Override
            public void onClick(View v) {
                if (CommonUtil.isURL(longURLEdt.getText().toString())) {
                                                                 //URL 的正则表达式验证
                    String contentStr = longURLEdt.getText().toString();
                    //跳转到生成的二维码显示页面
                    Intent intent = new Intent(GenerateURLActivity.this, UrlImageActivity.
class);
                    Bundle bundle = new Bundle();
                    bundle.putString("contentStr", contentStr);
                    intent.putExtras(bundle);
                    startActivity(intent);
                } else {   //若不能通过 URL 的正则表达式验证
                    //弹出一个自定义对话框提示 not valid url(不是有效的 URL 地址)
                     Dialog dialog = new MyGenerateUrlDialog(GenerateURLActivity.this, R.
style.MyScanDialog);
                    dialog.show();
                }
            }
        });
    }

    /**
     * 实现当 URL 编辑框获得焦点的时候(软键盘出现),显示光标,
     * 当编辑框失去焦点时(软键盘消失),隐藏编辑框光标
     */
    private void initLongURLEdt() {
        longURLEdt = (EditText) findViewById(R.id.edt_long_url);
        final View activityRootView = findViewById(R.id.linear);
            activityRootView. getViewTreeObserver ( ). addOnGlobalLayoutListener ( new
```

```
OnGlobalLayoutListener() {
            private int preHeight = 0;
            @Override
            public void onGlobalLayout() {
                int heightDiff = activityRootView.getRootView().getHeight() -
                activityRootView.getHeight();
                //在数据相同时,减少发送重复消息。因为实际上在输入法出现时会多次调用这
个 onGlobalLayout 方法。
                if (preHeight == heightDiff) {
                    return;
                }
                preHeight = heightDiff;
                if (heightDiff > 100) {
                    longURLEdt.setCursorVisible(true);
                } else {
                    longURLEdt.setCursorVisible(false);
                }
            }
        });

        //为 URL 输入框添加编辑框监听器,实现自动切换居中和居左显示
        longURLEdt.addTextChangedListener(textWatcher);
    }

    /**
     * 监听编辑框里输入内容的变化
     * 实现当 URL 的输入框中有输入了内容的时候,则将内容居中显示,
     * 当输入框中没有任何内容但输入框有获得焦点时,则将光标居左显示
     */
    private TextWatcher textWatcher = new TextWatcher() {
        @Override
        public void onTextChanged(CharSequence s, int start, int before, int count) {
        }

        @Override
        public void beforeTextChanged(CharSequence s, int start, int count, int after) {
        }

        @Override
        public void afterTextChanged(Editable s) {
            if (longURLEdt.getText().toString().trim() == null || longURLEdt.getText().
toString().trim().length() == 0) {
                longURLEdt.setGravity(Gravity.CENTER_VERTICAL);
            } else {
                longURLEdt.setGravity(Gravity.CENTER);
            }
        }
    };
}
```

代码文件：codes\11\QR where\cn\edu\hstc\qrwhere\activity\GenerateURLActivity.java

上面的程序中有几个需要注意的方法：

① 方法 initBackToGenerator()用于为返回按钮添加事件监听器，该事件监听器实现在结束自身 Activity 之前，检测页面中的软键盘是否是弹出状态，如果是，则将其隐藏。这样做的好处是，防止页面结束后，因为页面中的软键盘是弹出状态而造成上级页面的震动。读者可以注释掉隐藏软键盘的代码段，然后单击返回按钮，对比下效果。需要强调的是，在 QR where 的所有页面中，只要页面含有 EditText 编辑框，则在将该页面结束之前，都会使用该段代码检测软键盘并做处理。读者也可以将该段代码复用在其他应用中。

② 方法 initLongURLEdt 实现当 URL 编辑框获得焦点的时候(软键盘出现)，显示光标；当编辑框失去焦点时(软键盘消失)，隐藏编辑框光标。该方法也可以复用在有这个需求的应用中。

③ 程序中为 URL 输入框添加了编辑框监听器 TextWatcher。该监听器实现当 URL 的输入框中有输入了内容的时候，则将内容居中显示，当输入框中没有任何内容但输入框获得了焦点时，则将光标居左显示。TextWatcher 能监听编辑框中的内容变化，读者可以在一些特殊的需求中使用它。

11.5.2　开发根据 URL 地址生成二维码图片的页面 UrlImageActivity

在页面 GenerateURLActivity 中，用户输入了一串 URL 地址，若通过 URL 的正则表达式验证，则用户单击生成按钮时，会跳转到生成二维码图片的页面 UrlImageActivity，并通过 Bundle 对象将 GenerateURLActivity 页面中的 URL 地址传递到 UrlImageActivity 页面中。在 UrlImageActivity 页面中，程序将获得从 GenerateURLActivity 页面传过来的数据，根据数据生成二维码图片并显示在页面中。

UrlImageActivity 页面的布局比较简单，在页面左上角放置一个返回按钮供用户单击结束自身 Activity，页面右上角放置了一个按钮供用户单击后在页面底部弹出更多按钮，在这两个按钮的中间放置了一个 TextView 组件，用于显示根据长链接(网址)转换成的短链接，即页面标题，在标题的正下方放置了一个 ImageView 用以显示所生成的二维码图片，在 ImageView 的下方放置了一个 TextView 用于显示短链接。由于界面布局不复杂，在此不作展示。接下来，直接实现布局所对应的 Java 程序，即 UrlImageActivity 类的源码，如下所示：

```
package net.takewin.qrwhere.activity;

import java.io.File;
import java.util.ArrayList;

import net.takewin.qrwhere.R;
import net.takewin.qrwhere.encoding.EncodingHandler;
import net.takewin.qrwhere.entity.History;
import net.takewin.qrwhere.util.CommonUtil;
import net.takewin.qrwhere.util.Constant;
import net.takewin.qrwhere.util.HttpUtil;
import net.takewin.qrwhere.util.ShareHelper;
import android.app.Activity;
```

```
import android.app.AlertDialog;
import android.content.DialogInterface;
import android.content.DialogInterface.OnCancelListener;
import android.content.Intent;
import android.graphics.Bitmap;
import android.os.AsyncTask;
import android.os.Bundle;
import android.view.Gravity;
import android.view.View;
import android.view.View.OnClickListener;
import android.widget.ImageView;
import android.widget.TextView;

import com.google.zxing.WriterException;

public class UrlImageActivity extends Activity {
    private ImageView backImg, uploadImg;
    private TextView titleTxt, textTxt;
    private Bitmap qrCodeBitmap;
    private ImageView urlImageView;
    private String contentStr;
    private AlertDialog prompt;
    private AsyncTask<String, Integer, String> access;
    private String fileName = "";
    private SelectUploadTypePopupWindow menuWindow;

    @Override
    protected void onCreate(Bundle savedInstanceState) {
        super.onCreate(savedInstanceState);
        setContentView(R.layout.activity_url_image);
        initWidget();
    }

    private void initWidget() {
        //加载页面标题的TextView
        titleTxt = (TextView) findViewById(R.id.textview_title_url_image);
        //加载二维码图片下方显示URL短链接的TextView
        textTxt = (TextView) findViewById(R.id.textview_text_url_image);
        initBack();
        initShare();
        initUrlImageView();
    }

    /**
     * 实现返回按钮
     */
    private void initBack() {
        backImg = (ImageView) findViewById(R.id.back_to_generator_i);
        backImg.setOnClickListener(new View.OnClickListener() {
            @Override
            public void onClick(View v) {
```

```
                if (!fileName.equals("")) { //当用户单击过发送 Email 的按钮时,fileName 不为空
                    File file = new File(fileName);
                    CommonUtil.deleteFile(file.getParent());
                                                //删除发送邮件时所保存到 SDCard 的二维码图片
                    file = null;
                    fileName = "";
                }
                UrlImageActivity.this.finish();
            }
        });
    }

    /**
     * 右上角的按钮事件,单击后将会在底部弹出更多按钮
     * 1.Mail(发送邮件)
     * 2.Save to Album(保存至图库)
     * 3.Share to(分享至(微博、facebook 等))
     * 4.Cancel(取消)
     */
    private void initShare() {
        uploadImg = (ImageView) findViewById(R.id.action);
        uploadImg.setOnClickListener(new View.OnClickListener() {
            @Override
            public void onClick(View v) {
                //实例化 SelectPicPopupWindow
                menuWindow = new SelectUploadTypePopupWindow(UrlImageActivity.this,
                itemsOnClick);
                //显示窗口 设置 layout 在 PopupWindow 中显示的位置
                menuWindow.showAtLocation(UrlImageActivity.this.findViewById(R.id.linear_
image), Gravity.BOTTOM | Gravity.CENTER_HORIZONTAL, 0, 0);
            }
        });
    }

    /**
     * 获取 Bundle 中所带的数据(文本字符串),生成二维码图片
     */
    private void initUrlImageView() {
        urlImageView = (ImageView) findViewById(R.id.iv_qr_image);
        Intent intent = getIntent();
        Bundle bundle = intent.getExtras();
        contentStr = bundle.getString("contentStr");
        //根据字符串生成二维码图片并显示在界面上,第二个参数为图片的大小(350 * 350)
        try {
            //ZXing 生成二维码图片
            qrCodeBitmap = EncodingHandler.createQRCode(contentStr, 450);
            //将生成的二维码图片显示在页面 ImageView 组件
            urlImageView.setImageBitmap(qrCodeBitmap);
        } catch (WriterException e) {
            e.printStackTrace();
        }
```

```
        //调用异步任务
        acquire();
    }

    /**
     * 异步任务,访问地址,将长链接转换成短链接,操作数据库
     */
    private void acquire() {
        access = new AsyncTask<String, Integer, String>() {
            @Override
            protected void onPreExecute() {
                super.onPreExecute();
                //对话框
                prompt = HttpUtil.prompt(UrlImageActivity.this, "Loading...");
                prompt.setOnCancelListener(new OnCancelListener() {
                    @Override
                    public void onCancel(DialogInterface arg0) {
                        access.cancel(true);
                    }
                });
            }

            @Override
            protected String doInBackground(String... arg0) {
                String result = null;
                try {
                    //使用 HTTP GET 方式,传入长链接(网址),返回对应的短链接
                    result = HttpUtil.getDataByGet(contentStr);
                } catch (Exception e) {
                    e.printStackTrace();
                }
                return result;
            }

            @Override
            protected void onPostExecute(String result) {
                try {
                    if (result != null) {
                        titleTxt.setText(result); //将短链接显示在页面顶部 TextView 标题
                        textTxt.setText(result);  //将短链接显示在二维码图片的下方
                        handleDB();               //插入数据
                    }
                } catch (Exception e) {
                    System.out.println(e.getMessage());
                }
                prompt.dismiss();                 //取消对话框
            }
        };
        access.execute();
    }
```

```
    /**
     * 操作数据库,将数据插入 history 表
     */
    private void handleDB() {
        ArrayList<History> histories = new ArrayList<History>();
        String createTime = CommonUtil.getTime(); //获取当前时间
        History history = new History(createTime, titleTxt.getText().toString(),
        contentStr, "", "generate", "url");
        histories.add(history);
        MainActivity.dbManager.add(histories);
    }

    /**
     * 底部弹出按钮事件监听器
     */
    private OnClickListener itemsOnClick = new OnClickListener() {
        public void onClick(View v) {
            //单击选择框外任意位置,底部弹出按钮消失
            menuWindow.dismiss();
            switch (v.getId()) {
            case R.id.button1:
                //调用 sendEmail 方法发送邮件
                fileName = CommonUtil.sendEmail(UrlImageActivity.this, urlImageView,
                titleTxt.getText().toString());
                break;
            case R.id.button2:
                //调用 saveToAlbum 方法将二维码图片保存到系统图库中
                CommonUtil.saveToAlbum(UrlImageActivity.this, urlImageView);
                break;
            case R.id.button3:
                //调用第三方库 ShareSDK 分享至微博、facebook 等平台
                String share = "链接地址: " + contentStr + "." + Constant.SHAREFROM;
                String imagePathStr = CommonUtil.saveImage(urlImageView);
                ShareHelper shareHelper = new ShareHelper(UrlImageActivity.this,
                contentStr, share, imagePathStr, Constant.SHAREURL);
                shareHelper.showShare();
                break;
            case R.id.btn_cancel:
                //取消弹出按钮
                menuWindow.dismiss();
            default:
                break;
            }
        };
    };
}
```

代码文件: codes\11\QR where\cn\edu\hstc\qrwhere\activity\UrlImageActivity.java

上面的程序中有几个地方需要讲解一下:

① 方法 initUrlImageView 实现获取上级页面传过来的 Bundle 数据,即 URL 字符串,

然后调用工具类 EncodingHandler 的 createQRCode(String str, int widthAndHeight)方法,传入 URL 字符串以及所要生成的图片宽高,返回对应的二维码图片的 Bitmap 对象。在方法 createQRCode 中,生成二维码使用的是 ZXing 工具。因此,EncodingHandler 工具类是一个可以复用的工具类,读者可以在任意需要生成二维码图片的项目中调用该工具类,方便地实现功能。

② 单击页面右上角,会在底部弹出更多按钮,这些按钮自上而下排列。其实,这些排列好的按钮,整体上是一个事先自定义好的 PopupWindow。新建一个继承自 android.widget.PopupWindow 的类 SelectUploadTypePopupWindow,在该自定义类中加载对应的布局,其实,用户希望在底部弹出多少个按钮以及每个按钮对应的样式,都可以通过布局任意实现,本应用则是在一个 LinearLayout 布局中自上而下放置四个 Button 按钮并为各个按钮定义好相关样式和属性。在类 SelectUploadTypePopupWindow 中,程序只需加载布局中各个按钮并为其添加单击事件监听器即可。此处采用的是以构造函数参数的形式传入事件监听器对象,也就是上面程序中的 itemsOnClick 对象。至于 SelectUploadTypePopupWindow 类的实现,可以通过本书附带源码获得,并在其他项目中加以模仿利用。在需要调出 SelectUploadTypePopupWindow 类所定义的整个选择框的程序中,只需要创建出 SelectUploadTypePopupWindow 对象,并调用其 showAtLocation 方法即可。读者可以阅读上面程序中右上角按钮事件的方法 initShar()加以体会。

③ 上面程序所实现的底部弹出按钮的事件监听器中,为弹出的选择框中的第一个按钮绑定了单击事件,调用工具类 CommonUtil 中的 sendEmail()方法实现发送 Email。实际上,sendEmail()方法调用的正是普通的 Android API 来实现邮件的发送的。由于需要将二维码图片作为邮件的附件发送,根据 API 方法,程序需要在为邮件添加附件的代码实现之前,将页面中的二维码保存为图片文件,存放在 SDCard 中。因此,sendEmail()方法中调用了 saveImage 方法保存图片并返回文件路径。这也造成了用户单击发送邮件的按钮后,不管用户是否发送了该封邮件,都会在 SDCard 中保存了页面中所生成的二维码图片,造成了空间浪费,这就需要在页面销毁的时候,将该文件删除。因此,在实现返回按钮时,程序会判断是否有保存的文件,有则删除。

④ 底部弹框中的第二个按钮实现了将二维码图片保存至系统图库的功能。上面程序中调用的是 CommonUtil 工具类中的 saveToAlbum 方法来完成该功能。实际上,该方法也是调用普通的 Android API 来实现的。读者可以通过本书附带源码获得其源码,对该方法进行复用。事实上,整个 CommonUtil 工具类都是一些通用的实现,希望读者可以好好利用。

⑤ 底部弹框中的第三个按钮实现对应用的一键分享。单击该按钮,将会弹出一个选择框,在该选择框中显示了各大主流社交平台的图标,单击任意图标,实现快速将应用以及附带的其他文本信息分享至对应平台。例如,单击新浪微博的图标,将会弹出一个窗体生成一条微博,微博的内容为程序自动生成,该微博附带了 URL 地址的二维码图片,用户可以对该微博进行二次编辑,然后单击分享按钮,调用微博客户端或者网页版微博获取授权登录后发出该条微博。该按钮所实现的功能需要借助 Android ShareSDK,这是我们极力推荐的一个第三方产品。ShareSDK 是为 IOS、Android 的 App 提供社会化功能的一个组件,开发者仅需 10 分钟即可集成到自己的 App 中,它不仅支持国内外 40 多家的主流社交平台,帮助

开发者轻松实现社会化分享、登录、关注、获得用户资料、获取好友列表等主流的社会化功能，还有强大的社会化统计分析管理后台，可以实时了解用户、信息流、回流率、传播效率等数据，有效地指导移动 App 的日常运营与推广，同时为 App 引入更多的社会化流量。在其官方网站 http://www.mob.com/上可下载相关 SDK，在网站文档中心中可以进入 Android 集成文档的介绍页面，该页面详细介绍了如何获取 ShareSDK 的 AppKey、如何将下载的 SDK 快速集成到我们的项目中，一些必要的配置以及如何调用分享代码。通过文档中心的学习，可快速实现应用分享至各大主流社交平台的功能。建议读者去官网获取最新的 API。

⑥ 阅读上面的程序，可以发现程序通过自定义工具类 ShareHelper 的 showShare()方法实现了一键分享功能。程序在创建 ShareHelper 对象时通过其构造方法传入了必要的文本内容、链接地址、图片(二维码图片)地址等参数。showShare()方法正是实现分享的关键代码，实际上，这部分代码在 ShareSDK 的官网文档中心中会有介绍，读者学习后对其加以修改，可形成自己的自定义方法。读者可以通过阅读本书附带源码，加深该部分实现的体会。希望读者能够好好学习 ShareSDK 的知识，开发出更好的分享模块功能。

11.5.3 开发坐标拾取页面 GenerateLocationActivity

上面两个小节介绍了底部第三个菜单项对应的界面中的第一个按钮，单击其会跳转到 URL 编辑页面，输入正确的 URL 地址后，单击生成按钮，跳转到二维码生成页面，并在页面中处理将二维码图片通过邮件发送出去、保存到系统图库、一键分享至社交平台等功能。

界面中的第二个按钮所对应的功能则是跳转到联系人信息编辑页面，页面右上角提供了一个按钮供用户访问手机联系人应用，获取任意联系人数据，然后返回信息编辑页面，自动将该联系人的信息填充至页面对应的编辑框中。这部分的实现则是通过 ContentResolver 访问手机联系人应用的 ContentProvider 所暴露的数据，这与 9.3 节的内容相对应，在此不多作讲解，读者可以通过本书附带源码学习联系人信息编辑页面的代码。

同样，在页面中放置了一个生成二维码的按钮，单击该按钮则跳转到对应的二维码生成页面。在页面中根据上级页面所传递过来的联系人信息，调用通用的方法生成二维码图片。所有显示二维码图片的页面的功能大体相同，实现源码也大同小异。因此，对于 GeneratorFragment 界面中的第二个按钮生成联系人信息二维码则不多作介绍，相信读者通过阅读附带项目源码，配合源码中的注释，可以很好地理解其实现过程。

接下来，需要重点介绍的是 GeneratorFragment 界面中第三个按钮所对应的功能实现。单击界面中的第三个 TableRow，跳转到坐标拾取页面 GenerateLocatinActivity 中。该页面顶部中间放置了一个 TextView 组件用于显示经纬度信息，在其下方放置了一个生成按钮供用户单击跳转到生成二维码图片的页面，在生成按钮的下方则提供了高德地图组件，在页面底部从左至右放置了一个搜索按钮以及一个带有自动提示功能的编辑框 AutoCompleteTextView 组件。下面一起来看看 GenerateLocatinActivity 所对应的界面布局的实现源码：

```
<?xml version = "1.0" encoding = "utf - 8"?>
< LinearLayout xmlns:android = "http://schemas.android.com/apk/res/android"
    android:id = "@ + id/linear_contact"
```

```
android:layout_width = "match_parent"
android:layout_height = "match_parent"
android:background = "@drawable/back_hong"
android:orientation = "vertical" >

<! -- 顶部标题 -->
< RelativeLayout
    android:layout_width = "fill_parent"
    android:layout_height = "wrap_content"
    android:background = "@color/backhost"
    android:padding = "10dp" >

    < ImageView
        android:id = "@ + id/location_back_to_generator"
        android:layout_width = "wrap_content"
        android:layout_height = "wrap_content"
        android:layout_alignParentLeft = "true"
        android:layout_centerVertical = "true"
        android:src = "@drawable/navbar_back" />

    < TextView
        android:id = "@ + id/textview_title_generate_location"
        android:layout_width = "wrap_content"
        android:layout_height = "wrap_content"
        android:layout_centerHorizontal = "true"
        android:layout_centerVertical = "true"
        android:paddingBottom = "10dp"
        android:paddingTop = "10dp"
        android:text = "Location:"
        android:textColor = "@color/white"
        android:textSize = "18sp" />
</RelativeLayout >

< TextView
    android:layout_width = "fill_parent"
    android:layout_height = "0.5dp"
    android:background = "@color/gray" />

< ImageView
    android:id = "@ + id/imageview_location_btn"
    android:layout_width = "fill_parent"
    android:layout_height = "wrap_content"
    android:adjustViewBounds = "true"
    android:src = "@drawable/location_btn" />

< RelativeLayout
    android:layout_width = "fill_parent"
    android:layout_height = "fill_parent"
    android:background = "@color/white" >

    < LinearLayout
```

```
            android:id = "@ + id/linear_auto"
            android:layout_width = "fill_parent"
            android:layout_height = "wrap_content"
            android:layout_alignParentBottom = "true"
            android:orientation = "horizontal"
            android:paddingBottom = "20dp" >

            < ImageView
                android:id = "@ + id/searchButton"
                android:layout_width = "wrap_content"
                android:layout_height = "wrap_content"
                android:layout_gravity = "center_vertical"
                android:paddingLeft = "15dp"
                android:src = "@drawable/search_icon" />

            < AutoCompleteTextView
                android:id = "@ + id/keyWord"
                android:layout_width = "fill_parent"
                android:layout_height = "wrap_content"
                android:layout_marginLeft = "15dp"
                android:background = "@null"
                android:completionThreshold = "1"
                android:dropDownVerticalOffset = "1.0dip"
                android:hint = "Search..."
                android:imeOptions = "actionSearch"
                android:inputType = "text|textAutoComplete"
                android:maxLength = "20"
                android:popupBackground = "@color/black"
                android:singleLine = "true"
                android:textColor = "#000000"
                android:textSize = "18.0sp" />

        </LinearLayout >

        < com.amap.api.maps2d.MapView
            android:id = "@ + id/map"
            android:layout_width = "fill_parent"
            android:layout_height = "fill_parent"
            android:layout_above = "@id/linear_auto"
            android:paddingBottom = "5dp" />
    </RelativeLayout >

</LinearLayout >
```

代码文件：codes\11\QR where\res\layout\activity_generate_location.xml

上面的XML代码实现了坐标拾取页面的样式布局，为界面中的AutoCompleteTextView组件添加了android:imeOptions="actionSearch"的属性以及对应值，实现当用户单击底部的输入框时弹出的软键盘带有搜索按钮。读者还需要认识的是com.amap.api.maps2d.MapView这个组件，该组件需要读者在高德地图开放平台http://lbs.amap.com/上下载了相关SDK，并将Jar包集成在QR where中后，然后才可使用。

读者需要访问高德地图开放平台 http://lbs.amap.com/，然后免费注册开发者用户，单击导航栏中“开发”下的子级导航“Android 地图 SDK”和“Android 定位 SDK”，在这两个导航页面中下载相关的 Jar 包。

由于 QR where 需要使用到高德地图及其定位功能，所以需要在开放平台中下载 2D 地图显示包 AMap_2DMap_V2.x.x.jar、AMap_Services_V2.x.x.jar 以及定位包 AMap_Location_V2.x.x.jar。将下载好的三个 Jar 包复制到 QR where 项目目录中的 libs 文件夹下，即可在布局页面中使用 com.amap.api.maps2d.MapView 组件，程序在页面中即可使用该组件加载高德地图。

在开放平台的下载区中，除了相关的 Jar 下载链接之外，还提供了相关的 Demo 示例以及文档手册的下载。读者可以将相关的 Demo 示例下载到本地计算机中，然后将示例程序添加到 Eclipse 开发工具中，即可获得其源码程序。再根据示例程序中的功能模块的实现，复制这部分的代码到 QR where 中加以修改即可。例如需要创建一个基本 2D 地图，那么可以将 2D 地图示例工程的项目源码 Import 到 Eclipse 中，然后打开创建基本 2D 地图的 Activity，将创建地图的源码复制到自己的应用中并加以修改。

根据以上介绍的方法，摘取示例程序中的相关模块的实现代码，集成到本应用的坐标拾取页面 GenerateLocationActivity 中并加以修改。最终 GenerateLocationActivity 源码实现如下所示：

```
package net.takewin.qrwhere.activity;

import java.util.ArrayList;
import java.util.List;

import net.takewin.qrwhere.R;
import net.takewin.qrwhere.util.AMapUtil;
import android.app.Activity;
import android.app.ProgressDialog;
import android.content.Intent;
import android.location.Location;
import android.os.Bundle;
import android.text.Editable;
import android.text.TextWatcher;
import android.view.KeyEvent;
import android.view.View;
import android.view.View.OnClickListener;
import android.view.inputmethod.EditorInfo;
import android.widget.ArrayAdapter;
import android.widget.AutoCompleteTextView;
import android.widget.ImageView;
import android.widget.TextView;
import android.widget.TextView.OnEditorActionListener;
import android.widget.Toast;

import com.amap.api.location.AMapLocation;
import com.amap.api.location.AMapLocationListener;
import com.amap.api.location.LocationManagerProxy;
```

```
import com.amap.api.location.LocationProviderProxy;
import com.amap.api.maps2d.AMap;
import com.amap.api.maps2d.AMap.InfoWindowAdapter;
import com.amap.api.maps2d.CameraUpdateFactory;
import com.amap.api.maps2d.LocationSource;
import com.amap.api.maps2d.MapView;
import com.amap.api.maps2d.model.BitmapDescriptorFactory;
import com.amap.api.maps2d.model.Marker;
import com.amap.api.maps2d.model.MyLocationStyle;
import com.amap.api.maps2d.overlay.PoiOverlay;
import com.amap.api.services.core.AMapException;
import com.amap.api.services.core.PoiItem;
import com.amap.api.services.core.SuggestionCity;
import com.amap.api.services.help.Inputtips;
import com.amap.api.services.help.Inputtips.InputtipsListener;
import com.amap.api.services.help.Tip;
import com.amap.api.services.poisearch.PoiItemDetail;
import com.amap.api.services.poisearch.PoiResult;
import com.amap.api.services.poisearch.PoiSearch;
import com.amap.api.services.poisearch.PoiSearch.OnPoiSearchListener;

public class GenerateLocationActivity extends Activity implements InfoWindowAdapter,
TextWatcher, OnPoiSearchListener, OnClickListener, LocationSource, AMapLocationListener {
    private MapView mapView;                          //地图显示组件
    private AMap aMap;                                //地图实例
    private AutoCompleteTextView searchText;          //输入搜索关键字
    private String keyWord = "";                      //存放输入的 poi 搜索关键字
    private ProgressDialog progDialog = null;         //搜索时进度条
    private PoiResult poiResult;                      //poi 返回的结果
    private PoiSearch.Query query;                    //poi 查询条件类
    private PoiSearch poiSearch;                      //poi 搜索

    private ImageView back;                           //返回按钮
    private ImageView toLocationImage;                //跳转到生成二维码图片页面的生成按钮
    private TextView titleLocation;
                            //页面顶部标题,显示经纬度信息(Location:纬度度数, 经度度数)
    private double lat, lon;                          //存放纬度度数、经度度数

    private OnLocationChangedListener mListener;      //监听搜索后变化
    private LocationManagerProxy mAMapLocationManager;
    private boolean flag = true;
    private String locationCity = "";

    @Override
    protected void onCreate(Bundle savedInstanceState) {
        super.onCreate(savedInstanceState);
        setContentView(R.layout.activity_generate_location);
        mapView = (MapView) findViewById(R.id.map); //加载地图显示组件
        mapView.onCreate(savedInstanceState);         //此方法必须重写
        titleLocation = (TextView) this.findViewById(R.id.textview_title_generate_
        location);                                    //获取标题 TextView
```

```
        initMap();                                      //初始化地图
        initBack();                                     //返回按钮
        initToLocationImage();                          //生成按钮
    }

    @Override
    protected void onResume() {
        super.onResume();
        mapView.onResume();
    }

    @Override
    protected void onPause() {
        super.onPause();
        mapView.onPause();
        deactivate();
    }

    @Override
    protected void onSaveInstanceState(Bundle outState) {
        super.onSaveInstanceState(outState);
        mapView.onSaveInstanceState(outState);
    }

    @Override
    protected void onDestroy() {
        super.onDestroy();
        mapView.onDestroy();
    }

    /**
     * 初始化地图对象
     */
    private void initMap() {
        if (aMap == null) {
            aMap = mapView.getMap();
        }
        setUpMap();
    }

    /**
     * 返回按钮监听器
     */
    private void initBack() {
        back = (ImageView) this.findViewById(R.id.location_back_to_generator);
        back.setOnClickListener(new View.OnClickListener() {
            @Override
            public void onClick(View v) {
                GenerateLocationActivity.this.finish();
            }
        });
```

```
    }

    /**
     * 生成按钮监听器,将经纬度信息按照规则拼接成字符串,通过 Bundle 对象传递到生成二维
码的页面中
     */
    private void initToLocationImage() {
        toLocationImage = (ImageView) this.findViewById(R.id.imageview_location_btn);
        toLocationImage.setOnClickListener(new View.OnClickListener() {
            @Override
            public void onClick(View v) {
                Intent intent = new Intent(GenerateLocationActivity.this,
                LocationImageActivity.class);
                Bundle bundle = new Bundle();
                bundle.putString("nameStr", "Geo:" + lat + "," + lon + "?q=" + aMap.
getCameraPosition().zoom);
                bundle.putString("contentStr", "geo:" + lat + "," +lon + "?q=" + aMap.
getCameraPosition().zoom);
                intent.putExtras(bundle);
                startActivity(intent);
            }
        });
    }

    /**
     * 设置页面监听
     */
    private void setUpMap() {
        if (flag) {
            //自定义系统定位小蓝点
            MyLocationStyle myLocationStyle = new MyLocationStyle();
            myLocationStyle.myLocationIcon(BitmapDescriptorFactory.fromResource(R.
            drawable.location_marker));                //设置小蓝点的图标
            myLocationStyle.strokeColor(0);            //设置圆形的边框颜色
            myLocationStyle.radiusFillColor(0);
            myLocationStyle.strokeWidth(0.0f);         //设置圆形的边框粗细
            aMap.moveCamera(CameraUpdateFactory.zoomTo(20));
            aMap.setMyLocationStyle(myLocationStyle);
            aMap.setLocationSource(this);              //设置定位监听
            aMap.getUiSettings().setMyLocationButtonEnabled(true);
                                                      //设置默认定位按钮是否显示
            aMap.setMyLocationEnabled(true);   //设置为 true 表示显示定位层并可触发定位,
                                               false 表示隐藏定位层并不可触发定位,默认
                                               是 false
        }

        ImageView searButton = (ImageView) findViewById(R.id.searchButton);
        searButton.setOnClickListener(this);
        searchText = (AutoCompleteTextView) findViewById(R.id.keyWord);
        searchText.addTextChangedListener(this);     //添加文本输入框监听事件
        searchText.setOnEditorActionListener(new OnEditorActionListener() {
```

```
            @Override
            public boolean onEditorAction(TextView v, int actionId, KeyEvent event) {
                if (actionId == EditorInfo.IME_ACTION_SEARCH) {
                    searchButton();
                }
                return true;
            }
        });
    }

    /**
     * 单击搜索按钮
     */
    public void searchButton() {
        keyWord = AMapUtil.checkEditText(searchText);
        if ("".equals(keyWord)) {
            Toast.makeText(GenerateLocationActivity.this, "请输入关键字", 3000).show();
            return;
        } else {
            doSearchQuery("");
        }
    }

    /**
     * 开始进行poi搜索
     */
    protected void doSearchQuery(String scope) {
        if (progDialog == null) {
            showProgressDialog();                           //显示进度框
        }
        //第一个参数表示搜索字符串,第二个参数表示poi搜索类型,第三个参数表示poi搜索区
域(空字符串代表全国)
        query = new PoiSearch.Query(keyWord, "", scope);
        query.setPageSize(1);                               //设置每页最多返回多少条poiitem
        query.setPageNum(0);                                //设置查第一页

        poiSearch = new PoiSearch(this, query);
        poiSearch.setOnPoiSearchListener(this);
        poiSearch.searchPOIAsyn();
    }

    /**
     * 显示进度框
     */
    private void showProgressDialog() {
        if (progDialog == null)
            progDialog = new ProgressDialog(this);
        progDialog.setProgressStyle(ProgressDialog.STYLE_SPINNER);
        progDialog.setIndeterminate(false);
        progDialog.setCancelable(false);
        progDialog.setMessage("正在搜索:\n" + keyWord);
```

```
        progDialog.show();
    }

    /**
     * 隐藏进度框
     */
    private void dismissProgressDialog() {
        if (progDialog != null) {
            progDialog.dismiss();
            progDialog = null;
            titleLocation.setText("Location:");
        }
    }

    private String getSuggestCity(List<SuggestionCity> cities) {
        for (int i = 0; i<cities.size(); i++) {
            if (locationCity.equals(cities.get(i).getCityName())) {
                return locationCity;
            }
        }
        return cities.get(0).getCityName();
    }

    @Override
    public void onClick(View v) {
        switch (v.getId()) {
        /**
         * 单击搜索按钮
         */
        case R.id.searchButton:
            searchButton();
            break;
        default:
            break;
        }
    }

    @Override
    public void onPoiItemDetailSearched(PoiItemDetail arg0, int arg1) {
    }

    @Override
    public void onPoiSearched(PoiResult result, int rCode) {
        if (rCode == 0) {
            if (result != null && result.getQuery() != null) {  //搜索 poi 的结果
                if (result.getQuery().equals(query)) {         //是否是同一条
                    poiResult = result;
                    //取得搜索到的 poiitems 有多少页
                    List<PoiItem> poiItems = poiResult.getPois();
                                                //取得第一页的 poiitem 数据,页数从数字 0 开始
                    List<SuggestionCity> suggestionCities = poiResult.getSearchSuggestion-
```

```
                    Citys(); //当搜索不到poiitem数据时,会返回含有搜索关键字的城市信息
                    if (poiItems != null && poiItems.size() > 0) {
                        dissmissProgressDialog();            //隐藏对话框
                        aMap.clear();                        //清理之前的图标
                        aMap.moveCamera(CameraUpdateFactory.zoomTo(20));
                        PoiOverlay poiOverlay = new PoiOverlay(aMap, poiItems);
                        poiOverlay.removeFromMap();
                        poiOverlay.addToMap();
                        poiOverlay.zoomToSpan();

                        lat = poiItems.get(0).getLatLonPoint().getLatitude();
                        lon = poiItems.get(0).getLatLonPoint().getLongitude();
                        double lat_ = Math.round(lat * 1000) / 1000.0;
                        double lon_ = Math.round(lon * 1000) / 1000.0;
                        titleLocation.setText("Location:" + lat_ + ", " + lon_);
                    } else if (suggestionCities != null && suggestionCities.size() > 0) {
                        doSearchQuery(getSuggestCity(suggestionCities));
                    } else {
                        dissmissProgressDialog();            //隐藏对话框
                        Toast.makeText(GenerateLocationActivity.this, "对不起,没有搜索到
相关数据!", 3000).show();
                    }
                }
            } else {
                Toast.makeText(GenerateLocationActivity.this, "对不起,没有搜索到相关数
据!", 3000).show();
            }
        } else if (rCode == 27) {
            Toast.makeText(GenerateLocationActivity.this, "搜索失败,请检查网络连接!",
3000).show();
        } else if (rCode == 32) {
            Toast.makeText(GenerateLocationActivity.this, "key验证无效!", 3000).show();
        } else {
            Toast.makeText(GenerateLocationActivity.this, "未知错误,请稍后重试!错误码
为", 3000).show();
        }
    }

    @Override
    public void afterTextChanged(Editable s) {
    }

    @Override
    public void beforeTextChanged(CharSequence s, int start, int count, int after) {
    }

    @Override
    public void onTextChanged(CharSequence s, int start, int before, int count) {
        String newText = s.toString().trim();
        Inputtips inputTips = new Inputtips(GenerateLocationActivity.this,
                new InputtipsListener() {
```

```
                @Override
                public void onGetInputtips(List<Tip> tipList, int rCode) {
                    if (rCode == 0) {                         //正确返回
                        List<String> listString = new ArrayList<String>();
                        for (int i = 0; i < tipList.size(); i++) {
                            listString.add(tipList.get(i).getName());
                        }
                         ArrayAdapter<String> aAdapter = new ArrayAdapter<String>
(getApplicationContext(), R.layout.route_inputs, listString);
                        searchText.setAdapter(aAdapter);
                        aAdapter.notifyDataSetChanged();
                    }
                }
            });
        try {
            inputTips.requestInputtips(newText, "");
                //第一个参数表示提示关键字，第二个参数默认代表全国，也可以为城市区号
        } catch (AMapException e) {
            e.printStackTrace();
        }
    }

    @Override
    public View getInfoContents(Marker arg0) {
        return null;
    }

    @Override
    public View getInfoWindow(Marker marker) {
        View view = getLayoutInflater().inflate(R.layout.poikeywordsearch_uri, null);
        TextView title = (TextView) view.findViewById(R.id.title);
        title.setText(marker.getTitle());
        TextView snippet = (TextView) view.findViewById(R.id.snippet);
        snippet.setText(marker.getSnippet());
        return view;
    }

    @Override
    public void onLocationChanged(Location aLocation) {
    }

    @Override
    public void onProviderDisabled(String provider) {
    }

    @Override
    public void onProviderEnabled(String provider) {
    }

    @Override
    public void onStatusChanged(String provider, int status, Bundle extras) {
```

```
    }

    @Override
    public void onLocationChanged(AMapLocation aLocation) {
        if (mListener != null && aLocation != null) {
            locationCity = aLocation.getCity();
            mListener.onLocationChanged(aLocation);            //显示系统小蓝点
            lat = aLocation.getLatitude();
            lon = aLocation.getLongitude();
            double lat_ = Math.round(lat * 1000) / 1000.0;
            double lon_ = Math.round(lon * 1000) / 1000.0;
            titleLocation.setText("Location:" + lat_ + ", " + lon_);
            flag = false;
            deactivate();
        }
    }

    @SuppressWarnings("deprecation")
    @Override
    public void activate(OnLocationChangedListener listener) {
        mListener = listener;
        if (mAMapLocationManager == null) {
            mAMapLocationManager = LocationManagerProxy.getInstance(this);
            /*
             * mAMapLocManager.setGpsEnable(false);
             * 1.0.2版本新增方法,设置true表示混合定位中包含gps定位,false表示纯网络
定位,默认是true Location
             * API定位采用GPS和网络混合定位方式
             * ,第一个参数是定位provider,第二个参数时间最短是2000毫秒,第三个参数距离
间隔单位是米,第四个参数是定位监听者
             */
                    mAMapLocationManager. requestLocationUpdates ( LocationProviderProxy.
AMapNetwork, 3000, 10, this);
        }
    }

    @SuppressWarnings("deprecation")
    @Override
    public void deactivate() {
        mListener = null;
        if (mAMapLocationManager != null) {
            mAMapLocationManager.removeUpdates(this);
            mAMapLocationManager.destory();
        }
        mAMapLocationManager = null;
    }
}
```

代码文件: codes\11\11.2\QR where\cn\edu\hstc\qrwhere\activity\GenerateLocationActivity.java

上面的程序中有几个地方需要重点介绍:

① 上面的程序实现加载界面布局，初始化地图，为自动提示组件 AutoCompleteTextView 添加内容变化监听器，实现自动补全地理位置信息功能，并以下拉列表形式展示。即通过为 AutoCompleteTextView 对象添加文本输入框监听器，重写 onTextChanged 方法，在该方法中调用高德地图相关 API 实现联想查询。比如，在关键字输入框中输入潮州二字，则会弹出一个下拉列表，显示潮州路、潮州宾馆、潮州西湖等。用户直接选中任意一个则可直接补全关键字搜索框。这部分实现可以阅读 GenerateLocationActivity 类的 onTextChanged 方法进行理解。在以后的实际项目中，如果需要实现高德地图的关键字搜索框自动提示功能，则可以复用这部分的代码。

② 当用户单击关键字输入框左边的搜索按钮或软键盘右下角的搜索按钮时，将调用 searchButton()方法，检查输入框是否为空，如果为空，则提示用户；如果不为空，则调用 doSearchQuery 方法开始搜索地图。而 doSearchQuery 方法正是调用了高德地图的 API 进行 poi 搜索操作。这部分的代码也是参照官方平台所提供的示例程序中的实现代码而实现的。在实际项目开发中，应学会从示例程序中找到自己所需要的代码进行复用。poi 搜索的真正操作则是通过重写方法 onPoiSearched 实现的。该重写方法参照示例程序，添加进 QR where 所需要的特殊业务逻辑，比如，获取纬度度数以及经度度数，然后显示在页面顶部标题 TextView 中。如以下代码片段所示：

```
lat = poiItems.get(0).getLatLonPoint().getLatitude();
lon = poiItems.get(0).getLatLonPoint().getLongitude();
double lat_ = Math.round(lat * 1000) / 1000.0;
double lon_ = Math.round(lon * 1000) / 1000.0;
titleLocation.setText("Location:" + lat_ + ", " + lon_);
```

③ 上面所介绍的 poi 搜索的方法 onPoiSearched 中，实现了在地图中添加搜索位置的标记，当用户单击该标记，则在该标记上弹出一个窗体以显示地理名称以及街道信息，即作为标记的扩展信息。这部分的功能则是通过重写 getInfoWindow(Marker marker)方法实现的。该方法加载了标记扩展信息的显示窗体的自定义布局文件 poikeywordsearch_uri.xml，该布局从上而下放置了两个 TextView 用于显示地理名称以及街道信息。通过 marker 对象的 getTitle()方法获取地理名称以及 getSnippet()方法获取街道信息，然后将其显示在对应的 TextView 上。

在实际项目开发中，必须学会复用示例程序中的相关模块的代码。GenerateLocationActivity 类中关于高德地图的显示以及定位功能的实现都是参照示例程序中的代码，加入 QR where 特定的业务逻辑。在阅读理解 GenerateLocationActivity 类的源码时，需要结合示例程序相关模块的代码，以此学会复用示例代码的方法。

需要强调的是，为了使用高德地图，需要登录其开放平台，然后单击页面右上角的控制台，进入"我的 key"页面，该页面列出了登录账号所申请过的 key 以及 key 对应的应用名称。单击页面中的"获取 key"按钮，填写应用名称 QR where，选择绑定的服务平台，此处选择"Android 平台 SDK"，输入安全码 SHA1 以及 Package(单击页面中的查看 Android SHA1 与 Package 获取方式学习如何获取这两个参数)，然后单击"获取 key"按钮即可完成申请 key。

接着，需要在 app/src/main 目录下的 AndroidMainfest.xml 配置申请的 key 和相关

权限：

```
<?xml version="1.0" encoding="utf-8"?>
<manifest xmlns:android="http://schemas.android.com/apk/res/android"
    package="application.test.amap.com.myapplication" >

    <uses-permission android:name="android.permission.INTERNET" />
    <uses-permission android:name="android.permission.WRITE_EXTERNAL_STORAGE" />  权限
    <uses-permission android:name="android.permission.ACCESS_NETWORK_STATE" />
    <uses-permission android:name="android.permission.ACCESS_WIFI_STATE" />
    <uses-permission android:name="android.permission.READ_PHONE_STATE" />
    <uses-permission android:name="android.permission.ACCESS_COARSE_LOCATION" />

    <application
        android:allowBackup="true"
        android:icon="@mipmap/ic_launcher"
        android:label="My Application"
        android:theme="@style/AppTheme" >
        <meta-data
            android:name="com.amap.api.v2.apikey"          您申请的key
            android:value="请输入您申请的key" />
        <activity
            android:name=".MainActivity"
            android:label="My Application" >
```

程序中关于返回按钮、生成按钮的实现以及跳转后二维码图片的生成页面，皆与URL编辑页面以及生成URL二维码图片的页面相似，相信读者通过阅读源码，可以很快理解，此处不再赘述。

单击GenerateorFragment页面中的最后一个TableRow，跳转到WiFi信息编辑页面，此页面的实现以及WiFi二维码图片的生成页面皆与第一个TableRow类似，即11.5.1节介绍的GenerateURLActivity以及11.5.2节介绍的UrlImageActivity，在此亦不作赘述，读者可通过本书附带源码获得这部分的实现。

11.6 开发 MapResultActivity

在11.4节所介绍的HistoryFragment界面中，展示了存放在数据库中的数据，单击每个列表项，则跳转到对应的详情界面中，当然，用户也可通过底部第一个菜单项的扫描功能扫描二维码后跳转到相应的详情界面中。也就是说，从ScanFragment以及HistoryFragment这两个界面中同种类型所跳转到的详情页面是一样的。

对于URL、WiFi这两种类型的信息来说，所跳转到的详情页面都是将文本信息显示出来，并且显示了文本信息所对应的二维码图片，唯一区别是URL详情页面的底部提供了一个Open in Browser按钮供用户在手机浏览器中打开URL链接。这两种类型对应的详情页面都与11.5.2节所介绍的URL的二维码生成页面URLImageActivity的实现类似，此处不作展开介绍。

对于Contact这种类型的信息来说，跳转到的详情页面则同样在对应的TextView展示了联系人信息，页面中提供了发送短消息以及拨打电话的按钮，同时还提供了Create New Contact(创建新联系人)以及Add to Existing Contact(添加到现有联系人)这两个按钮供用户根据所传递过来的联系人信息调用联系人应用、创建新联系人或将传递过来的联系人信息添加到现有的任意联系人中。这些功能的实现，都是调用普通的 Android API 实现的，并没有任何特殊的业务逻辑，读者可以通过阅读附带源码，加以理解体会。由于上面所介绍的功能已经占据了界面相当大的一部分，故在同一个页面中展示二维码图片显得不太合理，故在页面底部提供了一个 QRCode 按钮供用户单击跳转到二维码图片显示页面。前面已经介绍过二维码图片的生成方法了，故此处不作展开介绍。

需要重点介绍的是扫描包含经纬度信息的二维码图片或从历史列表中单击 Location 类型的列表项后，所跳转到的详情界面 MapResultActivity。该页面获取上级页面中所传递过来的数据（包括经纬度度数），然后进行逆地理编码，即根据经纬度进行地图定位（GenerateLocationActivity 是根据搜索关键字进行地图定位）。这部分功能实现是通过参照示例程序中介绍 geocoder 地理编码功能的页面 GeocoderActivity 中关于逆地理编码的实现代码完成的。读者在理解 MapResultActivity 类源码的同时也需要参照示例程序中逆地理编码部分的源代码，体会复用源码过程。MapResultActivity 类源码如下所示：

```
package net.takewin.qrwhere.activity;

import java.util.ArrayList;
import java.util.List;

import net.takewin.qrwhere.R;
import net.takewin.qrwhere.entity.History;
import net.takewin.qrwhere.util.AMapUtil;
import net.takewin.qrwhere.util.CommonUtil;
import android.app.Activity;
import android.app.ProgressDialog;
import android.content.Intent;
import android.content.pm.PackageManager;
import android.content.pm.ResolveInfo;
import android.net.Uri;
import android.os.Bundle;
import android.view.View;
import android.widget.Button;
import android.widget.ImageView;
import android.widget.TextView;
import android.widget.Toast;

import com.amap.api.maps2d.AMap;
import com.amap.api.maps2d.AMap.InfoWindowAdapter;
import com.amap.api.maps2d.CameraUpdateFactory;
import com.amap.api.maps2d.MapView;
import com.amap.api.maps2d.model.BitmapDescriptorFactory;
import com.amap.api.maps2d.model.Marker;
import com.amap.api.maps2d.model.MarkerOptions;
import com.amap.api.services.core.LatLonPoint;
import com.amap.api.services.geocoder.GeocodeResult;
import com.amap.api.services.geocoder.GeocodeSearch;
import com.amap.api.services.geocoder.GeocodeSearch.OnGeocodeSearchListener;
```

```
import com.amap.api.services.geocoder.RegeocodeQuery;
import com.amap.api.services.geocoder.RegeocodeResult;

public class MapResultActivity extends Activity implements OnGeocodeSearchListener,
InfoWindowAdapter {
    //声明布局中的组件
    private TextView title, topAddress, topPoint, buttomLat, buttomLon, buttomCity, buttomAdd;
    private String titleStr, latStr, lonStr, qStr, content1;
    private ImageView back;
    private Button openMap, openQr;

    //高德地图相关
    private ProgressDialog progDialog = null;
    private GeocodeSearch geocoderSearch;
    private String cityName, addressName;
    private AMap aMap;
    private MapView mapView;
    private LatLonPoint latLonPoint;
    private Marker regeoMarker;

    @Override
    protected void onCreate(Bundle savedInstanceState) {
        super.onCreate(savedInstanceState);
        setContentView(R.layout.activity_map_result);
        mapView = (MapView) findViewById(R.id.map);
        mapView.onCreate(savedInstanceState);                    //此方法必须重写
        initView();
    }

    @Override
    protected void onResume() {
        super.onResume();
        mapView.onResume();
    }

    @Override
    protected void onPause() {
        super.onPause();
        mapView.onPause();
    }

    @Override
    protected void onSaveInstanceState(Bundle outState) {
        super.onSaveInstanceState(outState);
        mapView.onSaveInstanceState(outState);
    }

    /**
     * 显示进度条对话框
     */
    public void showDialog() {
```

```
        progDialog.setProgressStyle(ProgressDialog.STYLE_SPINNER);
        progDialog.setIndeterminate(false);
        progDialog.setCancelable(true);
        progDialog.setMessage("正在获取地址");
        progDialog.show();
    }

    /**
     * 隐藏进度条对话框
     */
    public void dismissDialog() {
        if (progDialog != null) {
            progDialog.dismiss();
        }
    }

    /**
     * 响应逆地理编码
     */
    public void getAddress(final LatLonPoint latLonPoint) {
        showDialog();
        //第一个参数表示一个Latlng,第二个参数表示范围多少米,第三个参数表示是火系坐标
          系还是GPS原生坐标系
        RegeocodeQuery query = new RegeocodeQuery(latLonPoint, 200, GeocodeSearch.AMAP);
        //设置同步逆地理编码请求
        geocoderSearch.getFromLocationAsyn(query);
    }

    /**
     * 加载布局中的各个View并初始化
     */
    private void initView() {
        initBack();                                         //返回按钮
        Intent intent = getIntent();
        titleStr = intent.getStringExtra("title");
        latStr = intent.getStringExtra("lat");
        lonStr = intent.getStringExtra("lon");
        if (intent.getStringExtra("q") != null && !intent.getStringExtra("q").equals("")) {
//地图缩放比例
            qStr = intent.getStringExtra("q");
        } else {
            qStr = "";
        }
        //根据获取的纬度以及经度,生成LatLonPoint对象,供逆地理编码以及逆地理编码回调使用
        latLonPoint = new LatLonPoint(Double.valueOf(latStr), Double.valueOf(lonStr));
        initTitleText();                                    //页面顶部标题TextView
        initTopText(); //悬浮在地图上面TextView,该TextView所在的最外层容器LinearLayout透明
        initButtomText();                                   //地图下方的各个TextView
        initMap();                                          //初始化地图
        initOpenMap();                       //在浏览器或谷歌地图APP中打开页面中的地图
        initOpenQr();                                       //跳转页面显示二维码图片
```

```
            getAddress(latLonPoint);                              //响应逆地理编码
            content1 = intent.getStringExtra("content1");
            if (intent.getStringExtra("flag").equals("scan")) {
                                              //如果是从 ScanFragment 界面中跳转过来的
                handleDB();                                       //插入一条新数据
            }
        }

        /**
         * 调用 API,在手机浏览器打开谷歌地图显示页面中的地理位置
         */
        private void initOpenMap() {
            openMap = (Button) this.findViewById(R.id.button_open_map);
            openMap.setOnClickListener(new View.OnClickListener() {
                @Override
                public void onClick(View v) {
                    String mapAdd = "http://ditu.google.cn/maps?hl=zh&mrt=loc&q=" +
latStr + "," + lonStr;
                    Intent intent = new Intent(Intent.ACTION_VIEW, Uri.parse(mapAdd));
                    intent.addFlags(Intent.FLAG_ACTIVITY_NEW_TASK & Intent.FLAG_ACTIVITY_
EXCLUDE_FROM_RECENTS);
                    intent.setClassName("com.google.android.apps.maps", "com.google.android.
maps.MapsActivity");
                    if (isIntentAvailable(intent)) {
                        startActivity(intent);
                        return;
                    }
                    intent = new Intent(Intent.ACTION_VIEW, Uri.parse(mapAdd));
                    startActivity(intent);
                }
            });
        }

        /**
         * 跳转页面,显示二维码图片
         */
        private void initOpenQr() {
            openQr = (Button) this.findViewById(R.id.button_open_qr);
            openQr.setOnClickListener(new View.OnClickListener() {
                @Override
                public void onClick(View v) {
                    Intent intent = new Intent(MapResultActivity.this, QrMapResultActivity.class);
                    Bundle bundle = new Bundle();
                    bundle.putString("title", titleStr);
                    bundle.putString("content1", content1);
                    intent.putExtras(bundle);
                    startActivity(intent);
                }
            });
        }
```

```
/**
 * 返回按钮
 */
private void initBack() {
    back = (ImageView) this.findViewById(R.id.back_map_result);
    back.setOnClickListener(new View.OnClickListener() {
        @Override
        public void onClick(View v) {
            MapResultActivity.this.finish();
        }
    });
}

/**
 * 获取上级页面中传递过来的数据项显示在顶部标题 TextView 中
 */
private void initTitleText() {
    title = (TextView) this.findViewById(R.id.title_map_result_TextView);
    title.setText(titleStr);
}

/**
 * 悬浮在地图上方的 TextView
 */
private void initTopText() {
    topPoint = (TextView) this.findViewById(R.id.top_point_map);
    topPoint.setText(latStr + "," + lonStr);
    topAddress = (TextView) this.findViewById(R.id.top_address_map);
}

/**
 * 在地图下面的各个 TextView
 */
private void initButtomText() {
    buttomLat = (TextView) this.findViewById(R.id.buttom_lat_textview);  //纬度
    buttomLon = (TextView) this.findViewById(R.id.buttom_lon_textview);  //经度
    buttomCity = (TextView) this.findViewById(R.id.buttom_city_textview);
                                                        //经纬度对应所在城市
    buttomAdd = (TextView) this.findViewById(R.id.buttom_add_textview);  //街道信息
    buttomLat.setText("Latitiude:" + latStr);
    buttomLon.setText("Longtitiude:" + lonStr);
}

/**
 * 初始化地图实例
 */
private void initMap() {
    if (aMap == null) {
        aMap = mapView.getMap();
         regeoMarker = aMap.addMarker(new MarkerOptions().anchor(0.5f, 0.5f).icon
(BitmapDescriptorFactory.defaultMarker(BitmapDescriptorFactory.HUE_RED)));
```

```
        }
        geocoderSearch = new GeocodeSearch(this);
        geocoderSearch.setOnGeocodeSearchListener(this);
        progDialog = new ProgressDialog(this);
    }


    /**
     * 操作数据库,插入数据
     */
    private void handleDB() {
        ArrayList<History> histories = new ArrayList<History>();
        String createTime = CommonUtil.getTime();
        History history = new History(createTime, titleStr, content1, "", "scan", "location");
        histories.add(history);
        MainActivity.dbManager.add(histories);
    }

    @Override
    public void onGeocodeSearched(GeocodeResult result, int rCode) {
    }

    /**
     * 逆地理编码回调
     */
    @Override
    public void onRegeocodeSearched(RegeocodeResult result, int rCode) {
        dismissDialog();
        if (rCode == 0) {
                if (result != null && result.getRegeocodeAddress() != null && result.
getRegeocodeAddress().getFormatAddress() != null) {
                    cityName = result.getRegeocodeAddress().getCity();
                    addressName = result.getRegeocodeAddress().getFormatAddress();
                    if (!qStr.equals("")) {
                    aMap.animateCamera(CameraUpdateFactory.newLatLngZoom(AMapUtil.
convertToLatLng(latLonPoint), Float.valueOf(qStr)));
                    } else {
                    aMap.animateCamera(CameraUpdateFactory.newLatLngZoom(AMapUtil.
convertToLatLng(latLonPoint), Float.valueOf(15)));
                    }
                    regeoMarker.setPosition(AMapUtil.convertToLatLng(latLonPoint));
                    regeoMarker.setTitle(result.getRegeocodeAddress().getCity());
                    regeoMarker.setDraggable(true);
                    regeoMarker.showInfoWindow();
                    topAddress.setText(cityName);
                    buttomCity.setText(cityName);
                    buttomAdd.setText(addressName);
                } else {
                    Toast.makeText(MapResultActivity.this, "对不起,没有搜索到相关数据!",
3000).show();
                }
```

```
        } else if (rCode == 27) {
            Toast.makeText(MapResultActivity.this, "搜索失败,请检查网络连接!", 3000).show();
        } else if (rCode == 32) {
            Toast.makeText(MapResultActivity.this, "key验证无效!", 3000).show();
        } else {
            Toast.makeText(MapResultActivity.this, "未知错误,请稍后重试!错误码为",
3000).show();
        }
    }

    @Override
    public View getInfoContents(Marker arg0) {
        return null;
    }

    @Override
    public View getInfoWindow(Marker marker) {
        //地图标记上的扩展信息
        View view = getLayoutInflater().inflate(R.layout.poikeywordsearch_uri_result, null);
        TextView title = (TextView) view.findViewById(R.id.title_result);
        title.setText(marker.getTitle());
        TextView snippet = (TextView) view.findViewById(R.id.snippet_result);
        snippet.setText(marker.getSnippet());
        return view;
    }

    private boolean isIntentAvailable(Intent intent) {
        List<ResolveInfo> activities = getPackageManager().queryIntentActivities(intent,
PackageManager.COMPONENT_ENABLED_STATE_DEFAULT);
        return activities.size() != 0;
    }
}
```

代码文件: codes\11\QR where\cn\edu\hstc\qrwhere\activity\MapResultActivity.java

11.7 开发第四个菜单项所对应的界面 SettingFragment

SettingFragment 对应第四个菜单项的界面。Setting 的中文含义为设置。该界面提供了两个可单击的列表项：About 和 Clear History。单击 About，弹出一个对话框，显示“Made In HSTC”；单击 Clear History，弹出一个询问对话框，询问“Are you sure?”。若用户单击 Confirm，则调用操作数据库的工具类清空作清空数据库表操作；若用户单击 Cancel，则取消对话框。SettingFragment 所对应的界面布局如下：

```
<LinearLayout xmlns:android = "http://schemas.android.com/apk/res/android"
    android:layout_width = "match_parent"
    android:layout_height = "match_parent"
    android:background = "@color/white"
    android:orientation = "vertical" >
```

```
<!-- 顶部标题 -->
<RelativeLayout
    android:layout_width="fill_parent"
    android:layout_height="wrap_content"
    android:background="@color/backhost"
    android:padding="10dip" >

    <TextView
        android:id="@+id/titleTextView"
        android:layout_width="wrap_content"
        android:layout_height="wrap_content"
        android:layout_centerHorizontal="true"
        android:layout_centerVertical="true"
        android:paddingTop="10dp"
        android:paddingBottom="10dp"
        android:text="@string/setting"
        android:textColor="@color/white"
        android:textSize="20sp" />
</RelativeLayout>

<RelativeLayout
    android:layout_width="fill_parent"
    android:layout_height="wrap_content" >

    <LinearLayout
        android:id="@+id/linearlayout_top"
        android:layout_width="fill_parent"
        android:layout_height="40dp"
        android:orientation="vertical" />

    <TextView
        android:id="@+id/textview_top"
        android:layout_width="fill_parent"
        android:background="@color/gainsboro"
        android:layout_height="0.5dip"
        android:layout_below="@id/linearlayout_top" />

    <ListView
        android:id="@+id/listview_setting"
        android:layout_width="fill_parent"
        android:layout_height="wrap_content"
        android:divider="@color/gainsboro"
        android:dividerHeight="0.5dip"
        android:layout_below="@id/textview_top" />

    <TextView
        android:id="@+id/textview_buttom"
        android:layout_width="fill_parent"
        android:background="@color/gainsboro"
        android:layout_height="0.5dip"
        android:layout_below="@id/listview_setting" />
```

```
    </RelativeLayout>

</LinearLayout>
```

代码文件：codes\11\QR where\res\layout\activity_setting.xml

上面布局中放置的ListView正是为了在界面中显示About列表项以及Clear History列表项。

界面Java程序SettingFragment类实现如下：

```
package net.takewin.qrwhere.fragment;

import java.util.ArrayList;
import java.util.List;

import net.takewin.qrwhere.R;
import net.takewin.qrwhere.util.MySettingListViewAdapter;
import net.takewin.qrwhere.view.MyAboutDialog;
import net.takewin.qrwhere.view.MyDialog;
import android.app.Dialog;
import android.os.Bundle;
import android.support.v4.app.Fragment;
import android.view.LayoutInflater;
import android.view.View;
import android.view.ViewGroup;
import android.widget.AdapterView;
import android.widget.AdapterView.OnItemClickListener;
import android.widget.ListView;

public class SettingFragment extends Fragment {
    private ListView settingListView;

    @Override
      public View onCreateView (LayoutInflater inflater, ViewGroup container, Bundle
savedInstanceState) {
        return inflater.inflate(R.layout.activity_setting, null);
    }

    @Override
    public void onActivityCreated(Bundle savedInstanceState) {
        super.onActivityCreated(savedInstanceState);
        initSettingListView();
    }

    private void initSettingListView() {
        List<String> listStrings = new ArrayList<String>();
        listStrings.add("About");             //关于
        listStrings.add("Clear History");     //清空数据库
        //获取界面布局中的ListView组件
        settingListView = (ListView) getActivity().findViewById(R.id.listview_setting);
        //设置适配器
            settingListView. setAdapter (new MySettingListViewAdapter (getActivity (),
```

```
listStrings));
        settingListView.setOnItemClickListener(new OnItemClickListener() {
                                                                        //列表项单击事件
            @Override
            public void onItemClick(AdapterView<?> arg0, View arg1, int arg2, long arg3) {
                if (arg2 == 0) {            //弹出关于的自定义对话框
                    Dialog dialog = new MyAboutDialog(getActivity(), R.style.MyAboutDialog);
                    dialog.show();
                } else if (arg2 == 1) {     //弹出询问是否清空数据库的自定义对话框
                                            //在对话框实现类 MyDialog 中操作数据库,清空表
                    Dialog dialog = new MyDialog(getActivity(), R.style.MyDialog);
                    dialog.show();
                }
            }
        });
    }
}
```

代码文件: codes\11\QR where\cn\edu\hstc\qrwhere\fragment\SettingFragment.java

上面程序中实现单击界面中的 About,弹出一个自定以显示对话框,该对话框中显示了提示消息“Made In HSTC”。从上面的 Java 程序上看,该自定义对话框是通过创建一个 Dialog 子类对象,然后调用其 show()方法实现的。如下代码片段:

```
Dialog dialog = new MyAboutDialog(getActivity(), R.style.MyAboutDialog);
dialog.show();
```

因此,我们需要知道 MyAboutDialog 类的实现:

```
package net.takewin.qrwhere.view;

import net.takewin.qrwhere.R;
import android.app.Dialog;
import android.content.Context;
import android.os.Bundle;
import android.view.View;
import android.widget.TextView;

public class MyAboutDialog extends Dialog {
    Context context;
    private TextView ok;

    public MyAboutDialog(Context context) {
        super(context);
        this.context = context;
    }

    public MyAboutDialog(Context context, int theme) {
        super(context, theme);
        this.context = context;
    }
```

```
    @Override
    protected void onCreate(Bundle savedInstanceState) {
        super.onCreate(savedInstanceState);
        //加载对话框布局
        this.setContentView(R.layout.dialog_about);
        //获取对话框中的 Ok 按钮
        ok = (TextView) findViewById(R.id.ok);
        //设置单击事件监听器
        ok.setOnClickListener(new View.OnClickListener() {
            @Override
            public void onClick(View v) {
                //取消对话框
                MyAboutDialog.this.dismiss();
            }
        });
    }
}
```

代码文件：codes\11\QR where\cn\edu\hstc\qrwhere\view\MyAboutDialog.java

上面 Java 类继承自 android.app.Dialog，重写 onCreate 方法，实现自定义一个对话框组件。在 onCreate 方法中，程序加载该自定义对话框的界面布局，获取布局中的按钮(此处为一个 TextVeiw)，然后为其添加单击事件监听器。该布局定义了对话框的样式，提示语 Made In HSTC 即是通过该布局实现。以下为自定义对话框对应的布局代码：

```
<?xml version="1.0" encoding="utf-8"?>
<LinearLayout xmlns:android="http://schemas.android.com/apk/res/android"
    android:layout_width="260dp"
    android:layout_height="match_parent"
    android:background="@drawable/segmented_selected_bg"
    android:orientation="vertical" >

    <TextView
        android:layout_width="fill_parent"
        android:layout_height="60dp"
        android:text="Made In HSTC."
        android:textSize="16sp"
        android:textColor="@color/black"
        android:gravity="center_vertical|center_horizontal"/>

    <TextView
        android:layout_width="fill_parent"
        android:layout_height="0.5dip"
        android:background="@color/lightgray" />

    <TextView
        android:id="@+id/ok"
        android:layout_width="fill_parent"
        android:layout_height="40dp"
        android:text="Ok"
        android:textColor="#0066FF"
```

```
        android:textSize = "20sp"
        android:gravity = "center_vertical|center_horizontal" />

</LinearLayout>
```

代码文件：codes\11\QR where\res\layout\dialog_about.xml

实际上，QR where 中所有需要用到弹出对话框的地方，都是通过自定义对话框布局，然后自定义对话框组件加载该布局，接着在对应的程序中创建该对话框的实例，调用 show()方法来实现的。包括单击 SettingFragment 界面中的第二个列表项 Clear History 所弹出的对话框亦是通过上述方式实现的。只不过在其对话框布局中放置的是两个按钮(此处为两个 TextView)：一个是 Cancel，另一个则是 Confirm，然后依次为这两个按钮添加单击事件监听器。为 Cancel 添加的监听器实现取消对话框，为 Confirm 添加的监听器实现调用数据库操作工具类 DBManager 的对象 MainActivity. dbManager 的 deleteTheTable(String tableName)方法清空 history 表数据。由于实现原理以及过程相同，此处不再赘述，读者可以通过附带源码获得实现代码。

11.8 QR where 运行效果图

前面已经将 QR where 的开发流程全部介绍完毕，当然，读者还是需要通过本书附带源码加深学习理解，部分工具类、二维码解析与生成、ShareSDK、高德地图等功能的实现，都是可以在其他 Android 项目中进行复用的，所以希望读者能对 QR where 的实现源码加以阅读体会。

为了读者更好地体会 QR where 的功能，下面将部分运行效果的截图贴出，如下所示：

① 打开软件，启动界面 MainActivity 默认打开底部第一个菜单项，如图 11.1 所示。

② 此时，单击右上角的按钮可打开或关闭手机自带闪关灯功能；打开左上角按钮，则调用系统相册，选中任意一张图片，返回 QR where 应用，解析返回的图片，若该图片中不包含二维码矩阵图，则会弹出一个对话框提示用户，如图 11.2 所示。

③ 若此时返回的图片中包含二维码矩阵图，则根据解析结果，判断对应的结果类型，跳转到相应的界面中，对于 URL、WiFi、其他普通文本类型则跳转到如图 11.3 所示的页面中(这几种类型的二维码显示页面相类似，这里以 WiFi 类型为例)。

④ 若返回的图片中所包含的二维码信息为 Contact 类型，则解析后跳转到的页面如图 11.4 所示。

⑤ 单击图 11.4 中手机号码右边的发送短信图标或拨打电话图标即可调用相应的系统应用，发送短信或拨出电话；单击页面中的 Send Message，则打开手机邮件应用发送邮件；单击页面中的最后一项 QRCode，则跳转到显示联系人二维码图片的页面；若单击页面中的 Create New Contact，则调用联系人应用的新增联系人模块并将页面中的联系人信息自动传递到联系人新增页面，如图 11.5 所示。

⑥ 单击如图 11.4 所示页面中的 Add to Existing Contact，则打开现有联系人列表，选中任意一个联系人(假如选择联系人列表中的“小红”)，跳转到该联系人修改页面，并将页面中的联系人信息传递至该页面中，如图 11.6 所示。

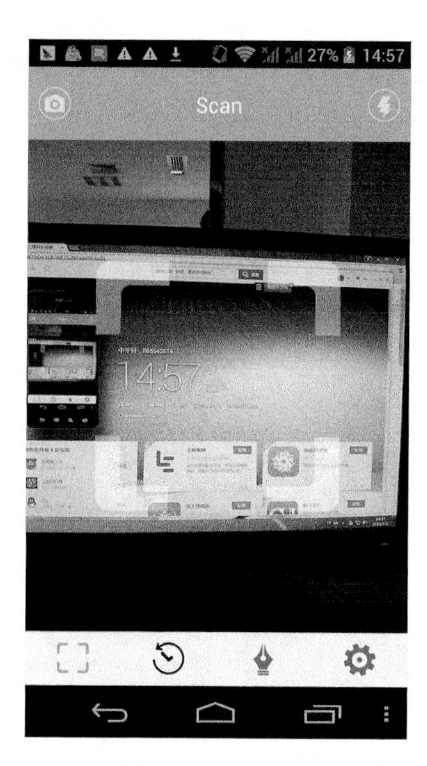

图 11.1 Scan 界面

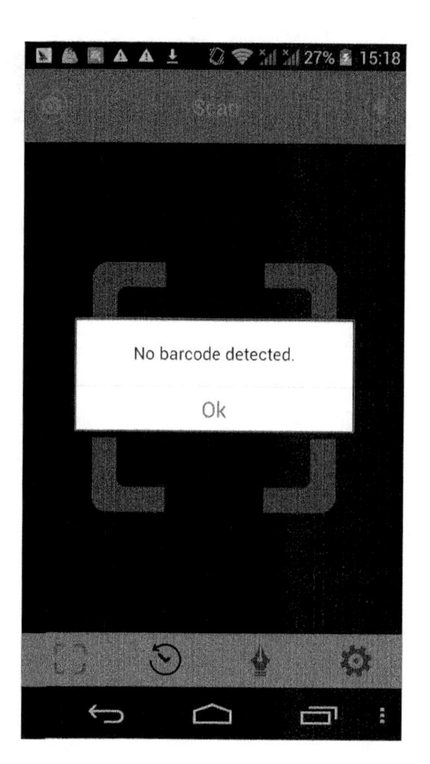

图 11.2 提示未检测到条形码

图 11.3 解析二维码(URL、WiFi、其他普通文本)后跳转到的页面

图 11.4 解析二维码(Contact 类型)后跳转的页面

图 11.5 新增联系人

图 11.6 添加至现有联系人

⑦ 若单击 Scan 页面左上角按钮返回的图片中的二维码矩阵信息为 Location 类型，则解析后跳转到的页面如图 11.7 所示。

⑧ 单击 Scan 页面左上角的图标，进入手机相册，选中一张图片，自动返回 Scan 页面并开始解析图片中的二维码，解析成功后进行页面跳转。由于这种操作实现的页面跳转与直接使用扫描框扫描二维码图片后的页面跳转相同，故此处不演示扫描框扫描二维码后的跳转过程。单击启动界面底部第二个菜单项，可以看到如图 11.8 所示的运行效果图。

⑨ 单击图 11.8 中的 Tab 页 Generate，可以看到如图 11.9 所示的运行效果图。

⑩ 单击 History 列表页面中的列表项，则根据类型跳转到对应的页面，这部分的跳转与解析二维码图片后的跳转页面类似，唯一的区别在于从历史列表跳转过去时，不再对 history 表执行插入新数据的操作。

⑪ 单击启动界面底部第三个菜单项，可以看到如图 11.10 所示运行效果图。

⑫ 单击如图 11.10 所示界面中的各个 TabRow，跳转到相应的信息编辑页面，并填写相应的信息，单击第一行，跳转到 URL 信息编辑页面，如图 11.11 所示。

⑬ 单击图 11.11 中的 Generate 按钮，跳转到下级页面，生成二维码并显示，操作数据库插入新数据，如图 11.12 所示。

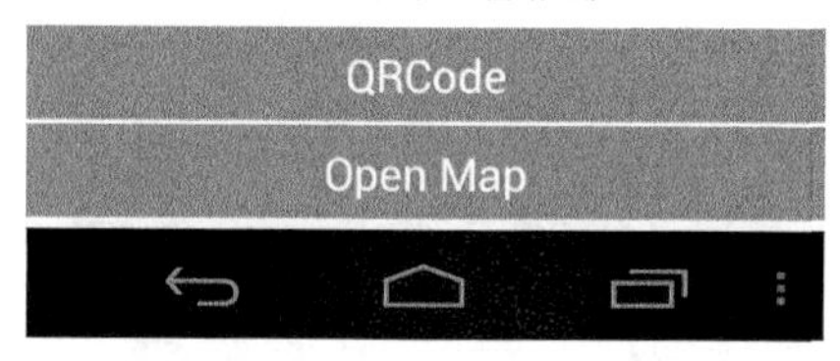

图 11.7　解析二维码(Location 类型)后的跳转页面

图 11.8　Scan 历史列表页面

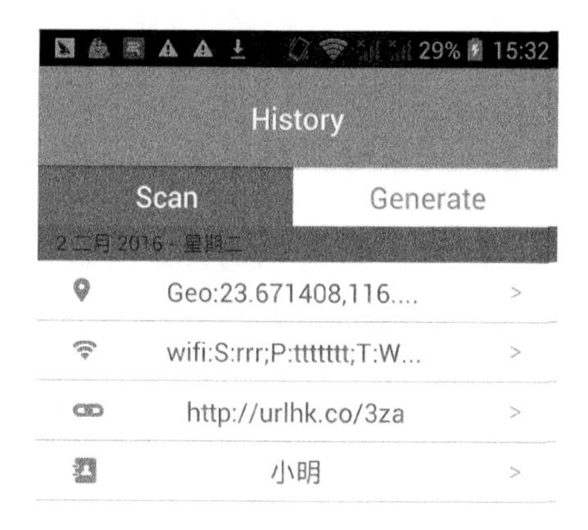

图 11.9　Generate 历史列表页面

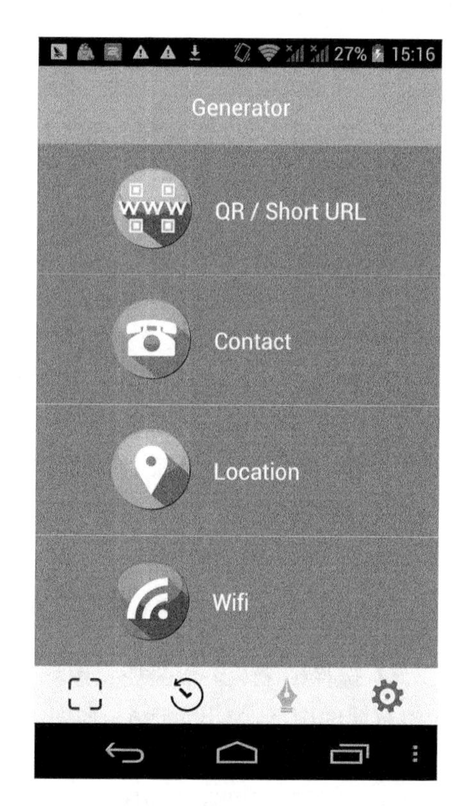

图 11.10　GeneratorFragment 界面

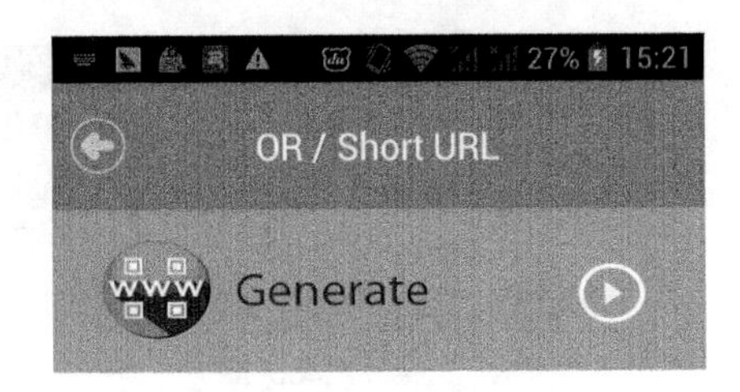

http://www.baidu.com

图 11.11 编辑 URL 信息

http://urlhk.co/3xG

图 11.12 生成 URL 二维码

⑭ 单击图 11.10 中的第二行，跳转到联系人信息编辑页面，编辑相应的信息，如图 11.13 所示。

⑮ 单击图 1.13 页面右上角的按钮，调出手机联系人应用，选中任意联系人，获取该联系人信息并返回页面，将对应信息填充到对应的编辑框中。

⑯ 单击图 1.13 页面中的 Generate 按钮，跳转到生成联系人二维码页面，如图 11.14 所示。

⑰ 单击图 11.10 中的第三行，跳转到坐标拾取页面，如图 11.15 所示。

⑱ 在坐标拾取页面中的关键字输入框中输入关键字，如图 11.16 所示。

⑲ 选中"潮州西湖"，单击搜索按钮进行搜索定位，可以看到如图 11.17 所示的效果图。

⑳ 单击页面中的 Generate 按钮，跳转到生成经纬度信息二维码的页面，如图 11.18 所示。

㉑ 单击图 11.10 页面中的第四行，跳转到 WiFi 信息编辑页面，如图 11.19 所示。

㉒ 在任意显示二维码图片的页面中，单击页面右上角按钮，都会在页面底部弹出一个选择框，如图 11.20 所示。

㉓ 单击选择框中的 Share to 按钮，则调用 Share 模块，弹出集成了各大社交平台的选择框，如图 11.21 所示。

㉔ 选择图 11.21 中的任意图标，则可一键分享至该平台，这里以分享至微博为例，单击新浪微博的图标，弹出对话框，如图 11.22 所示。

㉕ 单击启动界面底部第四个菜单项，跳转到设置页面，单击页面中的 Clear History，弹出确认对话框，可以看到如图 11.23 所示效果图。

图 11.13　联系人信息编辑页面

图 11.14　生成 Contact 二维码

图 11.15　坐标拾取页面

图 11.16　输入 poi 关键字

图 11.17　高德地图搜索结果页面

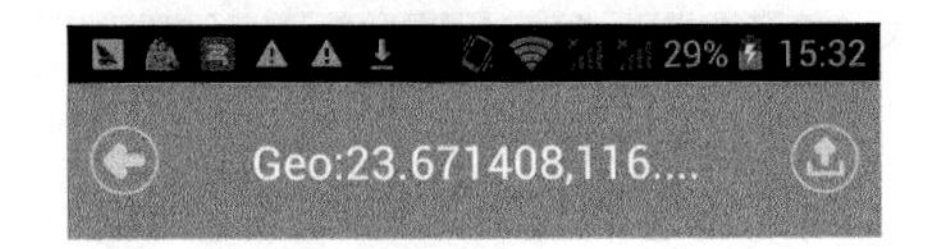

图 11.18　生成 Location 二维码

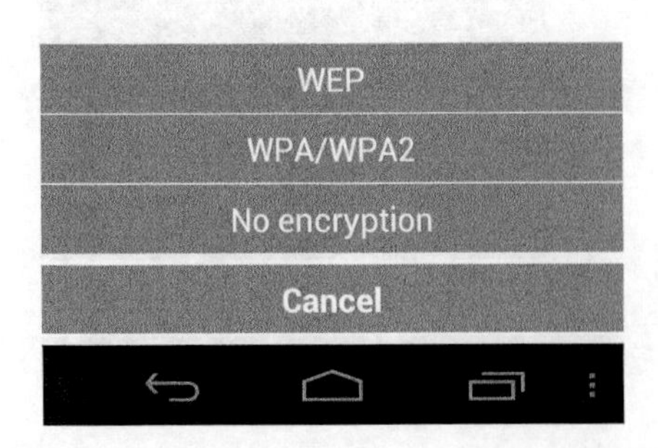

图 11.19　生成 WiFi 二维码

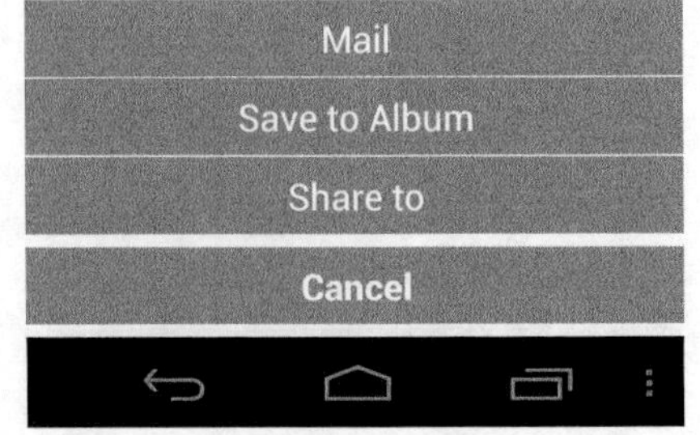

图 11.20　分享模块

图 11.21 弹出集成分享平台选择框

图 11.22 分享至微博平台

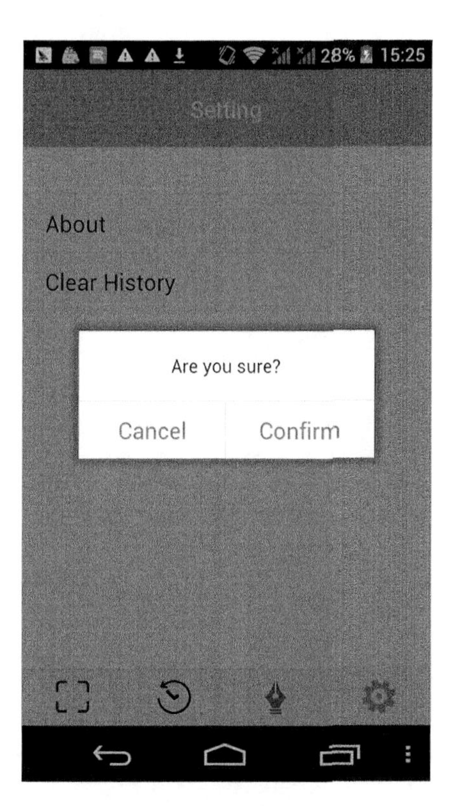

图 11.23 是否清空数据表的提示

㉖ 单击 Confirm 按钮，则将 QR where 的数据库表 history 清空，此时再返回 HistoryFragment，则已经看不到任何历史数据了。

11.9 本章小结

本章介绍了一款基于 Android 系统的二维码应用 QR where，该应用使用了 ZXing 框架以及 ZBar 框架解析和生成二维码，大量调用 Android API 操作系统相册、手机联系人应用，项目中还集成了一键分享平台 ShareSDK，使用高德地图进行地图显示以及定位，除此之外，项目还涉及 Android 自带的数据库 SQLite 的操作。通过该应用的实现，读者可以了解一些前面章节没有涉及的优秀的第三方产品，对项目中的工具类，底部选择框等实现源码，读者应该学会对其进行复用。

参 考 文 献

[1] 杨丰盛. Android 技术内幕[M]. 北京：机械工业出版社，2011.

[2] 郭霖. 第一行代码 Android[M]. 北京：人民邮电出版社，2014.

[3] Bill Phillips, Brian Hardy. Android 编程权威指南[M]. 北京：人民邮电出版社，2014.

[4] 任玉刚. Android 开发艺术探索[M]. 北京：电子工业出版社，2015.

[5] 张子言. 深入解析 Android 虚拟机[M]. 北京：清华大学出版社，2014.

[6] Bruce Eckel. Thinking in Java[M]. Prentice Hall, 2006.

图书资源支持

感谢您一直以来对清华版图书的支持和爱护。为了配合本书的使用，本书提供配套的资源，有需求的读者请扫描下方的"书圈"微信公众号二维码，在图书专区下载，也可以拨打电话或发送电子邮件咨询。

如果您在使用本书的过程中遇到了什么问题，或者有相关图书出版计划，也请您发邮件告诉我们，以便我们更好地为您服务。

我们的联系方式：

地　　址：北京海淀区双清路学研大厦 A 座 707

邮　　编：100084

电　　话：010－62770175－4604

资源下载：http://www.tup.com.cn

电子邮件：weijj@tup.tsinghua.edu.cn

QQ：883604（请写明您的单位和姓名）

资源下载、样书申请

书圈

用微信扫一扫右边的二维码，即可关注清华大学出版社公众号"书圈"。